Methods and Principles of Mycorrhizal Research

N. C. Schenck, editor
University of Florida

Published by
The American Phytopathological Society
St. Paul, Minnesota

Cover: Enlargement of Plate 6B (Page 234)

Library of Congress Catalog Card Number: 82-71988
International Standard Book Number: 0-89054-046-2

Printed in the United States of America

The American Phytopathological Society
3340 Pilot Knob Road
St. Paul, Minnesota 55121, USA

CONTENTS

AUTHORS AND THEIR AFFILIATIONS

Authors	Affiliation
M. F. Allen	Department of Wildlife Sciences, Utah State University, Logan, Utah 84322
M. F. Brown	Experiment Station Electron Microscope Facility, University of Missouri, Columbia, Missouri 65211
B. A. Daniels	Department of Plant Pathology, Kansas State University, Manhattan, Kansas 66506
J. M. Duniway	Department of Plant Pathology, University of California, Davis, California 95616
Larry Englander	Department of Plant Pathology-Entomology, University of Rhode Island, Kingston, Rhode Island 02881
J. J. Ferguson	Fruit Crops Department, University of Florida, Gainesville, Florida 32611
L. F. Grand	Department of Plant Pathology, North Carolina State University, Raleigh, North Carolina 27607
E. Hacskaylo	U.S.D.A. Forest Physiology Laboratory, Plant Industry Station, Beltsville, Maryland 20705
A. E. Harvey	U.S.D.A. Interior Forest and Range Experiment Station, Forestry Sciences Laboratory, Moscow, Idaho 83843
J. W. Hendrix	Department of Plant Pathology, University of Kentucky, Lexington, Kentucky 40546
M. C. Hirrel	Southeast Research and Experimental Center, University of Arkansas, Monticello, Arkansas 71655
R. S. Hussey	Department of Plant Pathology and Plant Genetics, University of Georgia, Athens, Georgia 30602
D. S. Kenney	Agricultural Chemicals Research, Abbott Laboratories, 36 Oakwood Road, Long Grove, Illinois 60047
E. J. King	Experiment Station Electron Microscope Facility, University of Missouri, Columbia, Missouri 65211
P. P. Kormanik	U.S.D.A. Forest Service, Forestry Laboratory, Institute for Mycorrhizal Research and Development, Athens, Georgia 30602
R. G. Linderman	U.S.D.A. Ornamental Research Laboratory, 3420 Orchard Street, Corvallis, Oregon 97330
Dale M. Maronek	Studebaker Nurseries, 1140 New Carlisle Road, New Carlisle, Ohio 45344
D. H. Marx	U.S.D.A. Forest Service, Forestry Laboratory, Institute for Mycorrhizal Research and Development, Athens, Georgia 30602
A.-C. McGraw	Department of Plant Pathology, University of Kentucky, Lexington, Kentucky 40546

J. A. Menge — Department of Plant Pathology, University of California, Riverside, California 92502

O. K. Miller, Jr. — Department of Biology, Virginia Polytechnic Institute and State University, Blacksburg, Virginia 24601

Randy Molina — Pacific Northwest Forest and Range Experiment Station, U.S.D.A., Forestry Sciences Laboratory, Corvallis, Oregon 97331

Stan Nemec — U.S.D.A. Horticultural Research Laboratory, 2120 Camden Road, Orlando, Florida 32803

J. G. Palmer — Center for Forest Mycology Research, U.S.D.A. Forest Products Laboratory, Madison, Wisconsin 53705

C. P. P. Reid — Department of Forest and Wood Sciences, Colorado State University, Fort Collins, Colorado 80523

L. H. Rhodes — Department of Plant Pathology, Ohio State University, Columbus, Ohio 43210

Jerry W. Riffle — U.S.D.A. Forest Service, Forestry Sciences Laboratory, University of Nebraska, Lincoln, Nebraska 68583

R. W. Roncadori — Department of Plant Pathology and Plant Genetics, University of Georgia, Athens, Georgia 30602

J. P. Ross — U.S.D.A., Plant Pathology Department, North Carolina State University, Raleigh, North Carolina 27607

G. R. Safir — Department of Botany and Plant Pathology, Michigan State University, East Lansing, Michigan 48824

N. C. Schenck — Plant Pathology Department, University of Florida, Gainesville, FLorida 32611

W. A. Sinclair — Department of Plant Pathology, Cornell University, Ithaca, New York 14853

H. D. Skipper — Agronomy and Soils Department, Clemson University, Clemson, South Carolina 29631

T. V. St. John — Natural Resource Ecology Laboratory, Colorado State University, Fort Collins, Colorado 80523

L. W. Timmer — Agricultural Research and Education Center, University of Florida, Lake Alfred, Florida 33850

James M. Trappe — U.S.D.A. Forestry Sciences Laboratory, 3200 Jefferson Way, Corvallis, Oregon 97331

Lidia S. Watrud — Monsanto Agricultural Products Co., 800 N. Lindbergh Blvd., St. Louis, Missouri 63166

Hugh E. Wilcox — Department of Enviornmental and Forest Biology, Syracuse University, Syracuse, New York 13210

S. H. Woodhead — Agricultural Chemicals Research, Abbott Laboratories, 36 Oakwood Road, Long Grove, Illinois 60047

PREFACE

The concept for METHODS AND PRINCIPLES OF MYCORRHIZAL RESEARCH was developed from discussions held by the Committee on Mycorrhizae of the American Phytopathological Society (APS). The proposal to prepare such a book was enthusiastically supported by this Committee (named below) and a subcommittee appointed to develop a tentative table of contents and possible authors. The contents and proposed authors developed by the subcommittee was modified and finally approved by the Committee on Mycorrhizae. The Committee on Publications for APS and the APS Council was approached about publishing METHODS AND PRINCIPLES OF MYCORRHIZAL RESEARCH. They approved this request for a book as a photocopy-ready publication. The book that you have before you is the result of a concerted effort of 39 authors to summarize the information and procedures dealing with most phases of research on these symbiotic fungi.

The purpose in preparing this book was to provide a single source of information on most aspects of mycorrhizae, with special reference to the methods and procedures used in their study. Hopefully, the book will provide the procedures or their sources that one would need to undertake most all projects with mycorrhizae, including identification, morphological studies, isolation, culture, methods of assay, and commercial application.

Considerable interest in mycorrhizae has developed in the past 15 years by researchers in many disciplines, but few individuals are trained in organized class room courses to work with these organisms. Few courses offer students actual "hands on" experience with mycorrhizae. Therefore, this book should provide these individuals with needed reliable information to prepare themselves for work with mycorrhizae. In addition, this book could be useful as a text for those wishing to initiate a course on mycorrhizae or for those instructors who wish to supplement their course offerings with a section on mycorrhizae. It most certainly would be a good resource tool for those already engaged in mycorrhizal research.

METHODS AND PRINCIPLES OF MYCORRHIZAL RESEARCH is divided into three sections. The first section covers the endomycorrhizae and predominantly deals with the vesicular-arbuscular mycorrhizae. The second section concerns the ectomycorrhizal fungi, including those referred to as ectendomycorrhizal fungi. The last portion of the manual covers areas that relate to both the endo- and ectomycorrhizae.

It is hoped that this book serves the purpose for which it was intended, a convenient, reliable, and useful source book on the procedures utilized with mycorrhizae.

N. C. Schenck
Editor

Members of the Committee on Mycorrhizae, American Phytopathological Society, that conceived the idea for the METHODS AND PRINCIPLES OF MYCORRHIZAL RESEARCH.

H. E. Bloss
J. W. Gerdemann
J. W. Hendrix
D. S. Kenney
R. G. Linderman
D. H. Marx
W. D. McClellan
J. A. Menge
S. Nemec
L. H. Rhodes
J. W. Riffle
G. R. Safir
N. C. Schenck
L. W. Timmer
S. H. Woodhead

Subcommittee that organized the METHODS AND PRINCIPLES OF MYCORRHIZAL RESEARCH.

S. Nemec
J. W. Riffle
N. C. Schenck
L. W. Timmer

INTRODUCTION

N. C. Schenck

During the latter half of the 19th century, several individuals noted the presence of fungi in plant roots without any apparent disease or necrosis. In 1885, Frank termed these fungus-root associations "mycorrhizae". He later distinguished between those that grew predominately intercellularly and developed extensively outside the root (ectomycorrhizae) and those that predominately developed within the root cell (endomycorrhizae) (2). Considerable speculation developed about the nature of these fungus-root associations. Some early studies suggested a possible pathogenic relationship, the fungus being parasitic on its companion host, while others suggested it was a mutually beneficial association. During the early 1900's, there were numerous reports on the occurence of fungi in mycorrhizal associations with plants, noting that ascomycetous, basidiomycetous and phycomycetous fungi were involved in these associations. In the 25 years from 1925 to 1950, most studies involved ectomycorrhizae, including the evaluation of growth improvement from ectomycorrhizae and inoculation with ectomycorrhizal fungi to establish pine in areas where it could not be grown previously (8).

In the period 1950 to 1960, the wide spread occurrence in soil, extensive host range, and growth benefits of endomycorrhizae, especially of the vesicular-arbuscular (VA) type, became apparent (1). Publications on VA mycorrhizae increased by over 100 papers from 1968 to 1973 (4) and the number has continued to increase substantially since then with 96 papers in 1979 alone (5). The interest in and publications on mycorrhizae in general has grown tremendously. Why this resurgence of interest in mycorrhizae whose existence was known for 100 years and whose presence was indicated in fossil records of some of the earliest land plants?

Several factors probably were responsible for this new interest. The identity of the fungal partner had been clarified and substantiated by pure culture studies for both the ecto- and endomycorrhizae. The ubiquitous nature of these fungi and their occurrence on roots of most plants has been established. The mycorrhizal condition of plants was shown to be the rule and not the exception in nature. This was especially true with the VA mycorrhizal fungi. In addition, procedures for pot culturing these fungi, for extracting their spores from soil, and for evaluating their incidence in roots had been established.

Perhaps the most important factor stimulating interest in mycorrhizae was the overwhelming published evidence indicating the growth promoting aspects of mycorrhizal fungi. Mycorrhizal fungi with their extramatrical hyphae increased the absorption of relatively immobile elements in soil (such as phosphorus, copper and zinc) by substantially extending the area of absorption beyond that for root hairs. The hyphae associated with mycorrhizal plants can ramify a greater soil volume and provide a greater absorptive surface than root hairs on a nonmycorrhizal plant. Plants with limited root hair development are frequently very dependent on mycorrhizae. Some plant species are so dependent on mycorrhizal fungi that they can not establish in nature or maintain normal growth without them (6). In addition, mycorrhizal plants were shown to have greater tolerance to toxic heavy metals, to root pathogens, to drought, to high soil temperatures, to saline soils, to adverse soil pH and to transplant shock than nonmycorrhizal plants. Because of these attributes, mycorrhizae are now considered important in the reestablishment of plants in inhospitable sites such as coal and copper mine wastes, borrow pits, and badly eroded locations.

Because of their widespread occurrence in nature and their numerous benefits to plants, researchers in several disciplines have become interested in studying mycorrhizae. Initially, scientists in only a few select disciplines were concerned with mycorrhizae, such as mycologists, plant pathologists, biologists and foresters. Now this group has been extended, to include agronomists, horticulturists, microbiologists, plant physiologists, soil scientists, ecologists, and many other disciplines.

The potential benefit of mycorrhizae to agriculture is more apparent than ever before. The need to increase food, fiber, and fuel production to keep pace with the increase in world population is crucial, especially in the lesser developed areas of the world (7). By the year 2000, the population of the world will double the number of people that have been on earth since the beginning of man's history (3). Challenges for the future are the utilization of mycorrhizae to i) increase production on previously marginal lands, ii) reduce the use of fertilizers that are becoming more expensive and in shorter supply, and iii) establish a sustainable agriculture less dependent on the energy-rich practices in current agriculture. If we are to be successful in captilizing on the potential benefits offered by mycorrhizae, considerably more knowledge must be generated by research in the evaluation and establishment of procedures for the use of mycorrhizae in commercial agriculture. We are hopeful that METHODS AND PRINCIPLES OF MYCORRHIZAL RESEARCH, in consolidating many of the procedures and present information on mycorrhizae, will serve a useful purpose in expediting such studies.

LITERATURE CITED

1. Gerdemann, J. M. 1968. Vesicular-arbuscular mycorrhizae and plant growth. Annu. Rev. Phytopathol. 6:397-418.
2. Kelley, A. P. 1950. Mycotrophy in plants. Chronica Botanica Co. Waltham, Mass. 223 pp.
3. Mayer, J. 1975. Agricultural productivity and world nutrition. Pages 97-108. In: A. Brown, T. Byerly, M. Bibbs, and A. San Pietro, eds. Crop productivity, research imperatives. Mich. Agr. Exp. Stn. and C. F. Kettering Foundation. Yellow Springs, Ohio. 399 pp.
4. Mosse, B. 1973. Advances in the study of vesicular-arbuscular mycorrhiza. Annu. Rev. Phytopathol. 11:171-196.
5. Nemec, S. 1982. Introduction to the symposium: Aspects of VA mycorrhizae in plant disease research. Phytopathology 72: (in press).
6. Rhodes, L. H. 1980. The use of mycorrhizae in crop production systems. Outlook on Agric. 10:275-281.
7. Ruehle, J. L. and Marx, D. H. 1979. Fiber, food, fuel, and fungal symbionts. Science 206:419-422.
8. Vozzo, J. A. 1971. Field inoculations with mycorrhizal fungi. Pages 187-196. In: E. Hacskaylo, ed. Mycorrhizae. U.S. Government Printing Office. Washington, D.C. 255 pp.

Methods and Principles of Mycorrhizal Research

TAXONOMY OF THE FUNGI FORMING ENDOMYCORRHIZAE

A. Vesicular-arbuscular Mycorrhizal Fungi (Endogonales)

James M. Trappe and N. C. Schenck

INTRODUCTION

Beniamino Peyronel (8, 9) in 1923 was the first to recognize that the vesicular-arbuscular mycorrhizal (VAM) fungi were members of the Endogonales, rather than chytrids, *Pythium* spp., or other fungi as suggested by earlier workers. Peyronel's discovery followed the revision of the Endogonaceae by Thaxter (11), who had not realized the mycorrhizal involvement of the family. The ensuing three decades witnessed little activity in taxonomic study of these fungi, because they were so rarely encountered. Gerdemann and Nicolson (5) then developed procedures for collecting spores from soil and described new species found by their techniques (7). The family was monographed in 1974 with segregation of the genus *Endogone* into seven genera (6); since then, steady discovery and publication of new taxa has added a genus and nearly doubled the number of species (13).

We now realize that the Endogonaceae are among the more common and widely distributed of the soil-borne fungi, and much new information about them has been obtained. Still, much mystery remains, and we continue developing hypotheses to clarify taxonomic concepts. To do this, similar individuals are grouped into species, a straightforward process for very distinctive groupings but an ambiguous one when extremely different spores are linked by a series of intergrading forms. The latter case occurs with some of the very common VAM fungi, e.g. *Glomus microsporum*-*G. fasciculatum*-*G. macrocarpum*. Unhappily, presently used criteria are inadequate to fix neat boundaries between intergrading taxa. Some of these complexes may be resolved by new techniques, but we must also be prepared to acknowledge that some may prove unresolvable in taxonomic terms. Nonetheless, we must continue to try, because nomenclature that reflects reality in speciation is necessary to scientific communication. An accurate species name permits us to make predictions about a fungus, and the literature now abounds with evidence that different VAM fungi can interact with hosts and environment in very different ways. Accurate species designations are required to avert chaos in VAM research.

The VAM Genera and Their Defining Characters

The Endogonales (Zygomycotina) as presently conceived consists of a single family, the Endogonaceae. The genera are separated at the first level on the manner of spore formation, as inferred from the morphology of spores and spore-bearing structures. Within the chlamydosporic taxa, further generic separation is based on sporocarp morphology. The type genus, *Endogone*, forms zygospores but is not known to produce VA mycorrhizae. *Acaulospora*, *Entrophospora*, *Gigaspora*, *Glaziella*, *Glomus*, and *Sclerocystis* have not been demonstrated to form zygospores and contain either proven or presumed VAM species.

Two additional genera have been tentatively placed in the Endogonales in the past. *Modicella* forms only sporangia and appears to be saprobic; we exclude it from the Endogonales, because it fits more comfortably with the Mortierellaceae (Mucorales). *Complexipes* is based on chlamydospores formed by certain Discomycetes, probably in the family Pyronemataceae (4).

Two kinds of spores are hypothesized for the VAM genera on the basis of spore morphology. No data are available on the cytology or sexual vs. asexual processes involved in formation of these spores; when such data become available, our concepts may change in some cases. The spores of *Acaulospora*, *Entrophospora*, and *Gigaspora* have been termed azygospores, i.e. parthenogenic zygospores, because they rather resemble the zygospores of *Endogone* spp., but no sexual orgin has been observed (2, 6). The spores of Glaziella, Glomus, and Sclerocystis are regarded as chlamydospores, i.e., specialized asexual resting cells.

The distinctive combination of characters that defines each VAM genus is summarized below. Each genus has been described in detail by Gerdemann and Trappe (6); this publication, out of print for several years, in now again available from the U.S. Forest Service, Pacific Northwest Forest and Range Experiment Station, 809 NE 6th Ave, Portland, OR 97232.

We avoid use of the term "vesicle", because in the VAM literature it has been applied to four quite different kinds of unrelated structures. To minimize confusion, we prefer to reserve the term only for the vesicles that form within roots as part of the vesicular-arbuscular mycorrhiza.

Acaulospora, (Fig. 1; Plate 1A).-- Azygospores bud laterally from the funnel-shaped stalk of a large, inflated hyphal terminus (6). By spore maturity, the funnel-stalked cell collapses; its amorphous remnants may cling to the spore but often detach to leave the spore with no sign of its origin except for a small pore with an inconspicuous rim. The spores are borne singly in soil or sometimes within roots (1, 6). Keys to the eight species presently known are provided by Trappe (13) and Walker and Trappe (15).

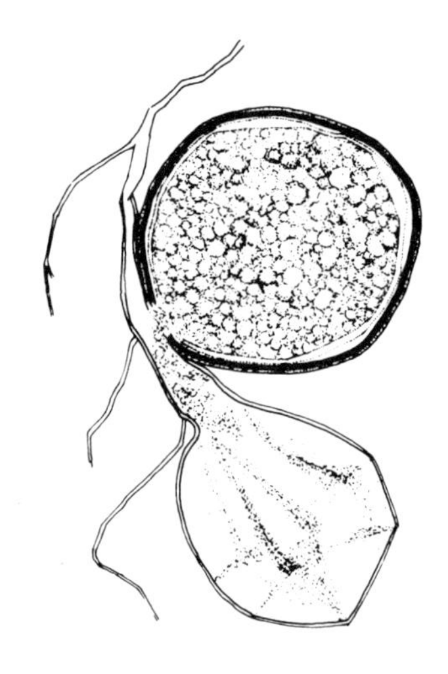

Fig. 1. Acaulospora laevis (cross section, x 100)

Entrophospora (Plate 1E)-- Azygospores are formed within the funnel-shaped stalk of a large, inflated hyphal terminus (2). As the spore enlarges, the stalk expands to produce a second swelling below the inflated hyphal terminus. The end result is a peanut or dumbbell-shaped cell with the spore in the basal, swollen part. Meanwhile, the walls of the spore-enclosing structure become greatly thickened. Ultimately the terminal swelling collapses and detaches, leaving the spore enclosed by the thick, appressed wall of the former stalk of the inflated hyphal terminus. The spores are borne singly in soil. Only one species has been described (2), but a second has been recently discovered (Schenck, unpublished data).

Gigaspora (Fig, 2: Plate 1B,C).--Azygospores bud from the bulbous, suspensor-like tip of a hypha. As the spore expands and approaches maturation, one to several septa form below the suspensor-like bulb, which usually remains attached to the spore and produces one to several, narrow, out-growing hyphae (6). The uppermost of these grows to the lower spore surface, where it apparently terminates without fusing with the spore. The spores are borne singly in soil. Keys to most of the 17 presently recognized species are provided by Trappe (13).

Fig. 2. Gigaspora calospora (cross-section, x 200)

Glaziella-- Chlamydospores, formed singly at hyphal tips, are scattered in the walls of large (1.5-5 cm diam), hollow, bright orange to red sporocarps (11). Known only from tropical lowlands, especially near the coast, this unique fungus forms sporocarps in the soil, but expansion of the sporocarps lifts them out of the soil. Often they are found lying loose on the ground. They float freely in water and may have evolved this sporocarp form for water transport by ocean currents. Only one species is presently recognized.

Glomus (Figs 3, 4; Plate 1F-H; Plate 2A,B).--Chlamydospores form at a hyphal tip, usually one per tip, but in G. fuegianum several spores emerge from a swollen hyphal tip (6, 11). By maturity, the spore contents are separated from the attached hypha by a septum (Fig.3) or by occlusion with deposits of wall material (Fig. 4). Two or more hyphae may be attached to spores of some species. Care must be taken not to confuse a basal protrusion of a Glomus spore (Fig. 3) with the suspensor-like bulb of Gigaspora spp. (Fig. 2). Spores of most Glomus species are borne singly in soil, but some of the same species may also form them in the cortex of roots (6, 10) or in sporocarps (6). A few species are known only from sporocarps. Most sporocarps of Glomus spp. are nonorganized conglomerations of spores. In G. radiatum the spores are randomly positioned but radiate from the sporocarp base, those nearest the base being the most mature. Most of the more than 40 species presently recognized are in Trappe's keys (13).

Fig. 3. Glomus mosseae (cross-section, x 200)

Sclerocystis (Fig 5; Plate 2C,D).--Chlamydospores form in sporocarps as a single, crowded layer of erect spores that surrounds the sides and top of a spore-free, central mass of tightly interwoven hyphae. The sporocarps may be borne singly in soil or fused together in crusts on organic debris or moss at the soil surface (the latter case appears to occur only in the wet tropics or in humid glasshouse conditions). All presently known species are included in the key by Trappe (13).

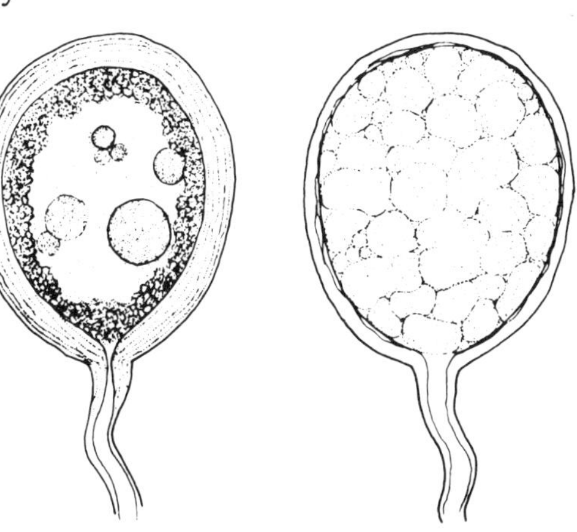

Fig. 4. Glomus fasciculatum, right spore young, left mature with laminated wall (cross-section, x 500).

Characters Important in Species Identification

Methods of observation.-- Firm identifications usually require numerous spores, although a single spore in prime condition can often be identified to species. To determine reaction to reagent or stains or to define variability in size, color, degree of maturation, etc., many spores from a single collection are needed. We like to have at least 20 and prefer 50 or more. Moreover, it is useful to crush spores to define wall structure; once crushed, a spore has limited value for other uses. Voucher collections for herbarium deposit increase in usefulness with increasing numbers of spores, and type collections of new taxa should contain as many spores as practicable.

Microscope mounts of spores from sporocarps are easy to make if the sporocarps are large enough to manipulate by hand. Cut the sporocarp in half with a razor blade along the axis of the point of sporocarp attachment, when such a point can be found. Intact spores can usually be teased from the exposed gleba directly onto

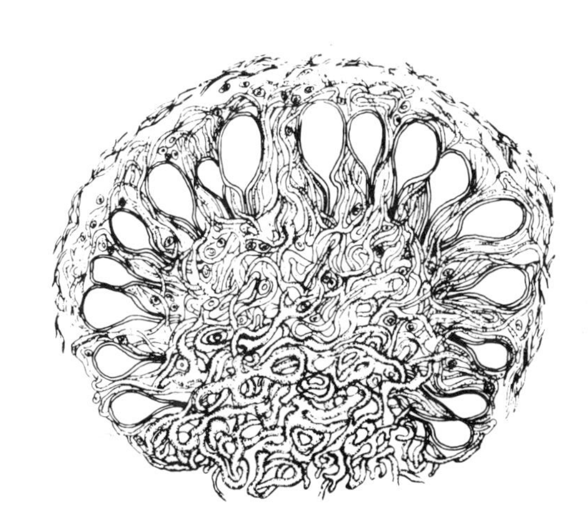

Fig. 5. Sclerocystis coremioides sporocarp (cross-section, x 100)

a microscope slide. Alternatively, thin slices of gleba can be cut from the exposed gleba; include the sporocarp surface in the slices for evaluation of peridial characters. Tiny sporocarps, such as those of Sclerocystis spp., are more difficult to section. With patience, decent sections can be obtained by holding them in a forceps and cutting razor blade slices under a stereomicroscope.

Individual spores can be picked up in a number of ways. If they are not suspended in fluid, touch them with the side of a dissecting needle (moisten the needle for dry spores), and they will cling to it. More often, the spores are suspended in fluid when they are to be manipulated. A simple tool for picking them out of fluid is a dissecting spade. To make a spade, heat the tip of a dissecting needle to redness, then hammer it flat. Make the tip especially thin. Reheat and bend the tip in a slight upward curve. Then reheat again and plunge it into cold water to temper the steel. Individual spores can be picked out of water ca. 5 mm deep by slipping the spade under them and lifting them (break them through the water surface tension abruptly). The technique is somewhat tricky but easily mastered with a little practice. In experienced hands it is very quick. A small-bore glass pipet is an easy alternative, (R. Ames, personal communication). Place a finger over the top hole of the pipet, then lower it into the water and over the spore. Release the finger, and the water rises into the pipet, dragging the spore with it. The higher the water level in the container, the more the flow into the pipet when the finger is released. Pipet tips on spring-action or thumbscrew suction control syringes are also excellent for taking up spores. C. Walker (personal communication) recommends the entomology tweezers made of thin gauge, springy stainless steel. These are designed for picking up insect specimens without damage and are excellent for spores. Unfortunately, commercial sources of these tweezers are difficult to find.

After nonsporocarpic spores have been extracted from soil (see Chapter 3), they should be examined stereomicroscopically. Collections from the field are usually a mix of several species. Several representive spores or sporocarps of each different appearing type can then be removed, placed on a drop of water on a microscope slide, and examined with a compound microscope. One can usually judge which of the groups seeming to differ by stereomicroscopy are indeed different taxa and which are simply variants or developmental stages of a single taxon. For example, some nearly colorless spores may prove to be young, thin-walled specimens of the same taxon that is also present as more colored, thick-walled, mature spores. Dull-surfaced spores may be senescent, with surfaces roughened by microbial degradation but otherwise clearly the same as similarly colored but bright and shiny spores. After these initial spores are evaluated, one can return to the steromicroscope and sort the rest of the spores, putting each type in a separate container of water (watch glasses or small evaporating dishes are appropriate if the spores are to be examined immediately; screw cap vials for storage in a refrigerator if examination will be delayed).

Standard mounting media for determining species can be water, lactoglycerine, or lactophenol. Avoid 3% KOH, a standard medium in mycology, because it induces unnatural swelling of the walls of some Endogonales. Two drops of the mounting medium can be put on a microscope slide, one drop near one end of the slide and one in the middle. A drop of Melzer's reagent can be put on the other end of the slide. A dozen or so spores can be selected from one of the sorted batches and placed in the drops of standard medium in the middle of the slide; attempt to include the smallest and the largest spores as judged by stereomicroscopic viewing. Three or four spores can be put into the end drop of standard medium and into the Melzer's reagent. Each drop is covered with a 20 x 20 mm cover slip. If the spores are exceptionally large, two or three drops may be needed to fill the space between the slide and the cover slip. The spores in the end spot of standard medium are percussion-crushed by gently bouncing a pencil, eraser side down, on the top of the cover slip. Avoid an excess of medium in the spot to be flattened by percussion crushing.

With these three mounts on a single slide, one can observe all the characters needed to key out the species. This can be done directly by starting in the key and observing each character in turn as called for by the key. Or, the spore characters can be recorded in full and then keyed from the notes. In the latter case, retain the slide in case the key calls for something not recorded.

Staining spores in cotton blue-lactophenol (or lactoglycerine) sometimes reveals spore structures difficult to see clearly in unstained specimens. To stain quickly, place several spores in a drop of cotton blue on a slide, add a glass (not plastic) cover slip, and flame to boiling by holding a lighted paper match under the cotton blue with the tip of the flame 1 cm below the slide. Remove the heat as soon as the medium begins to boil. When it reaches boiling, the mount may spit tiny droplets of cotton blue, so hold at arm's length when applying the heat. If the mount need not be examined for a day or so after preparation, the spores will usually stain satisfactorily in the interim without heating. After an extended period in cotton blue, most spores stain so intensely that structural details are obscured.

Sporocarps.-- These may range from rather large (*Glaziella*, some *Glomus* spp.--Plate 1F) to less than 0.5 mm in diam (some *Glomus* and *Sclerocystis* spp.--Fig. 4; Plate 2C,D). Important characters in addition to size are sporocarp surface color; nature of the peridium, ranging from absent to a loose tangle of surface hyphae, a tightly interwoven cutis, or a pubescence of erect hyphal tips; color and organization of the interior; and arrangement of spores. It is important to cut sporocarps along the axis of the point of sporocarp attachment, when such a point can be found, to accurately determine interior structure and spore arrangement.

Spore dimensions.--Spores are measured by use of an ocular micrometer in the eyepiece of a compound microscope. Spore length is along the axis of the point of spore attachment and width is at angles to that attachment. The two measurements are customarily presented length first, width second, but inconsistancies are common in the older literature and continue to appear occasionally. If spores are ornamented, measurements excluding the ornamentation are customarily recorded separately from the height of the ornamentation. Spore dimensions within a species and even within a single sievings can vary considerably, the more so with the larger spored species. Hence the range of dimensions should be recorded. Occasional spores sometimes exceed the dimensional range of the great majority of spores. These exceptional cases are recorded in parentheses, e.g. (26-) 30-55 (-61) µm.

Hyphal mantles.--Spores of some *Glomus* spp. are wrapped in interwoven hyphae. In some cases the mantles separate readily, but often the mantles are tightly adherent. This is an important character for species identification but it complicates study of the spore walls and attachment. Spores will pop out of nonadherent mantles when crushed by pencil percussion. If mantles adhere, more forceful percussion can be used to break spores apart, so that at least the broken ends of the walls can be observed. In either case, the hyphal attachment is difficult to find. One must then work patiently with large numbers of spores in hopes of finding hyphal attachments. Young spores in early stages of mantle development sometimes offer the best chance.

Spore ornamentation.--A distinct surface ornamentation delights the identifier of species of Endogonales, not only because of the beauty often displayed but also because it markedly narrows the choice of species. Ornamentation may be spines, warts, wrinkles, pits, a reticulum, etc., or a combination of two or more of these. The pattern, height, size of the elements (e.g. warts or spines) can all be diagnostically useful.

Biological or physical-chemical weathering can produce a dull roughening on the surface of old spores that were shiny-smooth when younger. The degradation can be mistaken for ornamentation, although it tends to be amorphous and erratic rather than distinctive or patterned. Weathered spores generally occur among spores in prime condition, and the two can generally be equated by characters other than surface if the collection contains numerous spores.

Spore walls.-- Walls are extremely important in taxonomy of the VAM fungi, but they can be confusing. All spores begin with thin walls (Fig. 4). As the spores mature, the walls may thicken to only 1 µm or so in some species or to more than 20 μm in others. Furthermore, the walls of some species differentiate into two to several distinct layers, and in a few of these species the outermost layer may flake away or erode just as inner layers are differentiating. Acaulospora and Gigaspora species commonly form one to several separable membranes within the primary spore wall. To compound the problem, the thickening of walls in some collections of Glomus spp. proceeds uniformily but in other collections is apparently intermittent. The latter circumstance results in a series of successively fused deposits which have been termed laminations (Fig. 4). When present in a spore wall, laminations are usually indistinct or discontinuous, but occasionally they appear as distinct and uniform as annual rings in a tree.

Assessment of spore wall structure is stepwise. First, intact spores should be examined with the high-dry microscope objective focused at optical cross section. Then, unless spores are opaque or extremely large, switch to oil immersion. The thickness, color, and any other distinctive feature should be recorded for each differentiated layer (but this generally is not useful for individual, fused laminations). A drawing is useful for complex differentiated wall layering. Repeat the steps with crushed spores, noting which layers are separable and differences from the observations of intact spores. Use the oil immersion objective. Then examine spores mounted in Melzer's reagent with the high-dry objective, specially noting distinctive color reactions of each layer. Before switching to oil immersion, crush the spores in Melzer's regent by percussion, then re-examine with the high-dry objective. If any distinctive color reaction is seen, examine with the oil immersion objective.

By these techniques, even the most complex spore walls can usually be accurately evaluated, although in some cases considerable time and patience is needed. If one is uncertain whether the specimens have truly differentiated wall layers or fused laminations of a single wall, the edge of the fracture of crushed spores will nearly always resolve the issue. Nonseparable differentiated layers will often break at slightly different spots, so the fracture edge will show angular discontinuities from one layer to the next (examine several spores). A single wall with fused laminations, in contrast, will generally fracture cleanly with no discontinuities between laminations.

Artifacts of spore wall structure sometimes result from parasitism. The invasion of a parasite through the spore wall, for example, can produce a pore; a single spore may have many such pores. These are random or erratic rather than in the regular pattern that would be expected if the pore were produced by the spore itself. Often the spore reacts to such parasitic penetrations by depositing inner wall material over them. The parasite then invades further, and the spore deposits additional wall material around it. This results in pored stalagmite-like growths from the inner surface of the spore wall.

Spore contents.-- Color of spore contents is occasionally useful in species identification. The spores of Glomus convolutum, for example, are filled with deep yellow oil globules, the source of the yellow color of the sporocarps (Plate IG). Spores of most species, however, have colorless contents. The form of contents of fresh, mature spores tends to differ between azygosporic taxa, which are filled with nearly uniform, small oil globules (Figs. 1, 2) and chlamydosporic taxa, which contain a wide range of oil globule sizes (Figs. 3, 4). Some chlamydospores, however, have "azygosporic-type" oil globules. With drying or freezing, the many oil globules within spores often coalesce into a single, large globule. The contents of senescing spores sometimes become crystalline or amorphous. Because of these variations, spore contents have limited taxonomic use. An exception is Glomus radiatum, which fills its spores with hyphae by maturity (6).

Hyphal attachments.-- The importance of hyphal attachment characters for generic determination has been outlined earlier in this chapter, but hyphal attachments are also useful at the species level. In Gigaspora, for example, the suspensor-like bulb is the same color as the spore in most species but a different color from the spore in a few species. Glomus species separate into two major categories of hyphal attachment closure; those closed by a septum (Fig.

3) vs those occluded by thickening of the spore wall (Fig. 4, Plate 2B). Some Glomus species are more readily differentiated by the hyphal attachment than by the spores themselves. For example, Glomus etunicatum spores sometimes resemble those of G. caledonicum, other times those of G. fasciculatum or G. macrocarpum. The hyphal attachment of G. etunicatum, however, is closed by a septum and is 5-8 um in diam, and the attached hypha is colorless, with only slightly thickened walls. In contrast, tne other three Glomus spp. have hyphal attachments occluded by spore wall thickening and routinely 10-20 µm in diam, and the attached hypha is pigmented and very thick-walled. Some Glomus species regularly have two or more attached hyphae. These may ultimately prove to be sexual spores, but for lack of evidence on sexuality and for their morphological similarity to chlamydospores of other Glomus spp., they are presently retained in Glomus.

The difference between the attachment of Gigaspora spp. and Glomus mosseae needs special mention, because many beginners confuse them. Compare figures 2 and 3. Gigaspora spp. bud from a suspensor-like bulb; the point of attachment is strikingly constricted, and a narrow pore penetrates from the bulb through the outer spore wall. Glomus mosseae has a basal protrusion of the spore itself, typically funnel-shaped as shown in figure 3 but sometimes cylindric; the point of attachment is not constricted and is closed by a septum with no visible pore.

Detailed descriptions of hyphal attachments and the attached hyphae have sometimes been neglected in the past, but now the characters of these structures are known to often have diagnostic value. Redescriptions and descriptions of new taxa should detail the diameter of the attachment and the nature of its closure. Descriptions of attached hyphae should include diameter at attachment plus changes in diameter below the attachment, color, wall thickness and layering, septation, presence of subsidiary hyphae or hyphal branching, and any other distinctive features observed.

Soil-borne auxiliary cells of Gigaspora spp.-- Hyphae of Gigaspora species produce fine, coiled hyphae terminated with small, specialized cells or clusters of cells (Plate 1D) termed "soil-borne vesicles" or "extramatrical vesicles" in the past. The function of these cells is unknown. To avoid confusion with other uses of "vesicle" in VAM terminology, we prefer to call these structures "soil-borne auxiliary cells" until a more specific, functional term can be coined. Gigaspora spp. differ in producing auxiliary cells singly vs in clusters of a few vs clusters of many. Young auxiliary cells are smooth and hyaline to subhyaline but when fully developed are mostly ornamented with knobs, warts, spines, or coralloid projections and, in some species, become brown walled. Gigaspora species can generally be identified by spore characters alone, but retrieval of auxiliary cells for confirmation of identification should always be attempted.

Manner of spore germination.-- Germination has been observed for only a relatively few species of Endogonales but is known to differ between species of a given genus. The inner and outer spore walls of some azygosporic species separate to form a cavity that becomes compartmentalized by cross walls. Each compartment may produce a germ tube through the outer spore wall. Other azygosporic species may simply germinate through the spore wall with no inner compartmentalization. Chlamydosporic species seem mostly to extend a germ tube from the hyphal attachment through the attached hypha, but a few species have been observed to germinate through the spore wall. As more data accumulate, the manner of spore germination will doubtless prove useful in reassessing genera and subgenera. It is useful in identifying taxa only when germinated spores are available.

Histochemical reactions.-- Histochemical reactions have assumed prominence in taxonomy of ectomycorrhizal fungi, but little has been attempted so far with VAM fungi. Melzer's reagent produces diagnostically useful reactions with some taxa, e.g. an intense reddening of inner membranes of Acaulospora spp. or presence of green to black granules and stains in hyphae of Glomus convolutum. Other reagents need to be tried, but at present we have no basis for suggesting which ones other than those which have proven useful with other groups of fungi.

Preservation of Voucher Collections

Type collections of new taxa are always deposited in a publically accessible herbarium, but we cannot emphasize too strongly the critical importance of similarly depositing voucher collections of VAM fungi collected in the field, grown in pot cultures, or used in experiments. We urge careful reading of the paper by Ammirati (3) on this subject.

Herbarium specimens of sporocarps can be air-dried and deposited in envelopes or packets. Tiny sporocarps or nonsporocarpic spores and hyphae are customarily preserved in vials of lactophenol, which will last indefinitely in tightly closed vials sealed with parawax and will prevent growth of contaminants. Unfortunately, spores sometimes change color with time in lactophenol. L-drying offers an excellent alternative for suitably equipped laboratories (12). As many spores as practicable should be deposited for each collection, but few spores are preferable to none. Auxiliary cells of *Gigaspora* species should be included with *Gigaspora* collections when available.

It is also helpful to include microscope slides of spores along with preserved specimens. One convenient method for preparing slides in polyvinyl alcohol lactophenol (PVL) is described by Walker (14). Dissolve 15 ml polyvinyl alcohol granules (50-75% hydrolyzed: 20-25 centipoise viscosity when in 4% aqueous solution at 20%°C) in 100 ml distilled water in a water bath at 80°C. This may take several days with occasional stirring. The resulting stock solution keeps well in dark bottles. To make PVL mounting medium, mix 56 parts of the stock solution with 22 parts each of lactic acid and liquid phenol. Fresh, rehydrated, or dry spores are placed on a microscope slide and covered with a drop of PVL. Arrange the spores at the center of the PVL with a needle and put on a glass cover slip. If an excess of PVL is present, it will often drag spores to the edge as it is flattened by the cover slip; better to use a minimal amount, then fill in vacant areas under the flattened cover slip by adding an excess of PVL from the side. If the spores are wet, the PVL sometimes becomes cloudy but will clear overnight. The slides are then allowed to dry and harden, a process which may take a week or more at room temperature but is hastened by keeping them on a warming table. PVL evaporates slowly even when solidified, so an initial excess at the edge of the cover slip is useful. Even so, gaps may appear under the cover slip during hardening and should be filled with fresh PVL. Once the mount is hardened and filled to the cover slip edge, seal with fingernail polish, porcelain cement, or a similarly impervious material. These slides will last indefinitely. Spores stained in cotton blue or Melzer's reagent can be similarly mounted in PVL, and the color reactions will be retained apparently indefinitely.

Data to accompany the herbarium-desposited specimens includes locality, date, and elevation of collection (or of original source of pot-cultured spores), associated hosts, history of pot culturing when applicable, color of spores and attached hyphae when fresh, and collectors' names. Other descriptive notes on the collection or photographs of the habitat or the specimens increase the value of the collection.

Any receptive public herbarium is suitable for deposit of collections of Endogonales. Normally the herbarium of the collectors' institution would be used. If the institution has no mycological herbarium, one option is the Oregon State University Mycological Herbarium, which has become a world depository for Endogonales. It also welcomes duplicates of collections placed in other herbaria.

LITERATURE CITED

1. Ames, R. N. and Linderman, R. G. 1976. *Acaulospora trappei* sp. nov. Mycotaxon 3:565-569.
2. Ames, R. N. and Schneider, R. W. 1979. *Entrophospora*, a new genus in the Endogonaceae. Mycotaxon 8:347-352.
3. Ammirati, J. 1979. Chemical study of mushrooms: the need for voucher collections. Mycologia 71:437-441.

4. Danielson, R. M. 1982. Taxonomic affinities and criteria for identification of the common ectendomycorrhizal symbiont of pines. Can. J. Bot. 60:7-18.
5. Gerdemann, J. W. and Nicolson, T. H. 1963. Spores of mycorrhizal Endogone species extracted from soil by wet sieving and decanting. Trans. Br. Mycol. Soc. 46:235-244.
6. Gerdemann, J. W. and Trappe, J. M. 1974. The Endogonaceae in the Pacific Northwest. Mycologia Mem. 5:1-76.
7. Nicolson, T. H. and Gerdemann, J. W. 1968. Mycorrhizal Endogone species. Mycologia 60:313-325.
8. Peyronel, B. 1923. Fructificatio de l'endophyte a arbuscules et a vesicules des mycorrhizes endotrophes. Bull. Soc. Mycol. France 39(2):1-8.
9. Peyronel, B. 1924. Specie de Endogone protructrici di micorize endotrofiche. Boll. Mens. R. Staz. Patol. Veg. Roma 5:73-75.
10. Schenck, N. C. and Smith, G. S. 1982. Additional new and unreported species of mycorrhizal fungi (Endogonaceae) from Florida. Mycologia 74:77-92.
11. Thaxter, R. 1922. A revision of the Endogonaceae. Proc. Amer. Acad. Arts. Sci. 57:291-351.
12. Tommerup, I. and Kidby, D. K. 1979. Preservation of spores of vesicular-arbuscular endophytes by L-drying. Appl. Environm. Microbiol. 37:831-835.
13. Trappe, J. M. 1982. Synoptic keys to the genera and species of zygomycetous (vesicular-arbuscular) mycorrhizal fungi. Phytopathology 72: (in press).
14. Walker, C. 1979. The mycorrhizast and the herbarium: The preservation of specimens from VA mycorrhizal studies. In: Program and Abstracts, 4th N. Amer. Conf. Mycorrhiza. Fort collins, Colorado.
15. Walker, C. and Trappe, J. M. 1981. Acaulospora spinsoa sp. nov. with a key to the species of Acaulospora. Mycotaxon 12:515-521.

B. Endomycorrhizae by Septate Fungi

Larry Englander

INTRODUCTION

Due to the limited number of studies and the diversity of both hosts and endophytes in this 'catch-all' group, it is more useful to present information which will familiarize the reader with the endomycorrhizal fungi rather than to attempt a taxonomic treatment of the fungi involved, especially since a number of these endophytes have not yet been identified, and for others, their symbiotic role has not yet been determined conclusively. For the convenience of presentation, the septate endomycorrhizae will be considered as three groups: arbutoid, ericoid, and orchid mycorrhizae.

ARBUTOID MYCORRHIZAE

This type of mycorrhiza has been reported on plants in a number of genera of Ericaceae, including *Arbutus*, *Arctostaphylos*, *Gaultheria*, *Leucothoe*, and *Vaccinium*, and on members of related families, the Pyrolaceae and Monotropaceae (6). It has been suggested (4) that a transition between ectomycorrhizae and endomycorrhizae exists in the arbutoid type of mycorrhiza, accounting for the term, ectendomycorrhiza, sometimes applied to this phenomenon. These mycorrhizae are characterized by a fungal sheath surrounding infected roots, by intracellular fungal penetration and, in some cases, development of an intracellular network of hyphae. Also, root dimorphism may occur, with infected roots remaining short.

It has been demonstrated that arbutoid mycorrhizae can be synthesized using several fungi which normally form ectomycorrhizae on forest trees (15). However, only a few fungal partners in naturally occurring arbutoid mycorrhizae have been identified thus far. Some genera of fungi determined to be arbutoid symbionts, principally through observations of the association of their sporophores with mycorrhizal roots, or by linking hyphae or rhizomorphs between sporophores and mycorrhizal roots, include *Amanita*, *Cortinarius*, and *Boletus* (5, 14).

ERICOID MYCORRHIZAE

Ericoid mycorrhizae occur throughout the fine-root system (hair roots) of ericaceous plants. Fungal strands or wefts develop on or in close proximity to the root surface. Epidermal and cortical cells are penetrated by these hyphae, and the fungus ramifies within each cell to form a coil or 'knot' of filaments occupying much of the volume within the cells. The stele is not invaded.

It is perhaps difficult to imagine how a mass of fungal hyphae, occupying up to 80% of the area within a root segment (11), might not be observed microscopically. Indeed, on healthy white hair roots in a region which is neither meristematic nor suberized (9,11), hyphal knots are readily detected (200-400X) in infected cells of root segments heated for 10 min at 90 C in lactophenol-0.05% trypan blue. However, one must take into account the fact that numerous patches of uninfected root cells may occur interspersed amongst heavily infected areas. Also, in less-than-prime root specimens, individual hyphal filaments comprising the knots may appear diffuse, and a positive determination is difficult.

The consensus of observations over the years has implicated slow growing, sterile dark mycelia as the ericoid endophytes. The controversy surrounding identification of the endophyte has been clarified considerably since the refutation of the theory that the fungus was a *Phoma* sp. which systemically invaded ericaceous plants (4,8). A major breakthrough was the successful isolation of a fungus from ericoid mycorrhizae, and the subsequent pure culture synthesis of typical ericoid mycorrhizae (8). Observation of the perfect state of the fungus led to its identification as *Pezizella ericae* Read (10).

P. ericae is a Discomycete. It exhibits a slow growth rate in culture. On malt agar, colonies are light to dark gray when viewed from above, with a dark brown to black color on the obverse side. Apothecia, borne on a short stalk, are at first concave, translucent to white, up to 1 mm in diam, becoming flattened and yellow to orange when mature (10). *P. ericae* has fruited consistently and abundantly for us in the greenhouse on the surface of soil-less (peat-perlite) mix in which inoculated Rhododendrons were grown for several months, and also directly on roots in pure-culture synthesized mycorrhizae in the lab (7).

Thus far, *P. ericae* is the only endophyte which has been both identified and successfully used to synthesize the ericoid mycorrhiza. There are indications that other fungi may also form this type of mycorrhiza. Several ultrastructural studies on roots with ericoid mycorrhizae have found Ascomycete and Basidiomycete hyphae, sometimes both in the same cell (1,9). The specific occurrence of simple clavate fruit bodies of *Clavaria* spp. within the rhizosphere of ericaceous plants has led several investigators to examine this association and, using techniques such as immunofluorescence and radioisotopes, researchers have concluded that this fungus is present in cells or is in close physiological relationship with roots having ericoid mycorrhizae (2,12). The lack of success in culturing *Clavaria*, and thus the inability to test synthesis of the ericoid mycorrhiza, precludes further clarification of the role of this fungus.

ORCHID MYCORRHIZAE

This type of mycorrhiza has been proposed as one of the most complex of symbiotic interactions, in which a dynamic balance is achieved between pathogenesis of host tissue and dissolution of the fungus. The fungi involved are members of the form-genus, *Rhizoctonia*, and mycelial characteristics common to this group simplify their recognition in culture. The perfect states of some of these fungi have been found, placing symbionts in the genera, *Ceratobasidium*, *Sebacina*, and *Tulasnella* (3,13). The complex taxonomy of the Rhizoctonias and limited information on orchid mycorrhizae precludes a systematic treatment of orchid symbionts.

LITERATURE CITED

1. Bonfante-Fasolo, P. 1980. Occurrence of a Basidiomycete in living cells of mycorrhizal hair roots of Calluna vulgaris. Trans. Br. Mycol. Soc. 75:320-325.
2. Englander, L., and Hull, R. J. 1980. Reciprocal transfer of nutrients between ericaceous plants and a Clavaria sp. New Phytol. 84:661-667.
3. Hadley, G. 1975. Organization and fine structure of orchid mycorrhiza. Pages 335-351 in: F. E. Sanders, B. Mosse, and P. B. Tinker, eds. Endomycorrhizas. Academic Press, London 626pp.
4. Harley, J. L. 1969. The biology of mycorrhiza. Leonard Hill, London. 334pp.
5. Largent, D. L., Sugihara, N., and Brinitzer, A. 1980. Amanita gemmata, a non-host-specific mycorrhizal fungus of Arctostaphylos manzanita. Mycologia 72:435-439.
6. Largent, D. L., Sugihara, N. and Wishner, C. 1980. Occurrence of mycorrhizae on ericaceous and pyrolaceous plants in northern California. Can. J. Bot. 58:2274-2279.
7. Moore-Parkhurst, S., and Englander, L. 1981. A method for the synthesis of a mycorrhizal association between Pezizella ericae and Rhododendron maximum seedlings growing in a defined medium. Mycologia 73:994-997.
8. Pearson, V., and Read, D. J. 1973. The biology of mycorrhiza in the Ericaceae. I. The isolation of the endophyte and synthesis of mycorrhizas in aseptic culture. New Phytol. 72:371-379.
9. Peterson, T. A., Mueller, W.C., and Englander, L. 1980. Anatomy and ultrastructure of a Rhododendron root-fungus association. Can. J. Bot. 58:2421-2433.
10. Read, D. J. 1974. Pezizella ericae sp. nov., the perfect state of a typical mycorrhizal endophyte of Ericacae. Trans. Br. Mycol. Soc. 63:381-419.
11. Read, D. J., and Stribley, D. P. 1975. Some mycological aspects of the biology of mycorrhiza in the Ericaceae. Pages 105-117 in: F. E. Sanders, B. Mosse, and P. B. Tinker, eds. Endomycorrhizas. Academic Press, London. 626 pp.
12. Seviour, R. J., Willing, R.R., and Chilvers, G.A. 1973. Basidiocarps associated with ericoid mycorrhizas. New Phytol. 72:381-385.
13. Warcup, J. H. 1975. Factors affecting symbiotic germination of orchid seed. Pages 87-104 in: F. E. Sanders, B. Mosse, and P. B. Tinker, eds. Endomycorrhizas. Academic Press, London. 626pp.
14. Zak, B. 1974. Ectendomycorrhiza of Pacific madrone (Arbutus menziesii). Trans. Br. Mycol. Soc. 62:201-204.
15. Zak, B. 1976. Pure culture synthesis of Pacific madrone ectendomycorrhizae. Mycologia 63:362-369.

MORPHOLOGY AND HISTOLOGY OF VESICULAR-ARBUSCULAR MYCORRHIZAE

A. Anatomy and Cytology

M.F. Brown and E.J. King

INTRODUCTION

Since vesicular-arbuscular (VA) mycorrhizal fungi do not induce distinctive alterations of root morphology, microscopic techniques play an essential role in the critical analysis of developmental patterns and anatomical features of VA mycorrhizal (VAM) infections as well as to merely confirm their presence and/or to determine the extent of infections in the root system. Preparative techniques used to study these associations vary greatly in complexity in relation to specific research objectives. These procedures range from basic root clearing and staining methods or histological techniques for light microscopy (1,16,17,19,22) which are considered in Section B of this chapter, to various techniques employed for ultrastructural studies (3,5,11,12,13,25). Procedures for transmission (TEM) and scanning electron microscopy (SEM) are provided in a later chapter of the manual. Consequently, in this presentation we have stressed the descriptive aspects of VAM infections rather than techniques, but have illustrated the structures involved with micrographs obtained by a number of correlative light and electron microscopic procedures.

External Morphological Features

VA mycorrhizal fungi are identified on the basis of morphological characteristics of various structures produced by the mycelial system in the soil and host as indicated in Chapter 1. Structures produced exterior to the root include sporocarps and/or ectocarpic chlamydospores in *Glomus* (Fig. 1-3) and *Sclerocystis*, azygospores in *Acaulospora*, and azygospores as well as soil-borne vesicles in *Gigaspora* (Fig. 4, 5) (9,10,21). The spores and sporocarps of VAM fungi are produced by an often extensive mycelial system extending from previously infected mycorrhizal roots into the rhizosphere but they do not always accompany successful VAM infections in the field (21). The external mycelium is dimorphic and consists of: 1) 'permanent,' coarse, thick-walled, generally aseptate hyphae which comprise a major portion of the mycelial phase and, 2) numerous fine, thin-walled and highly branched, lateral hyphae which become septate at maturity and are ephemeral in nature (18,20). Both components of the external mycelium of *G. fasciculatus* on soybean are shown in Fig. 3. Infections of host roots are initiated by hyphae emanating from sporocarps, chlamydospores, azygospores, internal vesicles or other viable components in root residues. Penetration of the root epidermis is often preceded by the formation of an appressorium. The frequency of penetration points may be as high as 20 per mm of root length (23) and, in some hosts, numerous closely spaced penetrations may be produced by "runner" hyphae which grow along the root surface (4,20).

Anatomical Features of VAM Infections

VAM infections are characterized by the production of thick-walled inter- or intracellular vesicles (Fig. 11, 12) that are believed to function as endophytic storage organs, and by the formation of intracellular, haustoria-like arbuscules (Fig. 7, 13-16). The arbuscles are considered to be the primary structures involved in the bidirectional transfer of nutrients between the fungal symbiont and host plant. Although generally considered with VAM endophytes, *Gigaspora* spp. typically form only arbuscules. Following penetration of the host epidermis, or in some instances root hairs, hyphae of the fungal symbiont grow inter- and/or intracellularly within the cortex. This differs in various hosts. Intercellular hyphae predominate in onion and maize (5,7). Longitudinally oriented hyphae of both types (Fig. 6, 7, 9) characterize infections in soybean (4) while in yellow poplar (Fig. 8, 10), the hyphal system is intracellular and composed of extensive loops and coils (7).

Arbuscules are produced progressively within cells of the inner cortex a short distance behind the advancing hyphal tips (Fig. 6) resulting in an age sequence of arbuscules from the leading ends of each 'infection unit' toward the oldest portion nearest the original point of penetration (5). Cox and Sanders (5) have indicated a maximum length of 5 mm for infection units in onion. Arbuscules develop in cortical cells close to the stele following penetration of the host wall by a lateral intracellular branch produced from a hypha within an adjoining cell (Fig. 7, 9, 10), or one in an adjoining intercellular space (Fig. 12). This lateral hypha, which becomes the arbuscular trunk, branches repeatedly in a dichotomous pattern (Fig. 13) and ultimately forms bifurcate terminal branches which may be less than 1.0 μm in diameter. When fully developed, arbuscules frequently occupy a large proportion of the host cell lumen (Fig. 7, 8, 10).

Fig. 1-5. SEM micrographs of Glomus and Gigaspora spp. Fig. 1. A sporocarp (S), probably containing a single spore, and an ectocarpic chlamydospore (C) of Glomus mosseae x54. Fig. 2. Chlamydospore of G. mosseae showing the distinctive funnel-shaped subtending hypha (arrow). x110. Fig. 3. Chlamydospores (C) and dimorphic hyphae of Glomus fasciculatus attached to a mycorrhizal soybean root. Stout, permanent hyphae (P), finely branched, ephemeral lateral hyphae (arrows). x150. Fig. 4. Basal portion of an azygospore of Gigaspora coralloidea illustrating the bulbous (B) and hyphal (arrow) suspensor-like cells produced by members of this genus. x200. Fig. 5. Echinulate soil-borne (accessory) vesicles of Gigaspora margarita. Vesicles of this species are borne in clusters on coiled hyphae (arrows). x440. Fig. 6. Leading edge of a cleared and stained (16) G. fasciculatus/soybean mycorrhizal infection unit. The youngest arbuscules (arrows) form a short distance behind the advancing tips (not shown) of longitudinally oriented hyphae (H). Mature, densely stained arbuscules (top center) are located toward the central portion of this developing infection unit. x375). Fig. 7. Incompletely developed (right center) and fully developed (left) arbuscules in cleared and stained preparation of G. fasciculatus/soybean. The immature arbuscule on the right has developed as a lateral branch of an intracellular hypha (IH) within an adjacent cortical cell. Host cell walls (W). x375. Fig. 8. SEM micrograph of G. mosseae/yellow poplar showing coiled, intracellular hyphae (arrows) in outer cortical cells and arbuscules (A) within cortical cells toward the interior of the root. Root epidermis (E). x525. Fig. 9. A 0.5 μm epoxy cross-section of a G. caledonius/soybean infection obtained as a light microscopic "survey" section prior to ultrathin sectioning for TEM. Three of the cortical cells contain arbuscules in an early stage of development, indicated by the densely stained, nonvacuolate cytoplasm within large and small order arbuscular branches (arrows). Intracellular hypha (IH), arbuscular trunk (T), host cell nucleus (N). x550. Fig. 10. Longitudinal paraffin section of G. mosseae/yellow poplar illustrating the coiled, intracellular hyphae (left center) typical of VA infections in this host, arbuscules which are fully developed but not yet deteriorating (A), and others in which portions of the arbuscule have aggregated into dense irregular clumps (arrows). Stele (S). x240. Fig. 11. Vesicles of G. fasciculatus in a cleared and stained soybean root. x150. Fig. 12. A 0.5 μm epoxy section of an immature intracellular vesicle (V) of G. caledonius in soybean. When fully developed, this vesicle would contain several large lipid globules and the cytoplasmic content seen here would be greatly condensed. Arbuscules in adjacent cells are fully developed, as indicated by the less intensely stained, vacuolate cytoplasm (compare with Fig. 9). Intercellular hyphae (H), arbuscular trunk (T), stele (S). x550.

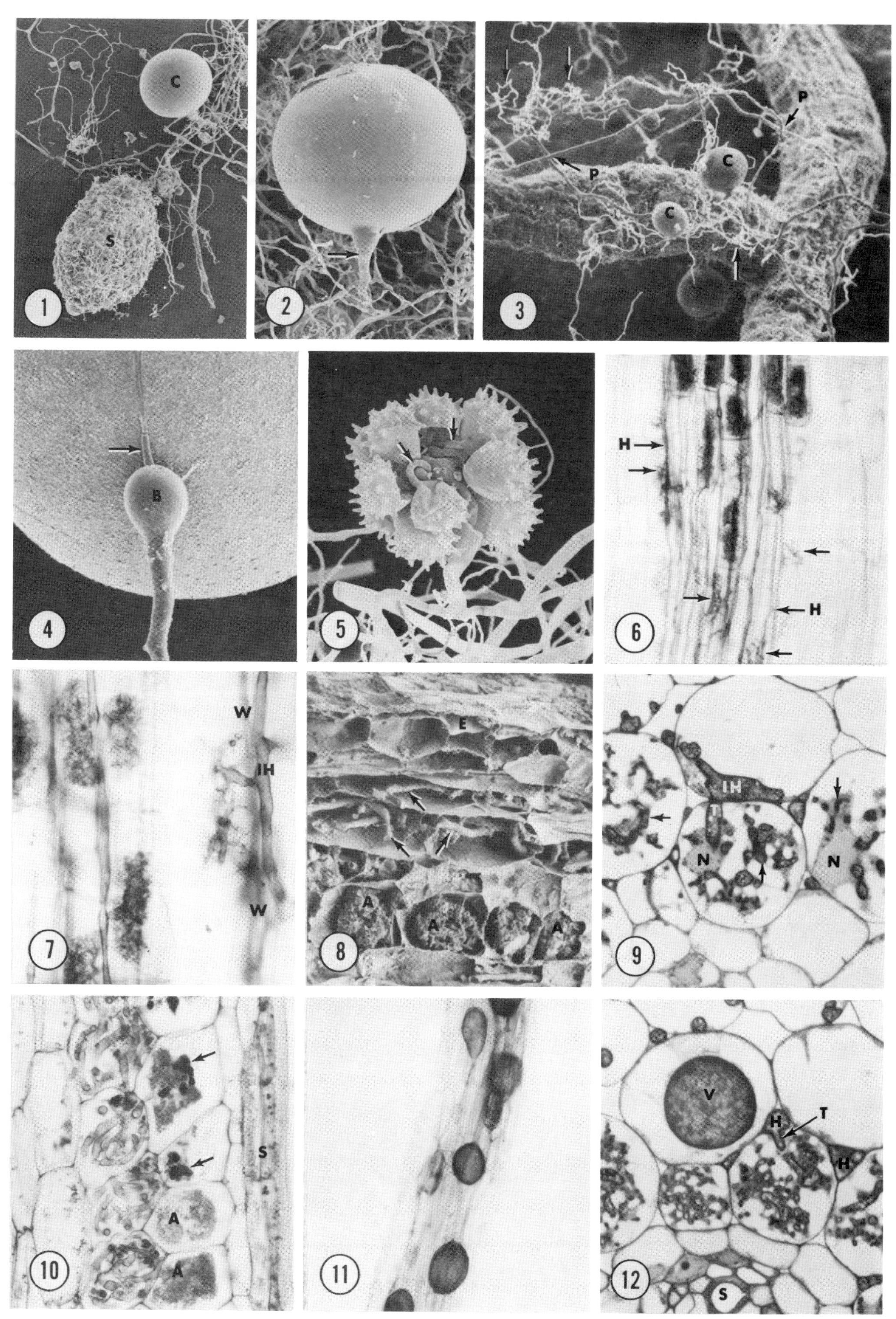
C
S
1
2
P
C
C
3
B
4
5
H
H
6
W
IH
W
7
E
A
A
A
8
IH
T
N
N
9
S
A
A
10
11
V
H
T
H
S
12

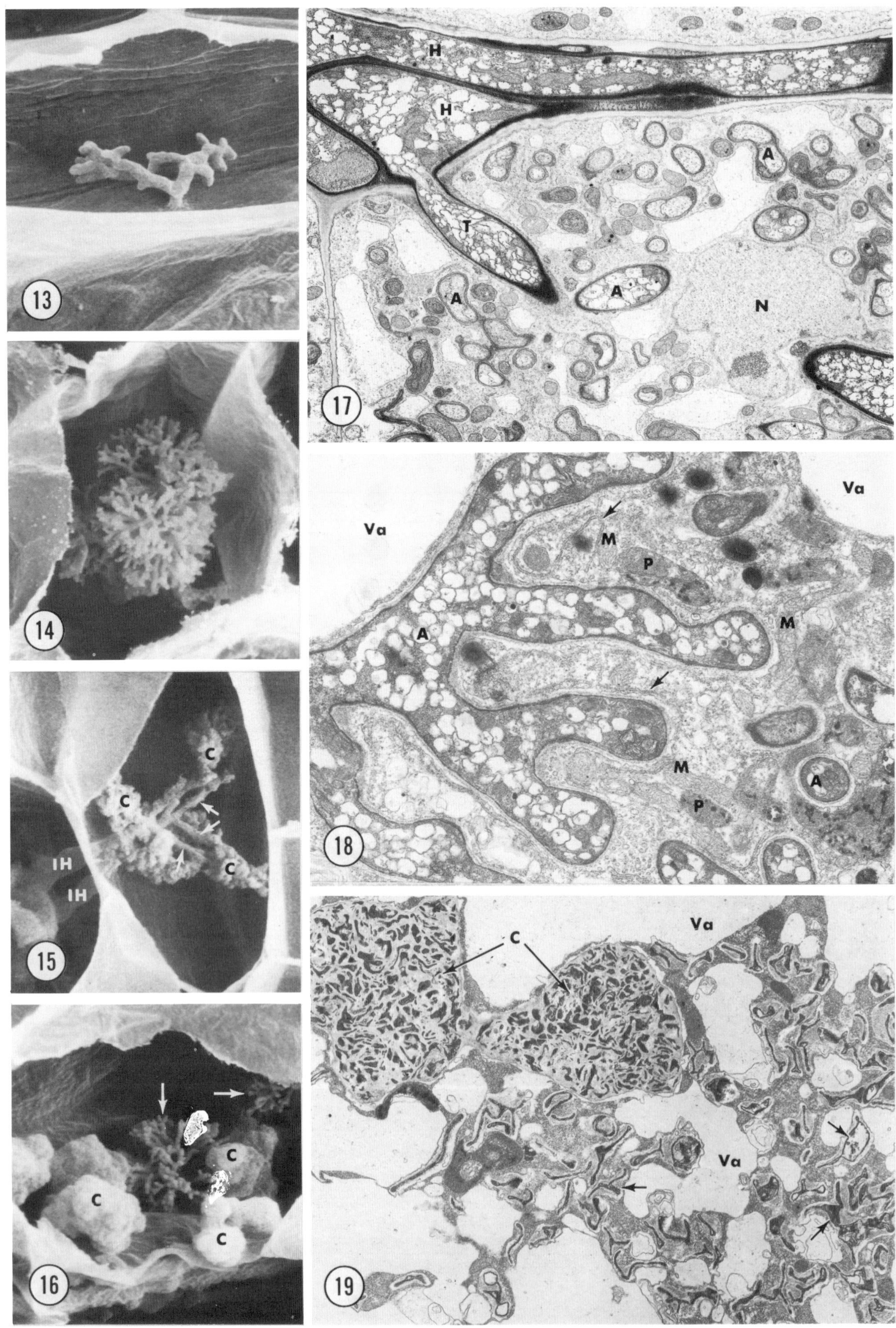
13
14
15
16
17
18
19
H
H
T
A
A
A
N
Va
Va
M
P
M
A
M
P
A
C
C
C
IH
IH
C
C
C
Va
C
Va

Arbuscules contain dense, nonvacuolated cytoplasm during early stages of development (Fig. 9) and vacuolation increases with maturity (Fig. 12, 17, 18). Disorganization of the fungal cytoplasm and vacuoles is subsequently accompanied by collapse of arbuscular branches and aggregation of them into dense, irregular masses (Fig. 10, 19). The breakdown of arbuscules may be initiated simultaneously in more than one portion of the branching system, but it proceeds toward the trunk until the entire arbuscule aggregates as a compact, lobed mass near the original point of penetration (5,15). These dense aggregations are composed of highly distorted and compressed arbuscular walls, remnants of entrapped host cytoplasm, and a substantial amount of interfacial material which is deposited exterior to the host plasmalemma. Interfacial material is present at all stages of arbuscular development, but it is most pronounced in association with deteriorated and aggregated portions of the structure. The host-fungal interface and variations in the interfacial material will be discussed in greater detail in a later chapter concerned with ultrastructural techniques. More than one arbuscule may develop concurrently within a single host cell (Fig. 15) and new ones may form in cells containing degraded arbuscules (Fig. 16) in some, perhaps most, hosts. The sequence of arbuscular development and breakdown in G. mosseae/yellow poplar mycorrhizae is shown by SEM in Fig.13-16 (12). Throughout this sequence, which is estimated to involve only 4 to 15 days (2,6) the arbuscule is surrounded by an intact host plasmalemma and an increased amount of host cytoplasm containing elevated numbers of host organelles (Fig. 17-19) (5,14,15,25 and others). The host nucleus becomes enlarged concomitantly with the first stages of arbuscular development (Fig. 9) and is often located among the arbuscular branches at later stages (Fig. 17) rather than along the host wall. The nucleus reverts to its normal size and position when the arbuscule has deteriorated fully and, concurrently, the number of host organelles returns to a level comparable to uninfected cortical cells.

Vesicles produced within infected roots, such as those shown in Fig. 11 and 12 may be inter- or intracellular, terminal or intercalary. They are often thick-walled and otherwise comparable to the chlamydospores produced by the species involved or they may be irregularly shaped and thin-walled, bearing little similarity to chlamydospores (8). The cytoplasm of vesicles in early stages of development is moderately dense, multinucleate and contains many

---

Fig. 13-16. SEM micrographs of arbuscules of G. mosseae in yellow poplar. Host cytoplasmic components have been removed to expose the arbuscules (12). Fig. 13. Early stage of arbuscular development showing dichotomous growth of the first branches. x2,000. Fig. 14. Fully developed arbuscule with numerous bifurcate terminal branchlets. x1,000. Fig. 15. Two arbuscules produced in the same host cell from separate intracellular hyphae (IH). The outer portions of the arbuscular branching systems have collapsed and aggregated into irregular clumps (C) while the main branches (arrows) remain intact. x1,000. Fig. 16. Residual clumps (C) of several fully deteriorated arbuscules and two small, but newly formed arbuscules (arrows) similar in structure to that shown in Fig. 14. x1,500. Fig. 17-19. TEM micrographs of arbuscules in host cortical cells. Fig. 17. G. caledonius/soybean showing the arbuscular trunk (T) produced from an intercellular hypha (H) and many vacuolate arbuscular branches (A) enclosed by host cytoplasm. Host nucleus (N). x5,200. Fig. 18. G. mosseae/yellow poplar. Portion of a vacuolate arbuscule (A) prior to deterioration and associated host organelles. Mitochondria (M), plastids (P), endoplasmic reticulum (arrows), host vacuole (Va). x8,700. Fig. 19. G. mosseae/yellow poplar. Portion of a deteriorating arbuscule with many collapsed branches throughout the lower right portion of the micrograph (arrows) and two clumps (C) of aggregated arbuscular branches comparable to those shown in Fig. 15. The clumps are composed of dense compacted arbuscular branches, host cytoplamic residues and lighter interfacial material. Host vacuole (Va). x5,700.

small lipid droplets as well as glycogen particles (11, M.F. Brown, unpublished data). The cytoplasm becomes more dense, apparently through condensation, and organelles become more difficult to define ultrastructurally as accumulation of lipids increases during maturation. At maturity, most of the volume of the vesicle is occupied by conspicuous lipid globules. Vesicles which form intracellularly are enclosed by host cytoplasm, only slightly greater in volume than uninfected cortical cells, the host plasmalemma and a compact layer of interfacial material (11,13). Typically vesicles are produced most abundantly in the outer cortical region of older infection units but, in some instances (Fig. 11), they appear to form rather prolifically without extensive prior development of arbuscules.

Although a substantial amount of research has been performed involving morphological studies of a small number of VAM fungal species and the infections they produce in certain hosts, most host-VAM endophyte combinations have not been examined comprehensively in this way. Additional critical studies at the light and ultrastructural levels would contribute significantly to an increased understanding of variations in developmental patterns and morphological characteristics exhibited by these symbionts.

Acknowledgements

We thank D.L. Pinkerton for performing the photographic work, D.A. Kinden for Figs. 8, 13-16, 18, 19, and J.A. White for reviewing the manuscript.

LITERATURE CITED

1. Bevege, D.I. 1968. A rapid technique for clearing tannins and staining intact roots for detection of mycorrhizas caused by Endogone spp., and some records of infection in Australasian plants. Trans. Br. Mycol. Soc. 51:808-811.
2. Bevege, D.I. and Bowen, G.D. 1975. Endogone strain and host plant differences in development of vesicular-arbuscular mycorrhizas. Pages 77-86 in: Sanders, F.E., Mosse, B. and Tinker, P.B. eds. Endomycorrhizas. Academic Press, London. 626 pp.
3. Carling, D.E., White, J.A., and Brown, M.F. 1977. The influence of fixation procedure on the ultrastructure of the host endophyte interface of vesicular-arbuscular mycorrhizae. Can. J. Bot. 55:48-51.
4. Carling, D.E. and Brown, M.F. 1982. Anatomy and physiology of vesicular-arbuscular and nonmycorrhizal roots. Phytopathology 72: (In press).
5. Cox, G. and Sanders, F.E. 1974. Ultrastructure of the host fungus interface in a vesicular-arbuscular mycorrhiza. New Phytol. 73:901-912.
6. Cox, G., and Tinker, P.B. 1976. Translocation and transfer of nutrients in vesicular-arbuscular mycorrhizas. I. The arbuscule and phosphorus transfer: a quantitative ultrastructural study. New Phytol. 77:371-378.
7. Gerdemann, J.W. 1965. Vesicular-arbuscular mycorrhizae formed on maize and tuliptree by Endogone fasiculata. Mycologia 57:562-575.
8. Gerdemann, J.W. 1968. Vesicular-arbuscular mycorrhiza and plant growth. Annu. Rev. Phytopathol. 6:397-418.
9. Gerdemann, J.W. and Trappe, J.M. 1974. The Endogonaceae in the Pacific Northwest. Mycol. Memoir No. 5. New York Botanical Garden and the Mycological Society of America. New York. 76 pp.
10. Hall, I.R. and Fish, B.J. 1979. A key to the Endogonaceae. Trans. Br. Mycol. Soc. 73:261-270.
11. Holley, J.E. and Peterson, R.L. 1979. Development of a vesicular-arbuscular mycorrhiza in bean roots. Can. J. Bot. 57:1960-1978.

12. Kinden, D.A. and Brown, M.F. 1975. Electron microscopy of vesicular-arbuscular mycorrhizae of yellow poplar. I. Characterization of endophytic structures by scanning electron stereoscopy. Can. J. Microbiol. 21:989-993.
13. Kinden, D.A. and Brown, M.F. 1975. Electron microscopy of vesicular-arbuscular mycorrhizae of yellow poplar. II. Intracellular hyphae and vesicles. Can. J. Microbiol. 21:1768-1780.
14. Kinden, D.A. and Brown, M.F. 1975. Electron microscopy of vesicular-arbuscular mycorrhizae of yellow poplar. III. Host-endophyte interactions during arbuscular development. Can. J. Microbiol. 21:1930-1939.
15. Kinden, D.A. and Brown, M.F. 1976. Electron microscopy of vesicular-arbuscular mycorrhizae of yellow poplar. IV. Host-endophyte interactions during arbuscular deterioration. Can. J. Microbiol. 22:64-75.
16. King, E.J., Schubert, T.S. and Brown, M.F. 1981. Techniques for developmental studies of VA mycorrhizae. Page 46 in: Program and Abstracts, Fifth North American Conference on Mycorrhizae. Universite Laval. Que. 83 pp.
17. Kormanik, P.P., Bryan, W.C. and Schultz, R.C. 1980. Procedures and equipment for staining large numbers of plant root samples for endomycorrhizal assay. Can. J. Microbiol. 26:536-538.
18. Mosse, B. 1959. Observations on the extra-matrical mycelium of a vesicular-arbuscular endophyte. Trans. Br. Mycol. Soc. 42:439-448.
19. Nemec, S. 1981. Histochemical characteristics of Glomus etunicatus infection of Citrus limon fibrous roots. Can. J. Bot. 59:609-617.
20. Nicolson, T.H. 1959. Mycorrhiza in the Gramineae. I. Vesicular-arbuscular endophytes, with special reference to the external phase. Trans. Br. Mycol. Soc. 42:421-438.
21. Nicolson, T.H. and Schenck, N.C. 1979. Endogonaceous mycorrhizal endophytes in Florida. Mycologia 71:178-198.
22. Phillips, J.M. and Hayman, D.S. 1970. Improved procedures for clearing roots and staining parasitic and vesicular-arbuscular mycorrhizal fungi for rapid assessment of infection. Trans. Br. Mycol. Soc. 55:158-161.
23. Rhodes, L.H. and Gerdemann, J.W. 1980. Nutrient translocation in vesicular-arbuscular mycorrhizae. Pages 173-195 in: Cook, C.B., Pappas, P.W., and Rudolph, E.D., eds. Cellular interactions in symbiosis and parasitism. Ohio State Univ. Press, Columbus. 305 pp.
24. Scannerini, S., Bonfante, P.F. and Fontana, A. 1975. An ultrastructural model for the host-symbiont interaction in the endotrophic mycorrhizae of Ornithogalum umbellatum L. Pages 313-324 in: Sanders, F.E., Mosse, B., and Tinker, P.B. eds. Endomycorrhizas. Academic Press, London. 626 pp.
25. Scannerini, S. and Bonfante-Fasolo, P. 1979. Ultrastructural cytochemical demonstration of polysaccharides and proteins within the host-arbuscule interfacial matrix in an endomycorrhiza. New Phytol. 83:87-94.

B. Histology and Histochemistry

Stan Nemec

INTRODUCTION

The apparently obligate characteristics of VA fungi have required that anatomical, morphological, and some physiological processes of the host-parasite relationship be examined with histological techniques. In a strict sense, the term histology deals only with the structural organization of the endophyte in cells and tissues. This section deals with an extension of this study to include, in part, histochemical analyses. Histochemical preparations enable the identification and localization of certain classes of chemical substances. The ultimate aim of histochemistry is to describe the dynamic organization of cells and tissues in terms of their structure, composition, and function (5).

Identification and Enumeration of Infection in Roots and Soil

Numerous simple to fairly complex techniques have been developed to study infection in roots. The simplest techniques involved no more than teasing open roots to examine them for infection (29, 33). Basic staining at first was done by immersing roots in preparations of aniline blue (29), cotton blue (12), or methylene blue (8) in lactic acid, or acid fuchsin in picric acid (31). This technique was improved by staining the fungus, usually with 0.01% acid fuchsin or aniline blue in lactophenol and clearing the roots in lactophenol (9, 11, 28, 34). This technique worked well with nonpigmented roots, but those plant roots containing tannins, other deposits, and heavily cutinized epidermal cells required a more extensive clearing treatment. To accomplish this, some researchers have resorted to extracting tannins with KOH (2, 21, 32, 36), bleaching roots in H_2O_2 (21, 32, 36), and clearing them with chloral hydrate (13) or sodium hypochlorite (2) prior to staining the fungus. The technique used most frequently in recent years is the one developed by Phillips and Hayman (32). Prior to staining the fungus in the root, they extracted tannins with KOH, or, in heavily pigmented roots, used H_2O_2 bleaching after a KOH treatment. Some variations of the Phillips and Hayman technique have been developed (22, 24). Roots processed by most of the above procedures are pressed between glass slides or a cover glass and slide and examined on the stage of a microscope. Because of the nonspecificity of the stains used in these studies, the technique selected to process roots should be determined primarily by the type of roots chosen and their intensity of pigmentation.

Staining the fungus has also aided in detection of it in soil. Kessler and Blank (19) readily recognized stained *Endogone* sporocarps in soil after the soil was soaked overnight in an acidified acid fuchsin solution.

Some researchers have used fixation, tissue dehydration, and sectioning techniques to study morphology in more detail. Tandy (35) processed fungus fructifications through formal-acetic acid-alcohol (FAA), followed by dehydration in a graded alcohol series, and infiltration in paraffin. Microtomed or freehand cut sections were routinely stained in ammoniacal Congo red. Marx et al. (26) examined *Endogone mosseae* Nicol. & Gerd. infection in citrus seedlings that were fixed, embedded in paraffin, sectioned, and stained.

Organic and Inorganic Components

In recent years, particular attention has been given to the mechanisms of phosphorus uptake by the plant and utilization of host photosynthate by the fungus. It has been pointed out that VA fungal hyphae and vesicles are rich in osmiophilic and lipid material (Table 1). The lipids in vesicles and hyphae appear to be chemically similar. Positive tests with OsO_4 (Plate 2, E), the Sudan dyes (Plate 2, F and G), and the oxazone component of Nile blue (Plate 2, H) suggest the presence of triglycerides, a form of stored energy, in these structures. Lipid droplets have been detected in arbuscules with the transmission electron microscope (3, 6). The composition of lipid in hyphae and vesicles appears to differ from the lipid detected in arbuscules by histochemistry (30). Acidic lipids, probably membrane related, were associated with arbuscules, and they were detected with the oxazin component of Nile blue (Plate 3, A), the gold hydroxamic (Plate 3, B), and the acid haematein (30) tests, all of which indicate the presence of phospholipids. Synthesis of lipid by the fungal endophyte has been suggested as an alternate storage sink for the plants' photosynthates (7, 16) and possibly a growth sink for the fungus.

Polyphosphate granules have been found in arbuscules, hyphae, and vesicles of VA fungi (Table 1). Callow, et al. (4) suggested that VA fungi synthesize polyphosphate vacuolar granules from soil phosphate, and that the granules are broken down in the arbuscules to inorganic phosphate for release to the host.

Carbohydrate-positive material was detected in vesicle and hyphal cell walls and in arbuscules by the periodic acid - Schiff reagent (Plant 3, C). Chitin appears to be the main carbohydrate-related material present in vesicle and hyphal walls (Table 1). However, in arbuscules, the carbohydrate materials appear to consist of glycogen particles (3, 17) and glycolipid material in arbuscule walls (30). Azure B has been used to show that arbuscules are rich in RNA (Plate 3, D), fast green indicated that basic proteins are a very common type in fungal cytoplasm (Plate 3, F), and the Nitroso test was used to demonstrate that vesicle and hyphal walls contain tannin compounds (Plate 3, G).

Enzymatic Components

Enzymatic localization studies in tissues require that the enzymes, because they are proteins, not be denatured in the processing of tissue. Most enzymes can be localized only in living, unfixed material (18). The preferred means of preparing sections for enzymatic analysis is to cut living material on cold microtomes. Cryostats are most suitable for this kind of work.

MacDonald and Lewis (25) reported the occurrence of acid phosphatase, glutamate dehydrogenase, succinate dehydrogenase, glyceraldehyde-3-phosphate dehydrogenase, glucose-6-phosphate dehydrogenase, and NADH and NADPH diaphorases in vesicles, arbuscules, and extra-matrical mycelium of *G. mosseae*. It was inferred that the fungus possesses an Emden-Meyerhof-Parnas system, a tricarboxylic acid cycle and a hexose monophosphate shunt. Alkaline phosphatase activity specific to VA mycorrhizae has been reported in onions and tobacco (1, 14).

Gianinazzi et al. (15) showed that strong alkaline α-naphthyl phosphatase and β-glycerophosphatase activities were localized within the vacuoles of the mature arbuscule and intercellular hyphae. They considered that the vacuolar alkaline phosphatase may, perhaps, be involved in the active mechanism of phosphate transport within hyphae of VA fungi. Cortical cells infected with arbuscules have been shown to respond by producing laccase activity (Plate 3, E). Catalase and peroxidase (Plate 3, H) activity in senescing arbuscules is of particular interest because it is possible that these enzymes are associated with α- and β-oxidation of fatty acids in the arbuscule wall (30).

Table 1. Chemical constituents detected in arbuscules, hyphae, vesicles, and chlamydospores of vesicular-arbuscular mycorrhizal fungi by histochemical techniques.

Chemical	Location in Fungus				Reaction or test reagent	Reference
	Arbuscule	Hyphae	Vesicle	Chlamydospore		
Lipid (oil)				+	Sudan IV	Mosse (27)
Lipid	+	+			Sudan IV or OsO_4	Cox et al. (7)
Lipid (fat)		+	+		Acriflavine	Holley and Peterson (17)
Lipid		+			Sudan IV	Cox and Sanders (6)
Lipid	+				Uranyl acetate and lead citrate	Kinden and Brown (20)
Lipid	+	+	+		Nile blue, Sudan black B, and others	Nemec (30)
Lipid	+				Uranyl acetate and lead citrate	Bonfante-Fasolo (3)
Protein	+	+	+		Mercuric bromphenol blue, fast green, and others	Nemec (30)
Chitin	-	+ (walls)	+ (walls)		KOH-IKI	Nemec (30)
Pectin	+(clumped)	-	-		Ruthenium red	Nemec (30)
Acidic mucopoly-saccharides	+	-	-		Alcian blue	Nemec (30)
Carbohydrates	+	+	+		Periodic acid-Schiff	Nemec (30)
Polysaccharides	+				Thiery reaction	Dexheimer et al. (10)
Polysaccharides	+				Thiery reaction	Holley and Peterson (17)
Glycogen	+				Uranyl acetate and lead citrate	Kinden and Brown (20)
Polyphosphate	+	+			Toluidine blue or lead sulphide	Cox et al. (7)
Polyphosphate		+	+		Toluidine blue	Ling-Lee et al. (23)
Polyphosphate		+	+		Toluidine blue	Nemec (30)
Nucleic acids	+	-	-		Azure B	Nemec (30)
Phenolics	-	+ (walls)	+ (walls)		Nitroso reaction, and ammoniacal silver nitrate	Nemec (30)

Note: +, positive color reaction; -, no color developed in fungus; or blank space indicates no evaluation.

LITERATURE CITED

1. Bertheau, Y. 1977. Études des phosphatases solubles des endomycorrhizes à vésicules et arbuscules. D.E.A. thesis. Université de Dijon, France.
2. Bevege, D. I. 1968. A rapid technique for clearing tannins and staining intact roots for detection of mycorrhizas caused by Endogone spp. and some records of infection in Australasian plants. Trans. Br. Mycol. Soc. 51:808-810.
3. Bonfante-Fasolo, P. 1978. Some ultrastructural features of the vesicular-arbuscular mycorrhiza in the grapevine. Vitis 17:386-395.
4. Callow, J. A., Capaccio, L. C. M., Parish, G., and Tinker, P. B. 1978. Detection and estimation of polyphosphate in vesicular-arbuscular mycorrhizas. New Phytol. 80:125-134.
5. Casselman, W. G. B. 1959. Histochemical technique. John Wiley and Sons, Inc. 205 p.
6. Cox, G., and Sanders, F. 1974. Ultrastructure of the host-fungus interface in vesicular-arbuscular mycorrhiza. New Phytol. 73:901-912.
7. Cox, G., Sanders, F., Tinker, P. B., and Wild, J. A. 1975. Ultrastructural evidence relating to host-endophyte transfer in a vesicular-arbuscular mycorrhiza. Pages 297-312. In: Sanders, F. E., Mosse, B., and Tinker, P. B. eds. Endomycorrhizas. Academic Press, London. 626 p.
8. Daft, M. J., and Nicolson, T. H. 1966. Effect of Endogone mycorrhiza on plant growth. New Phytol. 65:343-350.
9. Davis, R. M., and Menge, J. A. 1981. Phytophthora parasitica inoculation and intensity of vesicular-arbuscular mycorrhizae in citrus. New Phytol. 87:705-715.
10. Dexheimer, J., Gianinazzi, S., and Gianinazzi-Pearson, V. 1979. Ultra-structural cytochemistry of the host-fungus interfaces in the endomycorrhizal association Glomus mosseae/Allium cepa. Zeit für Pflanzenphysiologie 92:191-206.
11. Furlan, V., and Fortin, J. Andre. 1973. Formation of endomycorrhizae by Endogone calospora in Allium cepa under three temperature regimes. Naturaliste Can. 100:467-477.
12. Gallaud, I. 1905. Etudes sur les mycorrhizes endotrophes. Rev. gen de Bot. XVII.
13. Gerdemann, J. W. 1965. Vesicular-arbuscular mycorrhizae formed on maize and tuliptree by Endogone fasciculatus. Mycologia 57:562-575.
14. Gianinazzi-Pearson, V., and Gianinazzi, S. 1976. Enzymatic studies on the metabolism of vesicular-arbuscular mycorrhiza. I. Effect of mycorrhiza formation and phosphorus nutrition on soluble phosphatase activities in onion roots. Physiol. Veg. 14:833-841.
15. Gianinazzi, S., Gianinazzi-Pearson, V., and Dexheimer, J. 1979. Enzymatic studies on the metabolism of vesicular-arbuscular mycorrhiza. III. Ultra-structural localization of acid and alkaline phosphatase in onion roots infected by Glomus mosseae (Nicol. & Gerd.). New Phytol. 82:127-132.
16. Harley, J. L. 1975. Problems in mycotrophy. Pages 1-24. In: Sanders, F. E., Mosse, B., and Tinker, P. B. eds. Endomycorrhizas. Academic Press, London. 626 p.
17. Holley, J. D., and Peterson, R. L. 1979. Development of a vesicular-arbuscular mycorrhiza in bean roots. Can. J. Bot. 57:1960-1978.
18. Jensen, W. A. 1962. Botanical histochemistry. W. H. Freeman and Co. 408 p.
19. Kessler, K. J., and Blank, R. W. 1972. Endogone sporocarps associated with sugar maple. Mycologia 64:634-638.

20. Kinden, D. A., and Brown, M. F. 1975. Electron microscopy of vesicular-arbuscular mycorrhizae of yellow poplar. III. Host-endophyte interactions during arbuscular development. Can. J. Microbiol. 21:1930-1939.

21. Kormanik, P. P., Bryan, W. C., and Schultz, R. C. 1980. Procedures and equipment for staining large numbers of plant root samples for endomycorrhizal assay. Can. J. Microbiol. 26:536-538.

22. Lindermann, R. G., and Call, C. A. 1977. Enhanced rooting of woody plant cuttings by mycorrhizal fungi. J. Amer. Soc. Hort. Sci. 102:629-632.

23. Ling-Lee, M., Chilvers, G. A., and Ashford, A. E. 1975. Polyphosphate granules in three different kinds of tree mycorrhiza. New Phytol. 75:551-554.

24. Luedders, V. D., Carling, D. E., and Brown, M. F. 1979. Effect of soybean plant growth on spore production by _Glomus mosseae_. Plant and Soil 53: 393-397.

25. MacDonald, R. M., and Lewis, M. 1978. The occurrence of some acid phosphatases and dehydrogenases in the vesicular-arbuscular mycorrhizal fungus, _Glomus mosseae_. New Phytol. 80:135-141.

26. Marx, D. H., Bryan, W. C., and Campbell, W. A. 1971. Effect of endomycorrhizae formed by _Endogone mosseae_ on growth of citrus. Mycologia 63: 1222-1226.

27. Mosse, B. 1970. Honey-coloured, sessile _Endogone_ spores: II. Changes in fine structure during spore development. Arch. Mikrobiol. 74:129-145.

28. Mosse, B., and Hepper, C. 1975. Vesicular-arbuscular mycorrhizal infections in root organ cultures. Physiol. Plant Pathol. 5:215-223.

29. Neill, J. C. 1944. Rhizophagus in citrus. New Zealand J. Sci. Technol. 25:191-201.

30. Nemec, S. 1981. Histochemical characteristics of _Glomus etunicatus_ infection of _Citrus limon_ fibrous roots. Can. J. Bot. 59:609-617.

31. O'Brien, D. G., and McNaughton, E. J. 1928. The endotrophic mycorrhiza of strawberries and its significance. The West of Scotland Agric. College Research Bull. No. 1. 35 p.

32. Phillips, J. M., and Hayman, D. S. 1970. Improved procedures for clearing roots and staining parasitic and vesicular-arbuscular mycorrhizal fungi for rapid assessment of infection. Trans. Br. Mycol. Soc. 55:158-161.

33. Redhead, J. F. 1977. Endotrophic mycorrhizas in Nigeria: species of the Endogonaceae and their distribution. Trans. Br. Mycol. Soc. 69:275-280.

34. Schenck, N. C., and Schroder, V. N. 1974. Temperature response of _Endogone_ mycorrhiza on soybean roots. Mycologia 66:600-605.

35. Tandy, P. A. 1975. Sporocarpic species of Endogonaceae in Australia. Aust. J. Bot. 23:849-866.

36. Williams, S. E. 1979. Vesicular-arbuscular mycorrhizae associated with actinomycete-nodulated shrubs, _Cercocarpus montanus_ Raf. and _Purshia tridentata_ (Pursh.) D.C. Bot. Gaz. 140:115-119 (Suppl. S).

METHODS FOR THE RECOVERY AND QUANTITATIVE ESTIMATION OF PROPAGULES FROM SOIL

B. A. DANIELS and H. D. SKIPPER

INTRODUCTION

Over the years numerous attempts have been made to monoculture vesicular-arbuscular mycorrhizal (VAM) fungi on artificial media (15). These attempts, however, have met with little success. Thus, VAM fungi must still be collected from field soils, increased in the greenhouse on living host plants (nurse crops) and collected once again from these 'pot culture' soils. Clearly, the production and collection of propagules from soil remains an important, time-consuming and fundamental part of VAM research, both for inoculum preparation and population estimation.

Propagules of VAM fungi can consist of chlamydospores or azygospores, soil-borne vesicles and mycelium or infected root pieces. Used together, as they occur in soil, these propagules may be termed mixed inoculum as compared with spores which have been separated from soil and represent a 'purer' inoculum. Before a propagule recovery technique can be selected, the desired form of propagules must be determined. Either 'mixed' or spore inoculum may be used but there are advantages and disadvantages associated with each form.

Mixed inoculum has repeatedly given faster, more reliable infection than spore inoculum alone. However, if the inoculum is dried and stored for even 2 weeks, the infection rate drops, presumably because only spores survived (11). Thus, if mixed inoculum is not used under optimum conditions, the benefit (increased inoculum potential) of mixed inoculum is lost. Inoculum potential is defined here as "the energy of growth of a parasite available for infection of a host at the surface of the host organ to be infected" (8). Some VAM species such as Acaulospora trappei produce such small spores that they are easily lost or overlooked with our present extraction methods. Certain VAM species form no spores in nature, perhaps as an adaptation to clement climates (2). In these cases use of 'mixed' inoculum may result in better infection or a more accurate assessment of the mycorrhizal population contained in a given soil than might occur if only extracted spores were used.

In contrast, with extracted spores one can ensure that only the desired species is present, assess spore viability, and reduce the possibility that a plant pathogen or a VAM hyperparasite is carried in the inoculum. Thus the type of inoculum to be used and the extraction procedure should be carefully considered.

The use of spore inoculum has another advantage. It is far easier to quantify and, therefore, to standardize inoculum dosage. However, if the innate ability of species to increase plant growth is to be compared, equivalent inoculum potentials should be applied to each plant. Otherwise, the ability of VAM species to infect plants cannot be separated from their stimulating effect on plant growth once inside the plant. Simply equalizing spore numbers cannot guarantee equal inoculum potentials (4). For these types of comparisons inoculum potential can be assessed using most probable number methods and the inoculum dosage subsequently standardized (4).

Mixed inoculum is far more difficult to quantify. It has long been customary to inoculate plants by adding an equal volume of mixed inoculum into sterilized soil. The number of infective propagules and condition of the inoculum cannot thus be determined. If the mycorrhizal population in various soils or the growth-promoting ability of VAM fungal species is to be accurately compared, most probable number methods must again be used.

This chapter will describe the techniques currently in use by VAM researchers for propagule recovery and quantification in soil.

Recovery of Propagules from Soil

Numerous techniques are used to recover VAM propagules from soil. The most basic of these is wet sieving and decanting (10) to remove the clay and sand fractions of the soil while retaining spores and other similar-sized soil and organic matter particles on sieves of various sizes. This technique is relatively fast but further purification of the spores is usually necessary particularly if spore numbers in a soil are low. In addition, dense spores or sporocarps may be lost along with the sand fraction.

The use of elutriators to mechanize the sieving and decanting process for nematodes and some soil fungi is common if large quantities of soil are to be sieved and/or large quantities of spores are needed (3, 19). Mass production and extraction of VAM spores will be essential for applied phases. Tommerup and Carter (25) have suggested a dry separation technique for microorganisms from soil. Soil samples are comminuted, sieved and subsequently fractionated in a dry particle size analysis elutriator. The use of this technique for separation of VAM spores is new and needs further evaluation. Therefore, it will not be further detailed in this manual.

Most methods which further purify spores still require prior sieving and decanting. An exception is the plate method of Smith and Skipper (23) wherein small quantities of soil and water are mixed and examined directly. This method is fast but accurate only if spore numbers exceed 20 spores/g of soil. Its efficiency in a high organic matter or clay soil is untested. Similarly, the adhesion-flotation method (24) requires that soil be mixed in a graduated cylinder in water and poured through a separatory funnel where the spores adhere to the sides. This technique is useful only with small quantities of soil and spores which rapidly sink in water would be lost.

Following wet sieving and decanting, Mosse and Jones (16) demonstrated that spores could be collected by differential sedimentation on gelatin columns. While this technique indeed purifies the spores, it is a time-consuming process, suitable only for small quantities of soil, and some propagules may be lost. In the flotation-bubbling system (7), soil sievings are mixed in glycerol after which the spores remain in solution while heavier soil particles settle to the bottom of a glass column. This procedure is time-consuming but somewhat larger amounts of soil can be processed. Lighter debris, however, remains with the spores and heavier spores can settle out and be lost. Furlan et al (6) suggests that the flotation-bubbling technique be followed by a density gradient centrifugation to further purify the spores.

Density gradient centrifugation is now the most commonly used technique for VAM spore extraction. Its success, however, requires that a swinging bucket, horizontal head be used on the centrifuge. Sucrose gradients were suggested by Ohms (17), Ross and Harper (22), and Mertz et al (13). Furlan et al (6) demonstrated that radiopague media could also be used for gradients, reducing the osmotic shock which can occur if spores remain in sucrose too long. Gradients can also be made from 10, 20 and 30% Ficoll (Sigma Chemical Co., St. Louis, MO) when particularly clean preparations are desired (Daniels, unpublished). Each of these techniques requires prior sieving and decanting. However, density gradient centrifugation may offer the best compromise in choice of extraction techniques. Large or small quantities of soil may be processed rapidly with reasonable efficiency and little debris remains with the spores. Alternatively, a sucrose centrifugation techique which does not require gradients can be used to extract

VAM spores (12). While these extraction procedures vary in sensitivity, the spore number of certain species may be too low to be detected. In these cases, if all the species present in a soil are to be determined, host baiting techniques should be employed. A soil sample and/or roots from the test soil are added to a greenhouse potting mix and planted to a 'bait' plant. After 3 to 5 months the soil and roots from this pot can be sieved and the species present can be determined.

Wet Sieving and Decanting, Adapted from Gerdemann (9).

A. If the sievings themselves will be examined microscopically:

1. Mix a volume of soil (250 cc) in water (1000 ml) and allow heavier particles to settle for a few seconds.
   NOTE: If the number of spores contained in various soils is to be compared, the volume of soil sieved should be correlated to the dry weight of that soil sample.

2. Pour liquid through a coarse soil-sieve (500-800 m) to remove large pieces of organic matter. Collect the liquid which passes through this sieve. Wash the sieve in a stream of water to ensure that all small particles have passed through.

3. Resuspend the particles in the liquid which passed through the coarse sieve and allow heavier particles to settle for a few seconds.

4. Pass this suspension through a sieve fine enough to retain the desired spores, generally 38-250 m. (If total population is to be assessed, the finest available sieve should be employed.)

5. Wash the material retained on the sieve to ensure that all colloidal material passes through the sieve.

6. Transfer small amounts of the remaining debris to a petri dish and examine under a dissecting microscope.
   NOTE: A nest of varied-sized sieves may be used instead of steps 2, 3, and 4. If a nest of sieves is employed, care must be taken to ensure that the finest sieve does not overflow, losing sievings. If the bottom sieve does 'clog up' either a strong stream of water or patting the side of the sieve will often unclog it.

B. If the sievings are to be further extracted, the sieving technique may be shortened:

1. Mix a volume of soil in water and allow the heavier particles to settle for a few seconds.

2. Pour the liquid through a sieve fine enough to retain the desired spores. (If total population is to be assessed, the finest available sieve should be employed.)

3. Rewet the material which remained in the bucket and repeat steps 1 and 2.

4. Wash thoroughly the material retained on the sieve to ensure that all small particles pass through.

Flotation Bubbling (7).

1. Assemble a glass tube (55 cm X 6 cm i.d.) with a fritted disc (pore

size 4-5.5 m) 5 cm from the bottom which narrows to permit a compressed air connection. A screw clip is attached to the tubing through which the the compressed air passes to prevent leakage through the fritted disc.

2. Compressed air (12 psi, 0.84 kg/cm^2) is passed through the fritted disc into the column prior to adding one liter glycerol solution. Adjust the air pressure to 15 psi (1.05 kg/cm^2).

3. Add soil sievings to the glycerol and continue to bubble air through for 1-2 min.*

4. Tighten the screw clip and turn off the air supply. Allow suspension to settle 30 min.

5. Draw off supernatant with vacuum attached through a trap to a glass tube held just below the surface of the liquid.*

6. Rinse siphon tube by drawing through 100 ml of clean glycerol. Wash supernatant through a sieve fine enough to retain the desired spores.

*Soil particles adhering to the column sides should be washed down with glycerol.

<u>Density Gradient Centrifugation</u>.

A. First Alternative (17)

1. Add 10 ml of 50% sucrose to a clear 50 ml centrifuge tube.

2. Add a similar layer of 25% sucrose onto the 50% layer and add a layer of water above that using a hypodermic syringe with a curved glass tube extension to direct the liquid against the side of the tube.

3. Add a suspension of sievings and centrifuge 5 min at 3100 rpm (approx. 1100 X g).

4. Remove debris which collects in the middle layer to a fine sieve and wash with water.

B. Second Alternative (Developed in the laboratory of Dr. J. A. Menge, Univ. Calif., Riverside)

1. Transfer sievings to a blender and blend at high speed for 1 to 2 min. This frees any spores still attached to roots, contained in sporocarps or caught in the clay particles. Pour contents of blender through a fine sieve and wash the colloidal material through with a strong stream of water.

2. Using a manostat minipet deliver a 10 ml layer of 20% sucrose into a clear 50 ml centrifuge tube. Slowly, deliver a 10 ml layer of 40% and then a 10 ml layer of 60% sucrose into the bottom of the tube.

3. Add 10-15 ml of blended sievings onto the surface of the 20% sucrose.

4. Centrifuge 3 min at 3000 rpm (approx. 1,000 X g).

5. Remove debris which gathers at the 20-40% and/or 40-60% interfaces. Often the layer of spores is visible and can be removed without taking any of the debris which remained in solution.

6. Rinse spores on a fine sieve with a strong stream of water to remove sucrose.
NOTE: Ross and Harper (22) have suggested the use of 40, 50, 60, and 70% sucrose to form gradients and then centrifuging at 1500 rpm (approx. 250 X *g*) for 3 min.

C. Third Alternative (13)

1. Layer sievings on a large gradient (600 ml H_2O over 200 ml sucrose) in a 1 liter beaker and allow to settle by gravity.

2. Retrieve the spores and debris which collect at the sucrose-water interface by vacuum aspiration. Rinse in cold H_2O.

3. Centrifuge for 1-5 min at 1600 X *g* on a second gradient (15 ml H_2O over 20 ml sucrose in a 50 ml centrifuge tube).
NOTE: Mertz et al (13) suggest that roots and pot culture soil not be sieved until several months after plant death to allow root decay and minimize the number of root fragments in the preparation. If the sievings are blended as in Alternative 2 or if Ficoll is used for gradient preparation, root fragments will be minimized without the several month delay for root decay.

Sucrose Centrifugation (12).

1. Place a 100 to 500 cc soil sample on a 20-mesh screen and wash with water to remove large debris. Collect sieved soil and water in container.

2. Mix the sieved soil by stirring and allow mixture to stand for 30 sec. Decant through a 270 or 325-mesh sieve and collect the residue from the sieve in a beaker. Add water to the sieved soil and repeat decantation. Discard the remaining soil.

3. Transfer the collected residue into two 50-ml centrifuge tubes and centrifuge for 4 to 5 min at 1750 rpm (approx. 325 X *g*) in a horizontal rotor.

4. Decant the supernatant liquid carefully and resuspend pellet in a sucrose solution (454 g cane sugar/liter). Again centrifuge for 0.5 to 1 min.

5. Pour the supernatant (with spores) onto a 325-mesh sieve and rinse with water to remove the sugar.

Quantitative Estimation of Propagules from Soil

There are several methods by which the propagules in soil can be measured. Extracted spores in water can be pipetted into an eelworm counting slide (Hawksley & Sons Ltd., 12 Peter Rd., Lancing, West Sussex BN 158th, Great Britain) which is similar to a hemocytometer but has a 1 ml capacity. The slide is etched into rectangles or can be specially ordered with a rectangle divided into 30 parallel lines. Spores/ml can be calculated by counting the spores contained in a portion of the slide and multiplying (John Menge, personal communication). For large samples this is the most practical and easiest method yet described. For soils with high spore populations (20 spores/g) such as those in which pot cultures are to be compared, the plate method (23) is suitable. If species of mycorrhizae are to be compared or the viability and infectivity of spores is to be assessed, this can be done using the most probable number methods (4). This method may more accurately reflect the inoculum potential of spores than would simple spore counts.

If all mycorrhizal propagules in soil, i.e. spores, infected root pieces and mycelium are to be estimated the procedure is more complex. Moorman and Reeves (14) described a bioassay where soils were diluted with a 1:1:1 mixture of autoclaved perlite, vermiculite and sand. Surface-sterilized corn was planted in pots containing these soils and a number of replicate pots were harvested at 30, 60 and 90 days after planting. The relative amount of inoculum in the soils was assessed by observing the amount of mycorrhizal infection in the host plant roots. The assumption inherent in this technique is that all mycorrhizal species will develop equally well on a particular host. Estimation of propagules using most probable number techniques (20, 21) may avoid this assumption by only assessing whether or not infection occurs, not how much infection occurs.

Use of Most Probable Number to Assess VAM Population in Soils, Adapted from Porter (20) and Powell (21).

1. Make a ten-fold series of soil dilutions with the test soil to 10^{-3}, 10^{-4}, or 10^{-5} using steamed, autoclaved or fumigated soil as the diluent.

2. Place soil in small vials (20-100 cc), using 5 replicate vials per dilution.

3. Sow seeds or plant pregerminated seedlings of a test plant such as clover into each vial.

4. Grow plants in greenhouse or growth chamber for 6 weeks. Wash, clear, and stain roots (18), then examine microscopically. Determine whether infection is present or absent.

5. The most probable number of VAM propagules can then be calculated using Table VIII_2 (5) or Table 100-1 (1).

Plate Method For Population Studies (23).

1. Add one gram of soil to 9.0 ml of distilled water and shake vigorously.

2. Immediately pipet 1.0 ml in parallel lines onto a 9-cm, filter paper disc in a petri dish.

3. Spores can be counted (either wet or after air drying) under a dissecting microscope (7 to 30X).

SUMMARY

A number of procedures are available to extract and quantify VAM spores from soils. As noted, some such as elutriators or the dry sieving apparatus are suitable for large scale studies and others are limited to small quantities or to soils with high spore populations. For comparisons between or across soils, researchers, crops, etc., spore populations should be expressed on dry weight units of soil rather than volume.

LITERATURE CITED

1. Alexander, M. 1965. Most probable number method for microbial populations. In: C. A. Black, ed. Methods of soil analysis, Part 2, Chemical and microbiological properties. American Soc. Agron., Madison, Wis. pp. 1467-1472.
2. Baylis, G. T. S. 1969. Host treatment and spore production by Endogone. N. Z. J. Bot. 7:173-174.

3. Byrd, D. W., Jr., Barker, K. R., Ferris, H., Nusbaum, C. J., Griffin, W. E., Small, R. H. and Stone, C. A. 1976. Two semi-automatic elutriators for extracting nematodes and certain fungi from soil. J. Nematol. 8:206-212.
4. Daniels, B. A., McCool, P. M. and Menge, J. A. 1982. Comparative inoculum potential of spores of six vesicular-arbuscular mycorrhizal fungi. New Phytol. 89:385-391.
5. Fisher, R. A. and Yates, F. 1963. Statistical tables for biological, agricultural and medical research. Oliver and Boyd: Edinburgh, p. 146.
6. Furlan, V., Bartschi, H.and Fortin, J. A. 1980. Media for density gradient extraction of Endomycorrhizal spores. Trans. Br. Mycol. Soc. 75:336-338.
7. Furlan, V. and Fortin, J. A. 1975. A flotation-bubbling system for collecting Endogonaceae spores from sieved soil. Naturaliste Can. 102:663-667.
8. Garrett, S. D. 1970. Pathogenic root infecting fungi. Cambridge University Press, Cambridge, Great Britain, p. 9.
9. Gerdemann, J. W. 1955. Relation of a large soil-borne spore to phytomycetous mycorrhizal infections. Mycologia 47:619-632.
10. Gerdemann, J. W. and Nicolson, T. H. 1963. Spores of mycorrhizal *Endogone* extracted from soil by wet sieving and decanting. Trans. Br. Mycol. Soc. 46:235-244.
11. Hall, I. R. 1979. Soil pellets to introduce vesicular-arbuscular mycorrhizal fungi into soil. Soil Biol. Biochem. 11:85-86.
12. Jenkins, W. R. 1964. A rapid centrifugal-flotation technique for separating nematodes from soil. Plant Dis. Rep. 48:692.
13. Mertz, S. M., Heithaus, J. J. and Bush, R. L. 1979. Mass production of axenic spores of the endomycorrhizal fungus *Gigaspora margarita*. Trans. Br. Mycol. Soc. 72:167-169.
14. Moorman, T. and Reeves, F. B. 1979. The role of Endomycorrhizae in revegetation practices in the semi-arid West. II. A bioassay to determine the effect of land disturbance on Endomycorrhizal populations. Amer. J. Bot. 66:14-18.
15. Mosse, B. 1973. Advances in the study of vesicular-arbuscular mycorrhiza. Annu. Rev. Phytopathol. 11:171-196.
16. Mosse, B. and Jones, G. W. 1968. Separation of *Endogone* species from organic soil debris by differential sedimentation on gelatin columns. Trans. Br. Mycol. Soc. 51:604-608.
17. Ohms, R. E. 1957. A flotation method for collecting spores of a phycomycetous mycorrhizal parasite from soil. Phytopathology 47:751-752.
18. Phillips, J. M. and Hayman, D. S. 1970. Improved procedure for clearing roots and staining parasitic and vesicular-arbuscular mycorrhizal fungi for rapid assessment of infection. Trans. Br. Mycol. Soc. 55:158-161.
19. Phipps, P. M., Beute, M. K. and Barker, K. R. 1976. An elutriation method for quantitative isolation of *Cylindrocladium crotalariae* microsclerotia from peanut field soil. Phytopathology 66:1255-1259.
20. Porter, W. M. 1979. The 'most probable number' method for enumerating infective propagules of vesicular arbuscular mycorrhizal fungi in soil. Aust. J. Soil Res. 17:515-519.
21. Powell, C. L. 1980. Mycorrhizal infectivity of eroded soils. Soil Biol. Biochem. 12:247-250.
22. Ross, J. P. and Harper, J. A. 1970. Effect of *Endogone* mycorrhiza on soybean yields. Phytopathology 60:1552-1556.
23. Smith, G. W. and Skipper, H. D. 1979. Comparison of methods to extract spores of vesicular-arbuscular mycorrhizal fungi. Soil Sci. Soc. Am. J. 43:722-725.
24. Sutton, J. C. and Barron, G. L. 1972. Population dynamics of *Endogone* spores in soil. Can. J. Bot. 50:1909-1914.
25. Tommerup, I. C. and Carter, D. J. 1982. Dry separation of microorganisms from soil. Soil Biol. Biochem. 14: (In Press).

QUANTIFICATION OF VESICULAR-ARBUSCULAR MYCORRHIZAE IN PLANT ROOTS

P. P. Kormanik and A.-C. McGraw

INTRODUCTION

Vesicular-arbuscular (VA) mycorrhizal fungi have received considerable attention in recent years. Plant scientists have tried to clarify and describe the complex processes by which plant growth is affected by these symbiotic fungi. Until recently, the majority of workers involved in VA fungal research had extensive training in either plant pathology or mycology and were familiar with biological techniques used in assaying material for fungal colonization. These mycologically trained workers could easily adapt standardized mycological assay procedures to suit specific needs for research on VA mycorrhizal fungi. Currently, many plant scientists initiating studies involving VA mycorrhizal fungi have little on no formal training in mycological staining techniques, so there is a need for standard assay procedures that require minimum modification. Such procedures have an additional benefit over a myriad of modifications — scientists working with VA mycorrhizal fungi/plant interactions will have a reference with which to more readily compare research studies.

Whenever there is a mycorrhizal variable in an experiment, quantification is needed to determine the degree and intensity of root colonization of the plants. One cannot assume, even when endomycorrhizal inoculum from pot cultures are added to properly fumigated and well aerated soil, that plants grown in newly infested soils will become colonized. If field soil represents a part of the growth medium mixture, a VA mycorrhizae assay is desirable to determine if treatment response is affected by sporatic mycorrhizal colonization among treatments or plots, even when an endomycorrhizal variable is not an integral part of the plant nutrient study. The primary purpose of any assay for VA mycorrhizae is to establish whether roots are colonized and to determine the degree of development of mycorrhizae within the root system. In addition, depending on the purpose of a specific study, quantification of colonization may necessitate enumeration of the fungal morphological structures (vesicles, arbuscules, spores, mycelium, pelotons) present in the root system, particularly in studies describing physiological or ecological relationships of mycorrhizal development in relationship to plant growth.

The Phillips and Hayman (12) procedure for clearing and staining roots for rapid assay of mycorrhizal colonization represents a major breakthrough in VA mycorrhizal research. This gave plant scientists ready access to a technique that was specifically adapted to VA mycorrhizal research; it was less time consuming than embedding and sectioning and worked well for a wide range of host plants. This procedure quickly became the most commonly used clearing and staining procedure in mycorrhizal research. However, it had a serious drawback that was shared with other published procedures (4,6,11). The phenols or saturated chloral hydrate used in these staining and destaining procedures are hazardous. The fumes from these chemicals are highly toxic at room temperature and heating them — even under laboratory hoods — can result in adverse side effects. In many cases, especially in underdeveloped countries, and some field locations in technically advanced countries, facilities for confining or exhausting these toxic fumes are inadequate. Kormanik and others (9) published a procedure similar to that of Phillips and Hayman's (12) in which they eliminated the toxic phenols or chloral hydrates; their procedure has been equally effective in VA mycorrhizae clearing and staining for a wide range of host species. Either method, used with caution, is adequate for quantification of mycorrhizal colonization for most VA mycorrhizal assays.

METHODS

Any suitable assay procedure first requires proper selection of susceptible feeder roots. The primary site for VA mycorrhizae to develop is in the cortical region of the terminal feeder roots which is the most active site for nutrient uptake. As roots mature, the cortex ruptures and is sloughed off and, thus, mycorrhizae are seldom observed in older, less succulent plant roots. Fine terminal feeder roots are left in the soil if considerable caution is not taken during the excavation of plant root systems. If root systems are improperly excavated, roots undergoing secondary growth will constitute the greatest portion of the sample, resulting in a significant under estimation of the percentage of roots colonized by mycorrhizal fungi.

Root Sample Collections

Roots can be collected at any time of the year for mycorrhizal assay and need not be immediately processed in the laboratory. Regardless of whether the host plant is annual, perennial, herbaceous, or woody, the fine terminal feeder roots are the primary sites of VA mycorrhizal development. If reasonable care is exerted when roots are excavated, there will be sufficient terminal feeder roots attached to lower order roots that can be used in the sample. A representative sample of the entire root system can be obtained from four or five different portions of the root system and combined rather than obtaining one large sample from a single portion.

To preserve specimens, individual root samples of 1- to 2- grams (fresh weight) are placed in small plastic bags, vials, or reusable Tissue-Tek plastic capsules (Fisher Scientific Co., Pittsburgh, Pa.) with Formalin-Aceto-Alcohol (FAA) killing and fixing solutions. The standard FAA solution made with 50% alcohol with a v/v/v ratio of 90:5:5 is adequate. Tissue-Tek plastic capsules reduce handling considerably because the individual root samples need be handled only twice — once when the samples are placed in the capsule and once when they are placed in the destaining solutions. An additional benefit of these capsules is that treatment codes can be written on them with a No. 2 pencil when the samples are collected and remain visible during destaining. Samples immersed in FAA have been kept up to 2½ years before assay with no adverse effect to the penciled code or the root sample.

Samples collected and preserved in FAA are used primarily for assessing the degree and intensity of mycorrhizal development and are adequate for describing fungal morphological characteristics in the roots. These samples are not suitable for use in nutritional or biochemical analyses. However, tests for obtaining more complex and extensive nutrient or biochemical relationship data should be preceded by root assay of a representive sample for verification of the absence, presence, or degree of mycorrhizal colonization of the roots.

When plants are newly germinated from seed there is a time lag before mycorrhizae are well developed. It takes longer for colonization to occur if spores are the sole source of infesting material. In all probability, there is variability in the time required for different species of VA mycorrhizal fungi to successfully colonize a root system. Where precise measurements or treatments dictate knowing when roots are colonized, sub-sampling at weekly intervals would be advantageous.

Processing Root Samples

An autoclave is often used to heat the clearing and staining solutions because it reduces manhours and provides more consistent results. The clearing and staining schedule presented is that reported by Kormanik and others (9) deleting the toxic phenols from the staining and destaining solutions. A large number of different hosts and VA mycorrhizal fungal species combinations have been successfully assayed using this procedure. We recommend that phenols not be used in the staining and destaining solutions unless experience with specific root specimens show that elimination adversely affects specimen preparation. Information for using phenols in solutions is provided if one prefers or finds it necessary for optimum specimen preparation. Following is an outline of a clearing and staining procedure employing both an autoclave and a heating procedure for use under a ventilation hood. Also described is a "no heat" method, which can be used under only the most favorable conditions.

Clearing and Staining Specimens

1) Wash root specimens stored in capsules with FAA in tapwater; place in a glass beaker and cover with a 10% KOH solution. A 1,500-ml beaker will hold 45 to 50 capsules. Place petri plate tops on the specimens to keep them submerged. The capsules and KOH solution should not exceed 75 percent of the volume of the beaker to prevent boiling over in the autoclave. Place the specimens in an autoclave at 1.03×10^5 N/m^2 (15 psi) for 10 min (if an autoclave is not available, heat the specimens at 90°C for 1 h in a well-ventilated exhaust hood). The KOH solution clears the host cytoplasm and nuclei and readily allows stain penetration.

2) Pour off the KOH solution and rinse the capsules well in a beaker using at least three complete changes of tapwater or until no brown color appears in the rinse water. DO NOT agitate the capsules vigorously or loose caps may become detached.

3) Cover the capsules in the beaker with alkaline H_2O_2 at room temperature for 10 to 20 min or until roots are bleached. Alkaline H_2O_2 is made by adding 3 ml of NH_4OH to 30 ml of 10% H_2O_2 and 567 ml of tapwater. Three ml of regular household ammonia works well as the NH_4OH source. The alkaline H_2O_2 solution should be made up as needed; it loses its effectiveness even if stored overnight.

4) Rinse the capsules in the beaker thoroughly using at least three complete changes of tapwater to remove the H_2O_2.

5) Cover the capsules in the beaker with 1% HCl and soak for 3 to 4 min and then pour off the solution. DO NOT rinse after this step because the specimens must be acidified for proper staining.

6) Cover the capsules in the beaker with 0.01% acid fuchsin-lactic acid staining solution and autoclave for 10 min at 1.03×10^5 N/m^2 (15 psi). The lactic acid solution consists of 875 ml of laboratory grade lactic acid, 63 ml of glycerin, 63 ml of tapwater, and 0.1 g of acid fuchsin. (If an autoclave is not available, simmer the capsules in the beaker at 90°C in an exhaust hood for 10 to 60 min or until the roots are satisfactorily stained.)

7) After removing from the capsules, place the root specimens and the capsule top in glass petri plates for destaining and mycorrhizal assay. By retaining the capsule top with appropriate codes, the possibility of petri plate tops being inadvertently switched is eliminated. DO NOT rinse specimens after staining because the stain is readily removed from the fungal structure, requiring restaining. The destaining solution is the standard used in Step 6, but, of course, without the acid fuchsin component.

Several changes occur in the staining and destaining steps if lactophenol is used. The lactophenol solution for staining consists of 300 g of phenol, 250 ml of lactic acid, 250 ml of glycerin, and 300 ml of water. Acid fuchsin (0.1 g) is added to the staining solution, but excluded from the destaining solution. If lactophenol is used for staining, it must also be used for destaining. Specimens should be transferred to glycerin after several weeks to prevent excessive destaining. The staining solution made with lactophenol is seldom effective after the second use.

With or without phenols, this clearing and staining technique removes the cellular contents and makes the root opaque, but the VA mycorrhizal fungal structure stains bright red to light pink. During destaining the fungal structures are somewhat more distinct using the phenol solution for several weeks, but after a short period of time equal clarity is attained by the lactic acid solution.

Since the lactic acid staining procedure was first reported by Kormanik and others (9), other benefits of removing the phenols from these solutions have been observed (Kormanik, unpublished). The staining solutions have worked effectively for up to seven uses, after which they show no evidence of losing effectiveness. Since lactic acid is not volatile at autoclave temperatures, when staining becomes light additional stain can be added rendering it effective. Specimens have been held in the destaining solutions for up to 30 months with no significant loss of stain in the fungal structures. Activated carbon, Grade K-B, can be used to remove the stain from the lactic acid destaining solution which permits recycling. To clear the destaining solution, 10 g of activated carbon per liter of destaining solution is added and left standing overnight. The material is then filtered twice, first with Whatman No. 1 or 2 filter paper to remove the coarse material and larger carbon particles and second, with Whatman No. 42 filter paper to remove the finer carbon particles. This results in a clear solution.

Acid fuchsin is only one of the stains that has been found effective for VA mycorrhizae staining. Many workers prefer trypan blue, Sudan IV, or cotton blue. Of these stains, trypan blue (0.05% trypan blue in lactophenol) has been used extensively for staining many different host/fungus combinations. The use of trypan blue without phenols has not been widely tested. We do not know its effectiveness when used with the above lactic acid staining and destaining procedure. However, as is characteristic of any histological staining technique, slight modifications may be required for different root materials. The major modification will probably involve how long the clearing and staining solutions are heated, which is governed by the succulence of the roots. Too much heat applied to succulent roots results in flaccid or mushy root specimens. When this occurs, reducing time and pressure in the autoclave solves the problem.

No-heat Method.--The no-heat method works effectively only on succulent material. The solutions and procedures are the same as those used when heat is applied, but the time required for each step is greatly increased. If bleaching is required, up to 6 h may be required in unheated KOH. Two hours or longer may be required for the fungal structures to adequately absorb the stain.

Assessment of Colonization after Clearing and Staining

Recently, Giovannetti and Mosse (5) reported techniques for measuring mycorrhizae in roots following clearing and staining. These methods can be broken down into several distinct procedures. They are (i) visual assay, (ii) slide length (estimated length of colonized tissue in root segments of standard length mounted on slides), (iii) slide ± (presence or absence of colonization in the same root segments), and (iv) gridline intersect. These workers pointed out that there is no report comparing the different techniques and, thus, many researchers have developed or modified techniques to meet their own requirements. Their basic conclusion was that over the years the number of root segments per sample has tended to decrease while the number of observations per segment has tended to increase. These workers reported that only rarely have assessments been made based upon a visual assay of large samples containing unspecified numbers of roots.

Giovannetti and Mosse (5) reported that the gridline intersect method had the smallest standard error, followed closely by the visual method; root segments mounted on slides consistently resulted in the largest standard errors with either method. These authors feel that the gridline intersect method is, for most purposes, the most acceptable compromise when all the relative merits and shortcomings of the four systems are considered. The visual method, however, is quite rapid, easily mastered by the novice mycorrhizal researcher, and, because the assay is for a greater number of roots, may be biologically more meaningful to many plant science disciplines than the exacting, time-consuming measurements made by using a limited number of roots mounted on slides. We feel that the four common assay procedures reported by Giovannetti and Mosse (5) fall into two categories: systematic and nonsystematic. Their visual and slide ± methods are viewed as nonsystematic, while the gridline and slide length methods are considered systematic. We find the term "visual assay" misleading because all the procedures, whether systematic or nonsystematic, involved visual observations using either a dissecting or compound microscope. However, both nonsystematic procedures involve root scanning and subjectively give a ± rating to the roots without any attempt at being qualitative. The systematic procedures involve qualitative estimates of the percentage as well as the intensity of colonization within the root sample. For many mycorrhizal assessments, the percentage of roots colonized and the intensity of colonization within the root sample are readily determined by nonsystematic root scanning procedures (with a large, unspecified number of feeder roots). However, if exact measurements and enumeration of fungal components is required per unit root length, then observations can be made on subsamples located on a slide or gridline. For want of a better term, the latter procedure might be considered a "partial systematic procedure." If used properly, the gridline intersect method probably represents the most accurate assay procedure. We do, however, concur with Giovannetti and Mosse's (5) basic conclusion that it is not known which sampling procedure accurately measures the true level of root colonization.

A routine mycorrhizal assay is done on root samples in petri dishes under a dissecting microscope at 40 to 100X magnification. When a more detailed observation of fungal structures is desired, individual root segments must be mounted on a microscopic slide. Determination can then be made with a compound microscope at 100 to 250X magnification.

Once the clearing and staining is completed, it is necessary to differentiate between what is VA mycorrhizae and what is not. Plate 4A illustrates how root colonization appears under low magnification, Plate 4B shows the appearance of internal hyphae and arbuscules at higher magnification, and Plate 4C shows well-developed internal vesicles formed by a *Glomus* species. With careful root excavation and processing, external vesicles formed by VA mycorrhizal fungal species in the genus *Gigaspora* frequently are observed; these are illustrated in Plate 1D.

If possible, the first-time viewer of VA mycorrhizae colonization should seek the assistance of an experienced worker. Normally, VA mycorrhizal hyphae are thick and not smooth, hypae of non-VA mycorrhizae fungi in roots are usually thin and smooth, and frequently have septa at uniform spacing along the hyphal strands. However, older VA mycorrhizal hyphae can have septa. The arbuscules are distinct morphological features and not easily confused with anything else. Since arbuscules are delicate structures that are short-lived and readily absorbed, learning to recognize the coarse, nonabsorbed branches is important. One should not confuse these remnants (which can be septate) with non-VA mycorrhizae fungal structures. Identifying and separating VA and non-VA mycorrhizae fungal structures is frequently simplified because non-VA mycorrhizal fungi are not often encountered in VA mycorrhizal colonized roots in quantities that would significantly affect the root assay.

Nonsystematic Method

The evaluation of endomycorrhizal roots using the nonsystematic root scanning procedure can be expressed and quantified in different ways, depending upon the study objective. The primary objective of most studies is to determine the percentage of roots colonized as well as the intensity of colonization within the roots. To do this, roots are spread uniformly in a petri dish with dissecting needles to eliminate clumping and enhance light transmission. Within a sample, the number of susceptible feeder roots colonized determines the percentage of roots with mycorrhizae. The whole-root sample is carefully rotated on the microscope stage and an assessment of the colonized roots is made, usually in broad classes of percentage of colonization. A workable classification used at the Institute for Mycorrhizal Research and Development, USDA Forest Service, Athens, Georgia, is as follows: Class 1, 0-5%; Class 2, 6-26%; Class 3, 26-50%; Class 4, 51-75%; and Class 5, 76-100%. Normally a 3-min examination is sufficient to place the root sample in the appropriate class. When a sample is borderline between two classes, an assay of five or six random positions in the petri dish may be helpful to properly classify. Because of normal variation in percentage of roots colonized from a given root system, exact placement within a more narrow class range would be time consuming and, possibly, would not be biologically or statistically significant when comparing samples.

Many VA mycorrhizal researchers have found that the nonsystematic root scanning technique is the most rapid method for assessing the percentage of root colonization. Giovannetti and Mosse (5) reported that these subjective techniques gave reliable results when compared with more laborious methods, and that one could attain considerable proficiency with only a few hours' training.

Intensity of colonization is a separate assessment of colonization within roots. This assessment is as important as the percentage of roots colonized, but frequently is not attempted by VA mycorrhizal researchers. Kormanik and others (9) reported three categories for evaluating intensity of colonization. An intensity of 1 is assigned to roots with small colonization sites widely scattered along the roots; an intensity of 2 represents larger colonization sites more uniformly ditributed through the colonized roots, but rarely coalescing; and an intensity of 3 is given when the feeder roots are almost solidly colonized with few easily identified, isolated patches of colonization. A tentative intensity classification frequently can be obtained when percentage of roots colonized is determined, but a final value can be readily obtained by looking at five or six random microscope fields. Using this procedure, one must be constantly aware that only colonized roots are considered in the intensity evaluation because it is independent of the percentage of roots colonized.

While the root scanning nonsystematic procedure is adequate for many studies, it should not be used for quantifying morphological data (vesicles, mycelium, or arbuscules) on fungal infection per sé, but should be supplemented by assaying a subsample using a systematic procedure such as the slide or grid method. The nonsystematic procedure of assaying roots can be, and occasionally should be, checked against a systematic method to help reduce bias. A good time to do this is either when a new person is being trained in mycorrhizal assay or when changes are made in hosts and symbionts in a trial.

Systematic Methods

The most commonly used systematic procedure is the slide method, but the gridline intersect method is probably the most accurate if all theoretical requirements are met (5). Both of these methods are laborious. The stain retention in root segments, even to the naked eye, frequently indicates the degree of VA mycorrhizal colonization, and the accuracy of the assessment using the slide method depends entirely on the unbiased selection of root segments. Selection of root segments can be unintentionally biased.

Gridline Intersect Method.--This method can be used to estimate both the proportion of root length colonized and total root length in the sample. The root sample is spread out evenly in petri dishes that have gridlines marked on the bottom to form 1.27-cm squares. Many VA mycorrhizal researchers using the gridline intersect method ink the grid on a circular piece of acetate and place it on the bottom of the inverted plate top. The bottom portion of the plate containing the sample is then placed inside the top. The grid markings are distinct and the use of acetate is easier than marking grids on a large number of glass petri dishes.

The dimension of the grid squares is important for measuring the total length of roots. If only the percentage of root length colonized is to be determined, the gridlines serve only as a device for systematic selection of observation points. Vertical and horizontal gridlines are scanned and the absence or presence of colonization is recorded at each point where a root intersects a line. Good accuracy for percentage of root length colonized is obtained with this method if at least

100 gridline intersects are tallied. However, for an estimate of total root length, all gridline intersects, or intersects from a predesignated portion of the gridlines, must be counted. Provided that the exact grid spacing described by Giovannetti and Mosse (5) is used, the total number of root/gridline intersects equals the total length (expressed in centimeters) of roots in the petri dish. The theory behind this procedure is given by Newman (10) and Giovannetti and Mosse (5).

Slide Method.--Root segments, each approximately 1 cm long, are selected at random from a stained sample and mounted on microscopic slides in groups of 10. Giovannetti and Mosse's (5) work indicates that from 30 to 100 root segments from each sample should be used for this method. Length of cortical colonization is assessed (at 100 to 250X) in millimeters for each root segment, averaged for each of the 10 segments in a group, and expressed as a percentage of root length colonized. This method gives an assay based on total root length and takes into consideration the percentage of roots colonized, as well as the intensity of colonization. While time consuming, this method quantifies the mycorrhizal development quite well and, because a higher power microscope is used, the fungal structures are distinctive. This procedure may be of limited value when many specimens have to be assayed.

The slide procedure can be simplified by recording only the absence or presence of colonization in each root segment and expressing the results as a percentage of roots colonized. Giovannetti and Mosse (5) refer to this alternative procedure as the slide method ±. We feel this latter procedure is a nonsystematic assay and it is more time consuming than root scanning. This modified procedure appears to be less accurate because fewer roots would be assayed. Giovannetti and Mosse (5) reported that the modified procedure consistently resulted in the largest standard errors in their comparative tests of VA mycorrhizal assessment methods.

Mycorrhizal Assessment of Noncleared Roots

Regardless of which assay procedure is used, evaluating for VA mycorrhizae fungal colonization using clearing and staining methods is rather time consuming. Numerous attempts have been made to develop alternative procedures but none have been entirely satisfactory for a wide range of hosts or fungal structures, therefore, clearing and staining roots remains the primary method used. Colorimetric, chemical, and autofluorescence procedures offer considerable opportunities to improve the accuracy of assessments, as well as reduce the time spent making them; developing and perfecting alternative methods should be encouraged. Examples of these procedures follow, but details should be obtained from the references cited.

Colorimetric Assay.--Endomycorrhizal roots of some host plants develop a yellow pigmentation that diffuses into the water when roots are cut and also disappears rapidly when exposed to sunlight. Daft and Nicolson (3) attempted to assess mycorrhizal colonization of entire root systems by measuring the relative intensity of this yellow water-soluble pigment in tomato roots. The intensity of this yellow pigment was assessed visually by an ultraviolet mercury vapor lamp and was correlated with the intensity of mycorrhizae fungal colonization based on a systematic assay procedure that used paired root systems. The pigmentation assay gave good accuracy under low nutrient conditions when root colonization percentages were high, but basically it was a very subjective assessment. The yellow pigmentation was conspicuous under pot conditions in heavily colonized roots, was not detectable in field-grown plants, and occurred only in certain hosts.

Becker and Gerdemann (2) extracted this yellow pigment from onion roots by autoclaving root systems for 30 min at 121°C. After cooling, the roots were removed and the absorbance of the water extract was read at 400 nm against a water blank in a Bausch and Lomb Spectronic-20 spectrophotometer. If the absorbances could not be read immediately after cooling, the extract had to be kept in the dark to prevent the pigment from fading. They concluded that this colorimetric procedure was a valid alternative for root assessment for controlled short-term experiments, but not for long-term experiments where phenolic compounds of dying roots could interfere with the absorbance of the water extract.

Herrera and Ferrer (8) indicated that a direct measurement of the amount of endomycorrhizae fungal tissue present in roots can be obtained by clearing and staining roots with trypan blue in lactophenol and then eluting the absorbed stain from the fungal tissue. Supposedly the amount of stain eluted, measured colorimetrically, could be correlated with total colonization. It is possible that contamination by other microorganisms would interfere with measurements. This method has had limited testing.

Chemical Assay.--Chitin is a component in the hyphal walls of VA mycorrhizal fungi. Hepper (7) performed a chitin assay from both endomycorrhizal and nonmycorrhizal roots of several hosts to assess this compound as an assay method. In this procedure, chitin was converted to glucosamine and the absorbance at 650 μm was compared with purified glucosamine. The amount of glucosamine in the root samples was then correlated with the total weight of the root sample. It was also correlated with the percentage of mycorrhizal colonization when varying degrees of colonization were obtained by diluting endomycorrhizal and nonmycorrhizal roots. However, no good correlations were found when glucosamine was expressed per unit dry weight of the external mycelium of colonized roots. Hepper (7) concluded that the lack of correlation occurred because external mycelium from which the internal conversion factor was obtained was not a good measure of internal fungal biomass.

Ultraviolet-Induced Autofluorescence Assay.--Recently, Ames (1) reported a simple procedure for assaying arbuscules in fresh mycorrhizal roots by an ultraviolet-induced autofluorescence procedure.[1/] Colonization is assessed from fresh endomycorrhizal root segments with an epifluorescence microscope equipped with epifluorescent condensers, mercury lamps, exciter filters which passed wave lengths of 455-490 nm, and barrier filters which allowed wave lengths of 520-560 nm to pass to the viewer. Arbuscules autofluoresced in the root segments of all plants tested. Apparently, the species of VA mycorrhizal fungi, the host plant, and the conditions under which the symbiosis occurred did not adversely affect this phenomenon. However, only arbuscules autofluoresced under ultraviolet light; vesicles, spores, and hyphae within or exterior to the root did not. Although fluorescent intensity varied somewhat with the host, newly developed, highly branched arbuscules could be readily distinguished from the older collapsed arbuscules with their smaller, clumped appearance. Although only arbuscules can be detected with this procedure, it is potentially important because the same root segments can be used for determining the presence of VA mycorrhizae and can be used for either biochemical or nutritional assays. If further testing of this procedure reveals autofluorescence of arbuscules to be a reliable and consistent phenomenon, it could have a significant impact on the development of chemical or colorimetric assay procedures.

DISCUSSION

For the present, it appears that clearing and staining roots will remain the most common procedure used for endomycorrhizal root assays. Although it is time consuming when large numbers of samples must be processed, little specialized equipment is needed. Workers are discouraged from using toxic phenols or chloral hydrate in the staining and destaining solutions because suitable results can be obtained with less toxic materials. In our opinion, trials with different stains are not high priority since acid fuchsin and trypan blue have been proven effective for a wide range of hosts and VA mycorrhizae symbionts.

There has been no strong evidence in the literature favoring either the nonsystematic or the systematic procedures for assaying stained roots. The diverse root morphologies characteristic of the numerous endomycorrhizal host plants may,

---

[1/] Personal communication, Robert N. Ames, Natural Resources Ecology Lab., Colorado State University, Fort Collins, Colo.

in part, account for the fact that some investigators prefer a systematic procedure for root assays while others prefer a nonsystematic root scanning procedure. It appears, however, that a partial systematic root assessment is beneficial. In this procedure, large numbers of roots could be scanned for percentage of roots colonized and a subsample would be systematically assayed using the gridline intersect or slide method for determining intensity of colonization or enumeration of fungal components.

Further exploration in colorimetric, chemical, and autofluorescence procedures should be encouraged in an effort to develop more qualitative assay methods without the limitations so evident in published procedures. We need accurate, reliable, less time consuming procedures that will work on a wide range of host plants.

Finally, workers are cautioned on the excessive extension of data obtained from VA mycorrhizal assays. Our knowledge of this symbiotic relationship is not sufficient to state which threshold of root colonization will affect growth under different nutritional regimes. This obvious gap in knowledge probably accounts for poor correlations that have been observed by many researchers when growth is plotted against root colonization percentages. Thus, VA mycorrhizal assessments simply tell us what has occurred under the conditions of the test. We feel it is important to know this.

NOTE: The use of proprietary names in this publication does not constitute an endorsement by the U.S. Department of Agriculture or the Forest Service.

LITERATURE CITED

1. Ames, R. H., Ingham, E. F. and Reid, C. P. P. 1982. Ultraviolet-induced autofluorescents of arbuscular mycorrhizal root infections: An alternative to clearing and staining methods for assay infections. Can. J. Microbiol. March issue.
2. Becker, W. H. and Gerdemann, J. W. 1977. Colorimetric quantifications of vesicular-arbuscular mycorrhizal infection in onion. New Phytol. 78:289-296.
3. Daft, M. J. and Nicolson, T. H. 1972. Effect of <u>Endogone</u> mycorrhiza on plant growth. IV. Quantitative relationships between the growth of the host and the development of the endophyte in tomato and maize. New Phytol. 71:287-295.
4. Gerdemann, J. W. 1955. Relation of a large soil-borne spore to phycomycetous mycorrhizal infections. Mycologia 47:619-632.
5. Giovannetti, M. and Mosse, B. 1980. An evaluation of techniques for measuring vesicular-arbuscular mycorrhizal infection in roots. New Phytol. 84:489-500.
6. Hayman, D. S. 1970. Endogone spore numbers in soil and vesicular-arbuscular mycorrhiza in wheat as influenced by season and soil treatments. Trans. Br. Mycol. Soc. 54:53-63.
7. Hepper, C. 1977. A colorimetric method for estimating vesicular-arbuscular mycorrhizal infection in roots. Soil Biol. Biochem. 9:15-18.
8. Herrera, R. W. and Ferrer, R. L. 1978. Present knowledge about vesicular-arbuscular mycorrhizae in Cuba. Provisional Report No. 1. International Workshop of Tropical Mycorrhiza Research, International Foundation of Science. 425 pp.
9. Kormanik, P. P., Bryan, W. C. and Schultz, R. C. 1980. Procedures and equipment for staining large numbers of plant roots for endomycorrhizal assay. Can. J. Microbiol. 26:536-538.
10. Newman, E. I. 1966. A method of estimating the total length of root in a sample. J. Appl. Ecol. 3:139.
11. Nicolson, T. H. 1959. Mycorrhizae in the gramineae. I. Vesicular-arbuscular endophytes with special reference to the external phase. Trans. Br. Mycol. Soc. 42:421-438.
12. Phillips, J. M. and Hayman, D. S. 1970. Improved procedures for clearing and staining parasitic and vesicular-arbuscular mycorrhizal fungi for rapid assessment of infection. Trans. Br. Mycol. Soc. 55:158-161.

PRODUCTION OF ENDOMYCORRHIZAL INOCULUM

A. Increase and Maintenance of Vesicular-Arbuscular Mycorrhizal Fungi

J. J. Ferguson and S. H. Woodhead

INTRODUCTION

A scarcity of prime inoculum has limited the broad use of vesicular-arbuscular (VA) mycorrhizal fungi. The dynamics of inoculum production, the nature of host specificity, the activity of hyperparasites, and the absence of a regional or national inoculum bank recur as crucial factors affecting inoculum production. In addressing the topic of the increase and maintenance of VA mycorrhizal fungi, we will describe procedures and offer suggestions that may facilitate basic research and practical application of these beneficial microorganisms.

Starter Cultures

Source.--Although VA mycorrhizal starter cultures have been available from university and industrial sources in the past, no permanently designated, functioning inoculum bank has yet been established. The most accessible, most abundant yet taxonomically perplexing source of inoculum for starter cultures is the rhizosphere of endomycorrhizal plants in the field. Spores can be obtained by wet sieving and decanting rhizosphere soil samples taken from 0-15 cm depth and screened on soil sieves of various mesh sizes (18). With the aid of a dissecting microscope (20-70X), spores of mycorrhizal fungi can be removed from soil detritus with a microspatula or a Pasteur pipette fitted with a rubber bulb and subsequently used to establish pot cultures.

A modified method (17, 19, 20) can be used for soils that are not amenable to wet sieving and decanting because of high organic matter or low spore density. Roots and associated particles in the 147-833 μm size range obtained from the rhizosphere of the plant to be tested can be used to inoculate "trap plants." Monocots like sudan grass and bahia grass with rapidly developing fibrous root systems are ideal trap plants. When soil type precludes the use of these particular plants, any known endomycorrhizal host that is readily colonized by the desired mycorrhizal species and that can be easily grown in the greenhouse would suffice. Three to 4 mo after inoculation, soil from pot cultures can be sampled periodically to determine if new spores have matured. After spores or sporocarps have been extracted by wet sieving and decanting, species can be identified. Spores can be isolated from soil by density gradient centrifugation and can be surface disinfested (32). These spores can then be used to establish new pot cultures.

Inoculum should be placed in close contact with actively growing roots. When seeds are sown in pots, inoculum should be layered 2-3 cm below the seed. When seedlings are transplanted into pots, inoculum should be placed in a pot half filled with soil. Roots of transplanted seedlings can be put in direct contact with the inoculum, and the remainder of the soil added. A funnel inoculation technique (16) can be used to insure that roots contact inoculum.

Inoculum Preparation.--Various inoculum preparations, including soil inoculum (spores, soil, hyphae, and infected mycorrhizal roots) and spore inoculum (surface-disinfested spores that have been isolated by density gradient centrifugation) have been used in endomycorrhizal research. Soil inoculum is considered to be more rapidly infective than spore inoculum (12, 21), possibly because of the greater number of infective propagules (external hyphae, mycorrhizal root fragments and spores) and associated soil microflora that can favor germination of spores (26, 32, 38). Ferguson (12) reported that when both soil and spore inoculum at a similar range of spore densities were used to inoculate sudan grass, roots were infected more rapidly and subsequent production of spores was greater when soil inoculum was used than when spore inoculum was used. This was true especially at low spore density levels.

Air dried soil inoculum and fresh sievings (mycorrhizal roots, spores, hyphae, and associated microflora) have been successfully used in field and greenhouse experiments (12). Lyophilized mycorrhizal roots have also been reported to be an effective formulation (6, 24). This inoculum preparation is not as widely used and is generally considered to be less viable than other formulations. Procedures for the lyophilization of VA spores have, however, been proposed (49).

Surface-disinfested spores (32) have also been used to establish mycorrhizal infections. Spores from pot cultures or field soil may be obtained by wet sieving and decanting and can be separated from soil debris by using microspatulas or Pasteur pipettes as previously described. Alternatively, soil sievings suspended in water can be homogenized in a blender at low speed, layered onto the surface of a 60, 40, and 20% sucrose density gradient, and centrifuged for 5 min at 30,000 rpm on a clinical centrifuge. Mycorrhizal spores concentrated in bands at the density gradient interfaces can be extracted from the sucrose, rinsed in sterile water, surface-disinfested, and stored for short periods of time at 5°C in Ringer's Solution before use. Other methods to extract spores have been suggested (14, 33, 35, 47).

Species Variability.--Mycorrhizal species differ significantly in their ability to stimulate growth of host plants (10). This may be due to differing rates of infection, greater infectivity related to spore size, more external hyphae, more efficient nutrient uptake related to greater affinity for ions, or more rapid translocation of essential nutrient elements. Minimum spore densities required to insure rapid root colonization and host plant growth may vary, therefore, from species to species. *Glomus constrictus*, for example, has been shown to be considerably more infective and growth stimulative than either *G. fasciculatus* or *G. deserticola* at low spore densities (12).

Inoculum Density.--Mycorrhizal researchers have generally defined mycorrhizal infectivity or "inoculum potential" in terms of inoculum density: grams of soil inoculum or spore number per unit soil mass or area, plant, or pot. Although studies on the influence of mycorrhizal inoculum density on infection, sporulation, and host growth response have provided conflicting, often incomparable results (5, 7, 8, 12, 13, 15, 25, 42), the dynamics of colonization, sporulation, and growth enhancement in mycorrhizal systems suggest (i) the rate of onset and spread of infection appears to be a major factor affecting host stimulation and (ii) high inoculum densities are often associated with high levels of infection early in the growing season and high subsequent levels of sporulation. Although low spore concentrations can theoretically provide sufficient inoculum, we recommend an inoculum density of approximately 300-500 spores per pot (500g soil) and approximately 10,000 spores per m^2 of planting area to insure rapid colonization.

Horticultural Practices That Can Affect Inoculum Production

Pot Size.--Large plant size has been associated with increased levels of sporulation, probably because larger plants often have more extensive root systems, allowing greater mycorrhizal colonization and sporulation than do smaller plants (9, 43). Root mass can in turn be influenced by available soil or pot volume. Ferguson (12) reported that spore production per gram soil and per gram root dry weight increased with pot volume in mycorrhizal sudan grass. Plants in a 15,000 cm^3 pot produced 90X as many spores per pot as plants in a 750 cm^3 pot. A greater economy can therefore be obtained in greenhouse inoculum production by growing mycorrhizal seedlings in larger pots. Better soil aeration and root oxygen supply related to soil drainage may have influenced sporulation in large pots.

Soil Moisture.--Extreme fluctuations in soil moisture apparently can affect sporulation, probably by drastically affecting external hyphae rather than hyphae within the root cortex (41). The water regime most conducive to spore production appears to be that most favorable to plant growth: limited daily watering rather than heavy watering leading to waterlogged conditions or intermittent watering

(2, 40). Alternatively, water stress of mycorrhizal plants may induce senescence, thereby triggering early sporulation. Better soil moisture conditions may prevail in larger, deeper pots because of their better soil drainage. High soil moisture may also provide a favorable environment for the growth and reproduction of mycorrhizal hyperparasites, primarily phycomycetes, as well as some plant pathogens. Production of mycorrhizal inoculum on drought-hardy host species under low soil moisture regimes may provide a means of controlling these mycorrhizal parasites.

Soil Temperature.--It is generally accepted that environmental conditions that favor host plant growth also tend to maximize mycorrhizal infection and sporulation. Soil temperature both as a constant regime and as a stress-inducing principle may alter the physiology of mycorrhizal symbiosis to stimulate greater inoculum production by influencing root morphology and host plant nutrition and growth as well as the general ontogeny of mycorrhizal roots (27, 50). The existence of ecotypes, each with distinct physiological characteristics, including temperature dependent spore germination and infection can be inferred from the fact that particular VA mycorrhizal species (i) can be recovered more readily from summer than spring or winter crops (45), and (ii) can vary in optimal temperature requirements depending on the temperature of the environment to which the particular species is adapted (44). Quantitative and qualitative differences in mycorrhizal ontogeny have been associated with temperature regime. High soil temperatures generally favor infection and sporulation (12, 13, 22) whereas lower temperatures may favor arbuscule formation (22, 31, 45). Temporary cold treatments or cold shocks can also enhance infection and sporulation when these treatments are applied within the first 2 mo after inoculation (12). In order to enhance VA mycorrhizal infection and sporulation we recommend growing mycorrhizal plants at the maximum ambient air or soil temperatures consistent with both mycorrhizal species and host plant thermal habit.

Pruning.--Severe pruning or defoliation does not appear to increase and according to some reports, decreases infection and sporulation (2, 7, 41). Curtailment of translocated carbohydrate supply to the roots after pruning could affect sporulation. The physiological age of mycorrhizal plants and the sporulation cycle itself must also be considered. Pruning at an inopportune time (before the onset of sporulation), can affect inoculum production (12). Sanitary maintenance of greenhouse mycorrhizal pot cultures should involve periodic pruning done not on the basis of plant growth habit but rather with concern for the dynamics of spore production. We recommend mild pruning 3 to 4 mo after inoculation and thereafter as needed.

Light Effects.--The development of mycorrhizal infection depends to a considerable degree on the amount of light received by the host plant. Photon flux density (often referred to as light intensity) and photoperiod can strongly influence the development of VA mycorrhizal infection. Plants exposed to greater radiant energy in sunny rather than shady sites (37), during spring and summer rather than during autumn and winter (28, 51), at high elevations (11), or under various light regimes (7, 36, 46) had greater mycorrhizal infection than plants grown under lower radiant energy. Longer photoperiods stimulate greater mycorrhizal infection and growth response than do shorter photoperiods (3, 4, 22, 48). However, photoperiod appears to exert a greater effect than radiant energy (7, 22). Hayman (22), for example, reported that when mycorrhizal onions were grown under constant temperatures and light intensities but varying photoperiods infection was much less with 6-hr daylengths than with 12 or 18-hr daylengths. When light intensity alone varied, qualitative rather than quantitative differences were observed.

High light intensity appears to increase VA mycorrhizal infection and spore production (12, 15) probably by affecting the quantity of translocated photosynthate. In terms of greenhouse inoculum production, plants grown at higher light levels tend to have greater infection and sporulation than plants grown at lower light intensities (12).

Greenhouse light intensity is considerably lower than that of full sunlight. Mycorrhizal hosts, especially C_4 plants like sudan grass or Johnson grass that do not become light saturated at full sunlight (23), are usually grown under considerably less than saturated conditions. This suggests that higher radiant energy than that usually used under experimental conditions could further enhance inoculum production. High intensity discharge (HID) lamps can be used to provide high levels of radiant energy under controlled conditions. Both mercury vapor and metal halide lamps have been shown to enhance and hasten the onset of sporulation (12).

Maintenance of Vesicular-Arbuscular Mycorrhizal Pot Cultures

General Greenhouse Practices.--The proper maintenance of mycorrhizal pot cultures in greenhouses involves a number of cultural practices designed to (i) prevent contamination of cultures by pathogens and (ii) insure that cultures are not contaminated by adjacent cultures containing different mycorrhizal species. General greenhouse sanitation involves overall cleanliness, control of weed, insect, fungal and bacterial pests and rodents (mice will frequently dig up and eat newly planted grass seeds). Mycorrhizal cultures should be isolated as much as possible from experiments involving plant pathogens that can be carried on air currents, especially fungal spores and nematode eggs. The nozzles of watering hoses should not touch the ground. Contaminated nozzles may distribute pathogens throughout the greenhouse. All pots and tools should be sterilized before use by steaming, autoclaving, or with fire or a disinfectant. Wooden benches should be periodically painted with copper napthalate (1 part to 5 parts thinner) to further discourage growth of plant pathogens.

Additional precautions can be taken to prevent the establishment and restrict the multiplication of host plant pathogens in mycorrhizal pot cultures. These practices involve the traditional concept of crop rotation. Inoculum produced on monocotyledonous host plants can be used to inoculate dicotyledonous host plants. Further rotation schemes could involve the successive use of host plants from different plant families.

Containers.--Sudan grass, bahia grass and other fibrous rooted monocots are frequently used as mycorrhizal hosts. Because of their vigorous growth habit, roots often grow out of pot drainage holes onto the bench itself, causing cross-contamination of cultures. This can be prevented by using double clay pots, one inside the other. Roots grow circularly into the second outer pot but not onto the bench itself. As an added precaution against contamination from untreated greenhouse benches, double pots can be raised off the bed by placing individual pots on inverted clay saucers. Clay pots should be scrubbed clean of soil and organic matter and autoclaved if possible. Plastic pots should also be cleaned and then steamed. Low phosphorus sandy soil frequently used as planting medium for pot cultures will readily drain through pot drainage holes. Placing a piece of paper towel over the pot drainage hole before the sandy soil is put into the pot will prevent soil loss even under a heavy watering regime. In addition paper towels are porous enough to allow proper drainage. When filling pots with sandy soils, soil should be watered and allowed to compact under its own weight before additional soil is added. Water the soil when the pot is 1/3 full, 2/3 full, and after seeds or seedlings have been placed in the pots. Seedlings placed in dry, autoclaved, sandy soil will invariably lean when the soil is first watered.

Clay pots, plastic pots, and styrofoam cups can all be used. However, smaller clay pots (10 cm) and styrofoam cups should be placed inside steamed wooden planting flats to insure that pots do not tip over on the bench.

Fertilization.--Reports have indicated that high phosphorus concentrations in plants inhibit mycorrhizal infection (30, 39). Extractable soil phosphorus levels, zinc, manganese, copper, and percent organic matter have also been reported to be inversely correlated with mycorrhizal dependency in some systems (29). A coarse sand fertilized frequently with modified Hoagland's solution lacking phosphorus can provide a planting medium conducive to the development of mycorrhizal symbiosis. Low supplementary concentrations of specific elements may be needed for plants with special nutritional needs. When granular fertilizers are to be incorporated into a planting medium, fertilizer formulations devoid of phosphorus but containing micronutrients should be used.

Contaminating Microorganisms.--A number of contaminating nonpathogenic soil microorganisms are frequently found in pot cultures regardless of attempts to sterilize soil and pots beforehand. These organisms do not appear to inhibit mycorrhizal symbiosis. Oligochaetes (transparent, primitive round worms) may be mistaken for nematodes. However, these worms lack stylets and are considerably larger than nematodes. Mites, spring tails (Collembola), small beetles and other miscellaneous arthropods are common inhabitants of greenhouse cultures. Fern and moss gametophytes and sporphytes are also prevalent, especially in old cultures. The imperfect Chromelosporium state of Peziza ostracoderma occurs commonly on the surface of freshly sterilized soil in greenhouse pots or flats as yellow or cinnamon-colored, powdery colonies. The umber-brown apothecia of Peziza ostracoderma may be produced in old flats or pots (1). However, this fungus is saprophytic, and does not appear to affect mycorrhizal symbiosis. Papulospora, a fungal species belonging to the Form Class Mycelia Sterilia, produces reproductive structures that are similar in size, shape, and color to spores of some mycorrhizal species. These structures can be distinguished from mycorrhizal spores by their bulbous septate structure, each "spore" being in fact composed of irregular clusters of cells (bulbils) that are frequently pigmented.

Harvest and Storage

Whole Pot Culture.--Whole pot cultures can be harvested by pruning plants to the soil level, removing the compacted soil mass from the pot, and by chopping or otherwise segmenting the mycorrhizal roots. Razor blades, kitchen knives, and small saw blades have been used, but are quickly dulled by soil particles. Alternatively, soil adhering to roots can be shaken loose and saved. The fibrous roots can then be fragmented using the above mentioned tools, manual or motorized compost grinders or wood chipping machines. The roots and soil can be homogenized in a sterile fertilizer blender and examined for the presence of plant pathogens and mycorrhizal hyperparasites, cultural purity, spore number, and spore maturation.

Before storage, the soil mixture should be air dried to the point at which there is no free water. The actual soil moisture level will vary considerably with soil type. After drying, the culture should be packaged in plastic bags, sealed to prevent further drying, and stored at 5°C. Cultures have been successfully stored by this method for at least 4 yr.

Spores and Sporocarps.--Glomus epigaeus warrants special attention because of its unique ability to produce sporocarps on the soil surface of pot cultures in the greenhouse. The epigeous sporocarps can be collected from the soil surface without destroying the pot culture, which will continue to bear for years. G. epigaeus stored either as fresh sporocarps or as spores in sterile bentonite clay saturated with water maintained a high percent germination after 3 mo at 5°C (10). Care must be taken to avoid contamination of mycorrhizal pot cultures by the unharvested sporocarps of this species which can be blown about by greenhouse air currents.

A method of controlled drying of extracted spores (L-drying) before storage at 4°C has been reported as an effective method for preservation of G. caledonius, G. monosporus, Acaulospore laevis, and a Gigaspora sp. for short periods of time (49).

Literature Cited

1. Barron, G. L. 1968. The genera of Hyphomycetes from soil. Robert E. Krieger, New York. 363 pp.
2. Baylis, G. T. S. 1969. Host treatment and spore production by Endogone. N.Z.J. Bot. 7:173-4.
3. Boullard, B. 1957. Premierès observations concernant l'influence du photopériodisme sur la formation des mycorrhizes. Mem. Soc. Sci. Nath. Cherbourg 48:1.

4. Boullard, B. 1959. Relations entre la photopériode et l'abondance des mycorrhizes chez l'Aster tripolium L. (Composees). Bull. Soc. Bot. Fr. 106:131.
5. Carling, D. E., Brown, M. F., and Brown, R. A. 1979. Colonization rates and growth responses of soybean Glycine max plants infected by vesicular-arbuscular mycorrhizal fungi. Can. J. Bot. 57:1769-1772.
6. Crush, J. R., and Pattison, A. C. 1975. Preliminary results on the production of vesicular-arbuscular mycorrhizal inoculum by freeze drying. Pages 485-495 in: F. E. Sanders, B. Mosse, and P. B Tinker, eds. Endomycorrhizas. Academic Press, New York. 626 pp.
7. Daft, M. J., and El-Giahmi, A. A. 1978. Effect of arbuscular mycorrhiza on Plant growth. VIII. Effects of defoliation and light on selected hosts. New Phytol. 80:365-372.
8. Daft, M. J., and Nicolson, T. H. 1969. Effect of Endogone mycorrhiza on plant growth. III. Influence of inoculum concentration on growth and infection in tomato. New Phytol. 68:953-961.
9. Daft, M. J., and Nicolson, T. H. 1972. Effect of Endogone mycorrhiza on plant growth IV. Quantitative relationships between the growth of the host and the development of the endophyte in tomato and maize. New Phytol. 71:287-295.
10. Daniels, B. A., and Menge, J. A. 1981. Evaluation of the commercial potential of the vesicular-arbuscular mycorrhizal fungus, Glomus epigaeus. New Phytol. 81:345-354.
11. Dominik, T., Nespiak, A., and Pachlewski, R. 1954. Badanie mikotorfizmu zespolow roslinnych regla gornego ev Tatrach. Acta Soc. Bot. Pol. 23:487.
12. Ferguson, J. J. 1981. Inoculum production and field application of vesicular-arbuscular mycorrhizal fungi. Ph.D. Dissertation. University of California, Riverside. 117 pp.
13. Furlan, V., and Fortin, J. A. 1973. Formation of endomycorrhizae by Endogone calospora on Allium cepa under three different temperature regimes. Naturaliste Can. 100:467-477.
14. Furlan, V., and Fortin, J. A. 1975. A flotation-bubbling system for collecting Endogonaceae spores from sieved soil. Naturaliste Can. 102:663-667.
15. Furlan, V., and Fortin, J. A. 1977. Effects of light intensity on the formation of vesicular-arbuscular endomycorrhizas on Allium cepa by Gigspora calospora. New Phytol. 79:335-340.
16. Gerdemann, J. W. 1955. Relation of a large soil-borne spore to phycomycetous mycorrhizal infections. Mycologia. 47:619-632.
17. Gerdemann, J. W. 1961. A species of Endogone from corn causing vesicular-arbuscular mycorrhiza. Mycologia. 53:254-261.
18. Gerdemann, J. W., and Nicolson, T. H. 1963. Spores of a mycorrhizal Endogone species extracted from soil by wet sieving and decanting. Trans. Brit. Mycol. Soc. 46:235-244.
19. Gerdemann, J. W., and Trappe, J. M. 1974. The Endogonaceae in the Pacific Northwest. Mycol. Memoir No. 5. New York Botanical Garden and the Mycological Society of America. New York. 76 pp.
20. Gilmore, A. E. 1968. Phycomycetous mycorrhizal organisms collected by open-pot culture methods. Hilgardia. 39:87-105.
21. Hall, I. R. 1976. Response of Coprosma robusta to different forms of endomycorrhizal inoculum. Trans. Br. Mycol. Soc. 67:409-411.
22. Hayman, D. S. 1974. Plant growth responses to vesicular-arbuscular mycorrhiza. VI. Effect of light and temperature. New Phytol. 73:71-80.
23. Hesketh, J. D., and Moss, D. N. 1963. Variation in the response of photosynthesis to light. Crop Sci. 3:107-110.
24. Jackson, N. E., Franklin, R. E., and Miller, R. H. 1972. Effects of vesicular-arbuscular mycorrhizae on growth and phosphorus content of three agronomic crops. Proc. Soil Sci. Soc. Am. 36:65-67.
25. Johnson, P. N. 1977. Mycorrhizal Endogonaceae in a New Zealand forest. New Phytol. 78:161-170.
26. Manjunath, A., and Bagyaraj, D. J. 1981. Components of VA mycorrhizal inoculum and their effects on growth of onion. New Phytol. 87:355-363.

27. Marx, D. H., Bryan, W. C., and Davey, C. B. 1970. Influence of temperature on aseptic synthesis of ectendomycorrhizae by Thelephora terrestris and Pisolithus tinctorius on loblolly pine. For. Sci. 16:424-431.
28. Meloh, K. A. 1963. Untersuchungen zur Biologie der endotrophen mycorrhiza bei Zea mays L. und Avena sativa L. Arch. Mikrobiol. 46:369-381.
29. Menge, J. A., Jarrell, W. M., Labanauskas, C. K., Ojala, J. C., Huzar, C., Johnson, E. L. V., and Sibert, D. 1982. Predicting mycorrhizal dependency of Troyer citrange on Glomus fasciculatus in California citrus soils. Soil Sci. Soc. Am. J. 46:in press.
30. Menge, J. A., Labanauskas, C. K., Johnson, E. L. V., and Platt, R. G. 1978. Partial substitution of mycorrhizal fungi for phosphorus fertilization in the greenhouse culture of citrus. Soil Sci. Soc. Am. J. 42:926-930.
31. Minassian, V. 1979. Interactions between vesicular-arbuscular mycorrhizae and Tylenchulus semipenetrans in citrus. Ph.D. Dissertation. University of California, Riverside. 80 pp.
32. Mosse, B. 1962. The establishment of vesicular-arbuscular mycorrhizae under aseptic conditions. J. Gen. Microbiol. 27:509-520.
33. Mosse, B., and Jones, G. W. 1968. Separation of Endogone spores from organic soil debris by differential sedimentation on gelatin columns. Trans. Br. Mycol. Soc. 51:604-608.
34. Nicolson, T. H., and Schenck, N. C. 1979. Endogonaceous mycorrhizal endophytes in Florida. Mycologia. 71:178-198.
35. Ohms, R. E. 1957. A flotation method for collecting spores of a phycomycetous mycorrhizal parasite from soil. Phytopathology. 47:751-752.
36. Peuss, H. 1958. Untersuchungen zur Ökologie und Bedeutung der Tabakmycorrhiza. Arch. Mikrobiol. 29:112-142.
37. Peyronel, B. 1940. Prime osservazione sui rapporti tra luce e simbiosi micorrizica Lab. Chenousia Giordi. Bot. Alpino. Piccolo San Bernardo. 4:1.
38. Powell, C. L. 1976. Development of mycorrhizal infections from Endogone spores and infected root segments. Trans. Br. Mycol. Soc. 66:439-445.
39. Ratnayake, M. R., Leonard, T., and Menge, J. A. 1978. Root exudation in relation to supply of phosphorus and its possible relevance to mycorrhizal formation. New Phytol. 81:543-552.
40. Redhead, J. F. 1971. Endogone and endotrophic mycorrhizae in Nigeria. IV. IUFRO Cong. Sec. 24.
41. Redhead, J. F. 1975. Endotrophic mycorrhizas in Nigeria: Some aspects of the ecology of the endotrophic mycorrhizal association of Khaya grandifolia C.D.C. Pages 447-459 in: F. E. Sanders, B. Mosse, and P. B. Tinker eds. Endomycorrhizas. Academic Press, London. 626 pp.
42. Ross, J. P., and Harper, J. P. 1970. Effect of Endogone mycorrhiza on soybean yield. Phytopathology. 60:1552-1556.
43. Saif, S. R., and Khan, A. G. 1977. The effect of vesicular-arbuscular mycorrhizal associations on growth of cereals. Part 3. Effects on barley growth. Plant Soil. 47:17-26.
44. Schenck, N. C., Graham, S. O., and Green, N. E. 1975. Temperature and light effects on contamination and spore germination of vesicular-arbuscular mycorrhizal fungi. Mycologia. 67:1189-1192.
45. Schenck, N. C. and Schroder, V. N. 1974. Temperature response of Endogone mycorrhiza on soybean roots. Mycologia. 66:600-605.
46. Schrader, R. 1958. Untersuchungen zur Biologie der mycorrhiza der Erbsenmycorrhiza. Arch. Mikrobiol. 82:81-114.
47. Smith, G. W., and Skipper, H. D. 1979. Comparison of methods to extract spores of vesicular-arbuscular mycorrhizal fungi. Soil Sci. Soc. Am. J. 43:722-725.
48. Tolle, R. 1958. Untersuchungen über die Pseudomycorrhiza von Graminien. Arch. Mikrobiol. 30:285-303.
49. Tommerup, I. C., and Kidby, D. K. 1979. Preservation of spores of vesicular-arbuscular endophytes by L-drying. Appl. Environ. Microbiol. 37:831-835.

50. Wilcox, H. E., and Ganmore-Neuman, R. 1975. Effects of temperature on root morphology and ectendomycorrhizal development in Pinus resinosa Ait. Can. J. For. Res. 5:171-175.
51. Winter, A. G., and Meloh, K. A. 1958. Untersuchungen uber den Einfluss endotrophen Mycorrhiza auf die Entwicklung von Zea mays L. Naturwissenschaften 45:319.

B. HYPERPARASITISM OF ENDOMYCORRHIZAL FUNGI

J. P. Ross and B. A. Daniels

INTRODUCTION

Studies on the parasitism of vesicular-arbuscular mycorrhizal fungi (VAMF) have been few. Boosalis (1) pointed out that research reports on the biology of hyperparasitism in nature have been scarce and that most studies are based on dual cultures of parasite and host on synthetic media. Since he was basically referring to hyperparasites of plant pathogens, it should be apparent that studies revealing the biological significance of parasitism of obligate root parasites (VAMF) would be even more complicated and difficult to conduct than studies with fungus hosts capable of growth on artificial media. Boosalis also emphasized that most evidence of hyperparasitism in the soil has been circumstantial and research in this area has been largely ignored because of the complexity of the soil environment and lack of methods to deal with the problems. Fungal hyperparasites isolated from soil and shown to be hyperparasitic in dual cultures usually fail to exhibit significant parasitism when placed in the complexities of the natural soil. Studies of parasitism of soil fungi have been largely limited to fungi parasitizing other fungi. In complicated soil environments, however, parasites of VAMF could also include bacteria and actinomycetes (6).

Relevance

The importance of parasites of VAMF is difficult to determine. However, these hyperparasites may affect the ecological relationships and population dynamics of the VAMF (4) and successful culture for large scale inoculations of soil (2). Soils with high populations of parasites of VAMF may not allow high spore populations of the VAMF to develop and this, in turn, may result in poor mycorrhizal development and reduced agricultural production. Obviously, culturing VAMF for large scale inoculations would be seriously hampered by parasites. Stock cultures, derived from single spores, established on roots of plants in previously steamed or methyl bromide-fumigated soil would probably be parasite free. If experiments dealing with the effect of VAMF on host growth and development are conducted in soil previously exposed to sterilants (steam, methyl bromide, ionizing radiation, etc.) the relatively low populations of the harmful microorganisms may lead to over-estimation of beneficial effects of VAMF. Alternatively, the beneficial effects of VAMF may be underestimated in such experiments if hyperparasitized inoculum has been used.

Mycorrhizal Hyphal Types

Both hyphae and/or spores (sporocarps) of VAMF may be parasitized. Since VAMF hyphae function as the active mineral absorbing organ in the soil, the health of these hyphae can be very important to the functioning of the mycorrhizal system. Observations by the senior author have led him to postulate that there are two general types of VAMF hyphae. The relatively large, "thick-walled hyphae", sometimes pigmented, presumably serve as conduits and connective elements between the plant root and the more distal mineral-absorbing hyphae. The thick walls of these hyphae serve to insure their persistence. Some branches from these "thick-walled hyphae" may be "thin-walled ephermeral hyphae" (TEH). These are generally hyaline, probably serve as absorbing organs much like root hairs, are more vulnerable to the rigors of the soil environment (parasitism), and hence can be seen in various degrees of disintegration. The walls of the

TEH in many cases are difficult to observe even at 900X. The location of prior TEH branches are often evidenced by protuberances in the walls of the "thick-walled hyphae" where the "scar" was sealed over; thus the thick-walled hyphae may have a zig-zag appearance. The longevity of the TEH may be directly related to the nature and population level of harmful microorganisms (bacteria, actinomycetes) in the soil. The parasitizing or lysing ability of these microorganisms may play a large role in determining the beneficial effect of the mycorrhiza on the host plant.

Fungal Hyperparasites

Reports of research into the identification of specific hyperparasites of VAMF have been limited (2,4,5). In these studies, there were two basic techniques for working with the parasites. If the hyperparasite was obligate in nature, infection experiments were done using healthy-appearing spores or sporocarps as "bait", and spores bearing the parasite as inoculum. If the parasite could be cultured on artificial media, then its inoculum was produced in culture and used either in soil, sand, or agar along with healthy-appearing spores of the test host. Experiments dealing with organisms that may attack hyphae have either involved the establishment of pot cultures of VAMF on host roots with additions of the parasitic organisms to the soil (4), or placement of small pieces of agar containing the parasite on water agar near germinated spores of the VAMF (2). Microscopic observations of the VAMF hyphae after staining with cotton blue revealed whether parasitism occurred. These methods of establishing the parasitism of certain fungi under closely controlled conditions may not relate to the parasitic potential under field conditions.

Since chytridiaceous fungi have been identified by several workers as being parasitic on VAMF, these fungi are probably widely distributed parasites (2,4,5,7). These hyperparasites can be recognized by their zoosporangia adhering to the VAMF spores. A *Phlyctochytrium* sp. was found parasitizing chlamydospores of *Glomus macrocarpus* var. *geosporus* (Nicol. & Gerd.) Gerd. & Trappe obtained from plots in North Carolina (4), and a similar fungus of the same genus was found in California on spores of *Glomus fasiculatus* (Thaxter) Gerd. & Trappe, *G. epigaeus* Daniels & Trappe and *Gigaspora margarita* Becker & Hall (2). The chytrid *Rhizidiomycopsis stomatosa* Sparrow was found under natural conditions in Florida infecting *G. margarita*. Using the bait technique, *R. stomatosa* was found to infect five other species of *Gigaspora* and *Acaulospora laevis* Gerd. & Trappe (5,7). *Glomus* spp. have also been found under natural conditions with sporangia of *R. stomatosa* (5). These chytridiaceous fungi appear to attack only the various spore forms and initiate infections from zoospores which encyst on the spore wall, germinate, and penetrate; a rhizoidal system develops within the host.

An unidentified "Pythium-like" filamentous fungus was found in both spores and hyphae of *G. macrocarpus* var. *geosporus* inside and outside of soybean roots (4). Infected chlamydospores were smaller than noninfected spores and their contents, instead of being oil droplets of variable size, were rather uniform sized spores of the parasite, which germinated when the chlamydospore was ruptured in a water drop. The parasite grew well on cornmeal agar. The absence of cross walls in this "Pythium-like" fungus makes observation of the parasite in infected hyphae difficult. If spores of the parasite are absent, a sign of infection is the double-walled appearance of the *Glomus* hyphae which may still be evident; the hypha of the parasite may completely fill the lumen of the host hypha.

Humicola fuscoatra Traaen and *Anguillospora pseudolongissima* Ranzoni of the Fungi Imperfecti were found parasitizing chlamydospores in sporocarps of *Glomus epigaeus* and *G. fasiculatus* (2). Both parasites grow on potato dextrose agar or cornmeal dextrose yeast agar. Chlamydospores parasitized by *H. fuscoatra* contain hyphae and/or spherical spores whereas spores parasitized by A. pseudolongissima contain swollen sausage-shaped hyphae. Infection by these two hyperparasites may be limited to the spores of the host. Species of Glomus with dark pigmented walls (heavy melanization) were more resistant to H. fuscoatra and A. pseudolongissima than were species without melanization, and dark pigmented chalmydospores treated with demelanizing materials (H_2O_2 or $NaClO_3$) were more susceptible to parasitism than nontreated spores.

The use of chemicals (fungicides) to control parasites of VAMF has been reported (2), however, no definite chemical treatments have been devised to eliminate hyperparasites from cultures for commercial production of VAMF.

Nonfungal Parasites

The possibility of parasitism of VAMF by microorganisms other than fungi is worthy of consideration. The relatively low VAMF spore populations found under normal field conditions, compared with those formed under artificial experimental conditions using previously sterilized soil, could be caused by natural populations of bacteria or actinomycetes acting either as parasites or antagonists. Evidence of such biotic factors was recently reported (3). Whether these organisms are actual parasites or function as antagonists by producing substances which limit VAM fungal growth or sporulation is not known. However, if by their action, populations of VAM fungi are held at low levels, then perhaps under natural field conditions they are of great significance.

Results of recent experiments by the senior author have shown the presence of "bacteria-like" colonies on Glomus hyphae growing on soybean roots in previously sterilized soil (Fig. 1). Whether these colonies are parasitic in nature is unknown but their presence indicates that such colonies can exist on hyphae of VAMF. The presumed high susceptibility of the "thin-walled ephermeral hyphae" in a soil environment with microorganisms capable of colonizing them suggests the possibility of numerous kinds of interactions between the VAMF and beneficial, antagonistic, or parasitic microorganisms. At present, the importance of the interactions of VAMF hyphae and these soil microorganisms is unknown.

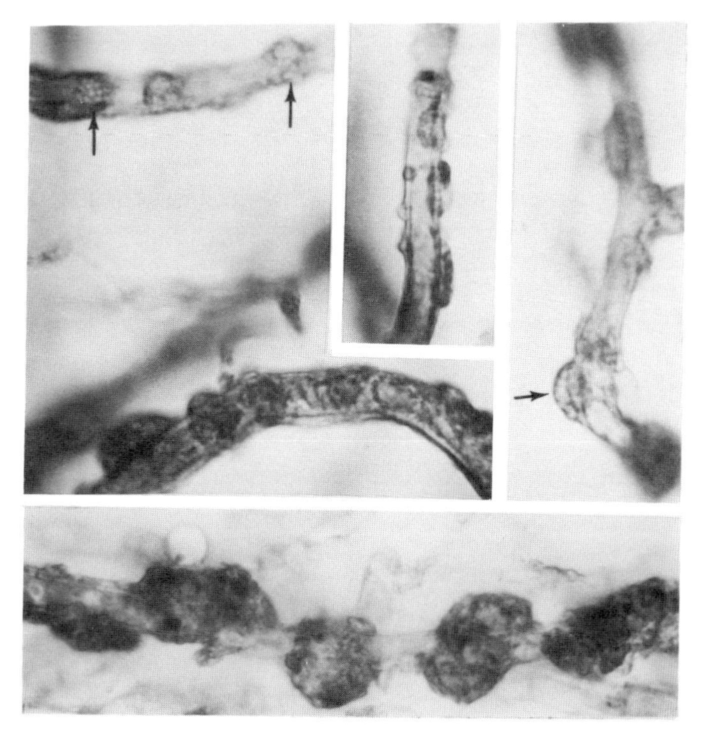

Figure 1. Glomus hyphae with clusters of "bacteria-like" structures. All photos at 900X.

Conclusions

The relative recency of the identification of various kinds of VAMF parasites make conclusions somewhat tenuous. Only future research directed into areas heretofore rather neglected will reveal the importance of VAMF parasites under natural conditions. Daniels and Menge (2) suggested that "hyperparasitism of VAM fungal spores would, in effect, cause disease in plants such as citrus and sweetgum". Although these authors suggested that the hyperparasites could be thought of as secondary plant pathogens, parasites of VAMF hyphae actively associated with host roots could cause a disease of the mycorrhizae and hence be classified as mycorrhizal pathogens.

LITERATURE CITED

1. Boosalis, M. G. 1964. Hyperparasitism. Annu. Rev. Phytopath. 2:363-376.
2. Daniels, B. A. and Menge, J. A. 1980. Hyperparasitism of vescular-arbuscular mycorrhizal fungi. Phytopathology 70:584-588.
3. Ross, J. P. 1980. Effect of nontreated field soil on sporulation of vesicular-arbuscular mycorrhizal fungi associated with soybean. Phytopathology 70:1200-1205.
4. Ross, J. P. and Ruttencutter, R. 1977. Population dynamics of two vesicular-arbuscular endomycorrhizal fungi and the role of hyperparasitic fungi. Phytopathology 67:490-496.
5. Schenck, N. C. and Nicolson, T. H. 1977. A zoosporic fungus occurring on species of *Gigaspora margarita* and other vesicular-arbuscular mycorrhizal fungi. Mycologia 69:1049-1053.
6. Sneh, B., Humble, S. J., and Lockwood, J. L. 1977. Parasitism of oospores of *Phytophthora megasperma* var. *sojae*, *P. cactorum*, *Pythium* sp., and *Aphanomyces euteiches* in soil by oomycetes, chytridiomycetes, hyphomycetes, actinomycetes, and bacteria. Phytopathology 67:622-628.
7. Sparrow, F. K. 1977. A Rhizidiomycopsis on azygospores of *Gigaspora margarita*. Mycologia 69:1053-1058.

PROCEDURES FOR INOCULATION OF PLANTS WITH VESICULAR-ARBUSCULAR MYCORRHIZAE IN THE LABORATORY, GREENHOUSE, AND FIELD

J. A. Menge and L. W. Timmer

TYPE OF INOCULUM

The type of vesicular-arbuscular (VA) mycorrhizal inoculum used depends greatly on the nature of the investigation. For growth response studies in the greenhouse as well as the field, mixed inoculum consisting of infected roots, spores, and mycelium may be satisfactory, is readily obtainable, and is usually very effective. Mixed pot culture inoculum generally has greater inoculum potential than purified spores or root material (7, 11). However, mixed inoculum is more difficult to quantitate and to make comparisons between individual experiments and between tests of various workers. For more precise work, it is often desirable to use inoculum which is readily quantitated.

Several investigators have compared infectivity and growth response with components of VA mycorrhizal inoculum. Hall (11) found that inoculation with 50 *Glomus* spores or 2 g of chopped, infected root fragments per plant gave equal growth responses, however, with *Acaulospora* sp., root fragments produced more growth than did spores. With onions, Manjunath and Bagyaraj (27) obtained a greater growth response with spores of *Glomus* than with root fragments in sterilized soil, however in unsterilized soil, both types of inocula were equally effective. Inoculation with spores of *G. mosseae* increased growth of clover more than inoculation with infected root fragments (37). In axenic culture, Hepper (17) found that new plants seeded into cultures from which all spores and infected roots had been removed still became infected, thus indicating that external mycelium can also serve as a source of infection. Next to mixed pot culture inocula, chlamydospores or azygospores are generally the most satisfactory inocula.

QUANTITATIVE AND QUALITATIVE EVALUATION OF INOCULUM

Spores for inoculum may be recovered by wet sieving, sucrose centrifugation, or other techniques described by Daniels and Skipper (Chap. 3). Spore numbers are relatively easily quantitated by counting in a nematode counting chamber or other suitable device. Quantification of root fragment inoculum is more difficult to achieve with accuracy. Roots must be cleared and stained and the percent infection determined by methods described by Kormanik and McGraw (Chap. 4). Percent infection serves only as an indication of the inoculum potential of root fragments.

Many papers report differences in responses of various plant species to different mycorrhizal species, but give no assurance that inoculum used was equally viable. Often much can be learned about the quality of inoculum by microscopic examination of spores at high power. Old and thick-walled, broken, or parasitized spores that may not germinate can often be detected. If a high percentage of spores are not viable, spore counts are not meaningful.

Viability stains have been little used, but may be helpful in evaluating the quality of inoculum. Daniels (personal communication) tested triphenyl tetrazolium chloride and rose bengal and found it difficult to differentiate living from dead spores. However, she found that when spores were placed in 1% aqueous thionin for 30 min, living spores stained deep blue, whereas dead spores accumulated stain only in the walls. Some experience was required to use the technique effectively. Stained spores germinated well when placed in soil.

Spore germination tests may also be helpful in evaluation of inoculum, but require more time. Spores can be disinfested using 0.5% NaOCl or other techniques (Watrud, Chap. 8A) and plated on agar with low nutrient levels (5, 16, 18). Germination usually begins in a few days and continues for up to 3 weeks. In the past, it was difficult to get consistent germination of many species (10), but it is now possible to obtain a high percentage germination with the proper combination of surface sterilization procedures, metal-free agar, and low nutrient levels.

Measurement of inoculum potential is the only critical way of comparing inoculum quantitatively. One of the most effective methods for evaluation of inoculum potential is the most probable number technique (Daniels and Skipper, Chap. 3). Several dilutions of inoculum with sterile soil are planted with seed. After a few weeks, roots are removed and stained to detect infection. An estimate of the population of spores in the original inoculum can be obtained from the greatest dilution at which infection occurs. This technique has the advantage of measuring infective propagules of the fungus rather than just the numbers of living spores. The disadvantage is that time is required to complete the evaluation and the effectiveness of the inoculum may diminish if improperly stored.

GREENHOUSE INOCULATION PROCEDURES

For many greenhouse studies of plant growth response and similar experiments, inoculum of any type may be mixed thoroughly with the soil. In other cases, it may be desirable to add inoculum next to the roots as plants are being transplanted. For other purposes, spores may be suspended in a viscous, nontoxic medium and plant roots dipped prior to planting. Carboxymethyl cellulose at 1% in water and a number of other materials of about the same viscosity have been effective for inoculation of citrus seedlings (S. Nemec, personal communication).

For more precise studies, it is often necessary to assure that a single plant or root becomes infected from a single or relatively few spores. For example, in the establishment of pot cultures, the use of only a few spores reduces the possibility that the cultures may be contaminated with plant pathogens, hyperparasites, or other mycorrhizal species. To accomplish this, it is necessary to confine a spore close to the root or assure that roots grow near spores. Several techniques are effective. Spores may be pipetted directly onto roots when transplanting seedlings. However, spores applied in this manner become dislodged from the roots easily. Alternately, spores can be pipetted onto a small piece of filter paper and the paper wrapped around rootlets to keep spores in place. The filter paper will decompose within 2 to 3 weeks.

The funnel technique (Fig. 1) is commonly used to inoculate plants in the greenhouse. The funnel forces roots to grow near the spores and assures infection even when few spores are used. Funnels can be formed from aluminum foil and simply peeled away to allow access to roots or for easy transplanting of the inoculated seedling.

The buried slide technique may be useful in following mycorrhizal infection from various inocula or under different soil conditions. Powell (29) coated slides with agar, placed spores or root fragments on one end of the slide and onion seed on the other and buried the slides. Onion roots grew down the slide and became infected. The infection process was followed microscopically by drying recovered slides and staining with trypan blue.

LABORATORY INOCULATION PROCEDURES

Frequently, it is desirable to establish axenic cultures of mycorrhizal fungi (Watrud, Chap. 8A and Allen and St. John, Chap. 8B). Hepper (17) described three techniques to establish axenic cultures: 1) the agar slant method, 2) the paper substrate technique, and 3) the Fahraeus slide method (Figs. 2A, B, C). With the agar slant method, a surface sterilized seed is added to the top of the slant and allowed to germinate. Later pregerminated spores are placed on the roots and a small amount of water added to the bottom of the slant (Fig. 2A). In the paper substrate technique, the seed is placed under a paper strip supported by a microscope slide in a test tube of nutrient solution. After the seed has germinated,

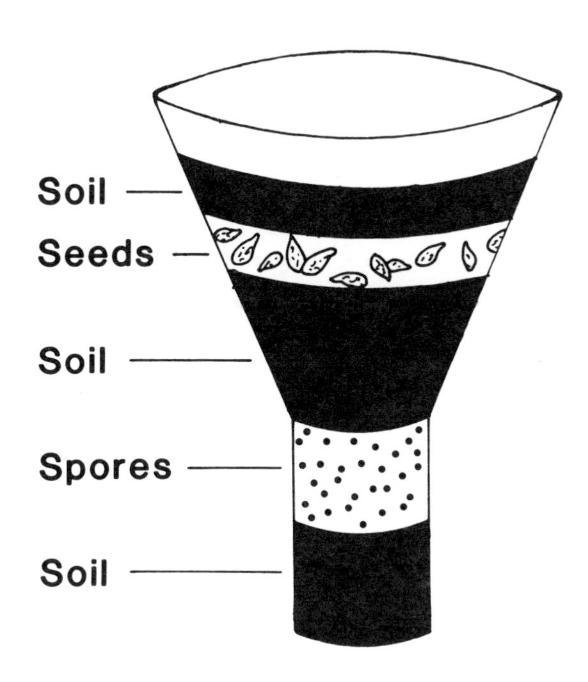

Fig. 1. The soil funnel technique for inoculation of plants with mycorrhizal fungi. After the seeds germinate, roots are forced to grow into the area where inoculum is present.

some nutrient solution is removed and spores are placed in contact with the roots (Fig. 2B). The Fahraeus slide technique is similar except roots are allowed to grow between the microscope slide and a cover slip fixed at 800 μm from the slide surface (Fig. 2C).

FIELD INOCULUM PREPARATION

Literature describing inoculum especially prepared for large scale field use is rare. However, most field inoculations are carried out with inoculum prepared in one of the following ways:

Granular.--Mixed inoculum of VA mycorrhizal fungi used for field inoculation usually consists of ground, crudely produced pot culture inoculum containing plant roots, mycorrhizal spores, and growth media such as perlite, peat moss, vermiculite, sand, or soil. Inoculum of this type can be ground using wood-chipping machines or hammer mills, or chopped finely using hand tools. This inoculum is usually air dried to about 5-20% moisture and resembles granular fertilizer in its consistency. Freeze-dried root material has been used to inoculate field-grown plants, but large amounts of this inoculum are difficult to obtain and may be less effective than pot culture inoculum (4, 7). Natural field soil containing mycorrhizal fungi has been successfully used as field inoculum (1, 2). However, spores are not concentrated in this material and large amounts are often required. Because of the danger of inadvertently introducing unwanted pathogens, the use of natural field soil as mycorrhizal inoculum is not recommended for commercial inoculations.

Pellets.--Pellets containing mycorrhizal inoculum have been used successfully to inoculate plants with mycorrhizal fungi (12, 13). Mycorrhizal pellets were formed by mixing 20 parts mycorrhizal inoculum (finely ground, roots, soil, and spores from a pot culture), 1 part sedimentary-loess clay (mean particle size, 16 μm) and 1 part tertiary sedimentary clay (mean particle size, 2-6 μm). Water was added until the mixture was malleable and could be rolled into pellets. One

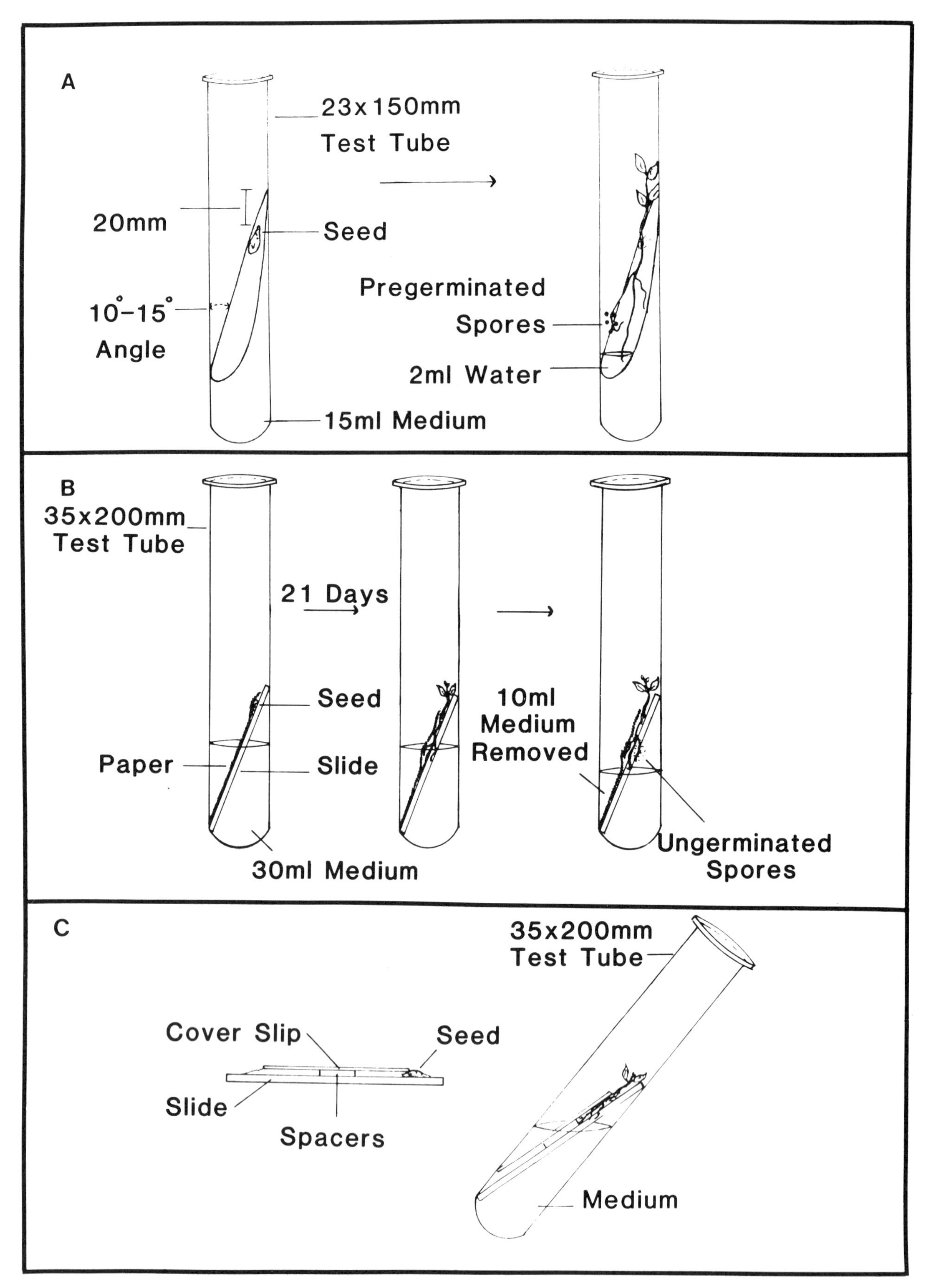

Fig. 2. Methods for inoculation of plants with mycorrhizal fungi in culture. A) the agar slant technique; B) the paper substrate method; C) the Fahraeus slide technique. See text for procedures.

pellet weighing 1.4 g was adequate to successfully infect white clover plants after storage for up to 28 days.

Jung et al (20) reported a unique method of producing pelleted inoculum of VA mycorrhizal fungi which is called polymer entrapped microorganisms. This method (6) requires mycorrhizal inoculum plus the following solutions: 1) phosphate buffer, pH 7 (1 g KH_2PO_4 and 2 g Na_2HPO_4/liter); 2) 238 g/liter of acrylamide in phosphate buffer; 3) 12.5 g/liter of N, N'-methylenebisacrylamide in phosphate buffer (excess N,N'-methylene-bisacrylamide is eliminated by filtration); 4) 126 mg of ammonium persulfate in 1 ml of distilled water (this solution must be prepared just before use from ammonium persulfate kept in a tightly closed vial); and 5) 76 μℓ N, N, N, N'-tetramethylethylene diamine solution (Merck). The mycorrhizal inoculum is imbedded by mixing the mycorrhizal inoculum in 150 ml of phosphate buffer, 50 ml of acrylamide solution, 50 ml of N, N'-methylenebis-acrylamide solution, 1 ml of ammonium persulfate solution, and 76 μℓ of N, N, N, N'-tetramethyl-ethylene diamine. Gel formation should be complete in 10 min. The solidified gel containing mycorrhizal inoculum can then be fragmented with a knife into 20-50 cm^3 pieces and washed under running tap water for 24 hr. After being washed, the gel, still in its wet state, can be fragmented into blocks of desirable size. The inoculum can be applied wet immediately as a pellet or dried 1-3 days at 25-28 C. This material can then be added granular or ground further into a powder. Mycorrhizal inoculum prepared in this manner was stored for several months at room temperature without loss of viability, infectivity, or efficiency.

FIELD INOCULATION PROCEDURES

Mycorrhizal inoculum will probably lead to successful mycorrhizal infection in the field when viable inoculum is placed in the root zone of actively growing plants which are not heavily fertilized. Various methods proposed for placement of mycorrhizal inoculum are illustrated in Fig. 3. Placement of inoculum in pot studies and even in fumigated field soil is not as critical as proper placement of mycorrhizal fungi in non-sterile field soil. Here, growth responses depend entirely on rapid infection and colonization of host roots by mycorrhizal fungi, which are superior to indigenous strains. The quantity of infection by the introduced endophyte depends first on placement and then on inoculum potential (Daniels and Skipper, Chap. 3), inoculum density (7), type of inoculum (11, 27), and on environmental factors. Table 1 illustrates the amount of pot culture inoculum and the number of spores which provided adequate infection with different placement in

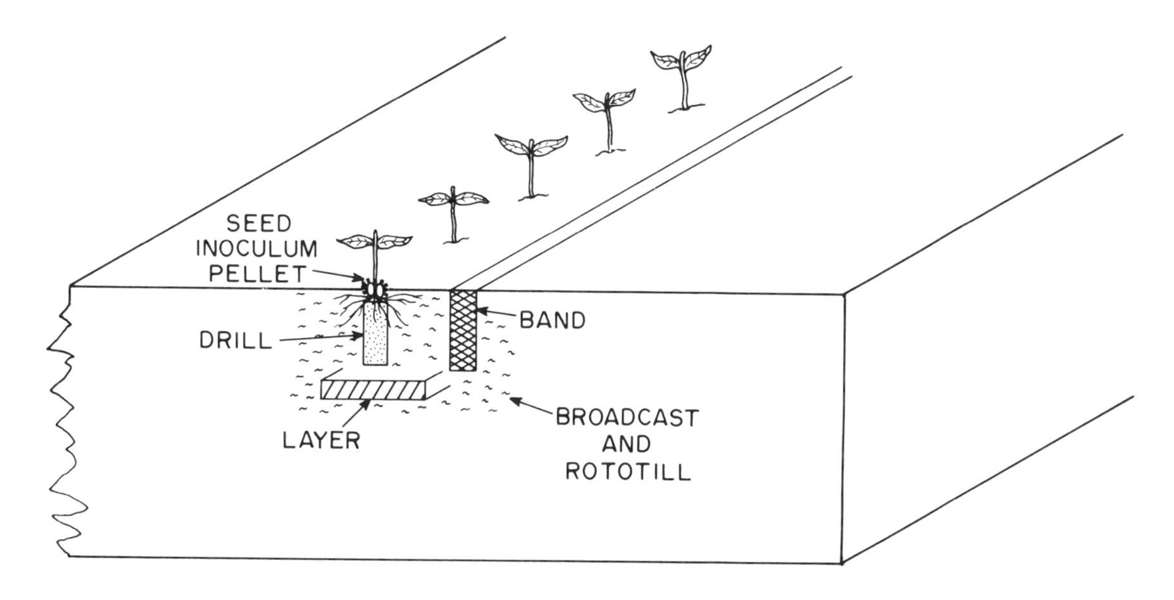

Fig. 3. Various methods proposed for placement of mycorrhizal inoculum for field inoculation.

Table 1. Quantities of VA mycorrhizal inoculum of *Glomus* spp. used to successfully inoculate crops under field conditions

Crop	Mycorrhizal species	Inoculation method	Amount of mixed culture inoculum (soil, roots, spores)	No. of chlamydospores	Reference
soybean	*G. macrocarpus*	mixed into soil	533 g/m^2	29,333/m^2	34
citrus	*G. fasciculatus*	layer under seed	300 g/linear m	--	23
citrus	*G. fasciculatus*	seed pellet			14
peach	*G. fasciculatus*	pad under seed	5 g/plant	--	24
potatoes	*G. macrocarpus*	layer under seed	3000 g/m^2	32,500/m^2	2
citrus	*G fasciculatus*	mixed into soil	666 g/m	--	38
onions, alfalfa, barley	*G. mosseae* *G. caledonius*	pad under seed	10-20 g/plant	--	28
Lotus pedunculatus	*G. fasciculatus*	soil pellet under seed	1.4 g/plant	28/plant	13
barley	*G. mosseae* *G. caledonius* *G. fasciculatus*	pad under seed	20 g/plant	--	3
citrus	*G. fasciculatus*	broadcast mixed into soil	300 g/m^2	4,800/m^2	7
citrus	*G. fasciculatus*	band	52 g/linear m	5,200/linear m	7

a variety of crops. These figures provide only a rough estimate of inoculum quantity and probably do not adequately reflect inoculum quality or inoculum potential.

Inoculum placement.--Placing inoculum in layers or pads beneath seeds (Fig. 3), so that roots will penetrate the inoculum, appears to be one of the most desirable methods of inoculating plants with mycorrhizal fungi. Jackson et al (19) studied several different methods to inoculate corn and found that layering inoculum 5-cm under the seed was superior to placing inoculum around the seed or banding the inoculum alongside the seed. Of several means of inoculating white clover with VA mycorrhizal fungi, sowing seed 2 cm above pelleted mycorrhizal inoculum was the most effective method for obtaining mycorrhizal infection (4). Menge et al (25) found that layering inoculum below citrus seed or applying inoculum in a band beside established seedlings was superior to pelleting seed with inoculum. Furthermore, a layer of inoculum 2.5 cm below the citrus seed was superior to a layer placed 5 cm below the seed. Placing inoculum in layers or pads beneath seeds has been effectively used for peach (24), soybean (34), white clover (30), onion (28), barley (3, 28), citrus (23), and potatoes (2). Unfortunately, most experiments using placement of this kind are done on a small scale by hand or using manual planters. Large application of this type can be done by applying inoculum through commercial tractor-drawn seeders or fertilizer banders prior to planting seed.

Banding or side dressing.--Banding or side dressing with VA mycorrhizal inoculum next to seedlings or seeds (Fig. 3) can be an effective way to apply mycorrhizal inoculum, especially if the quantity of inoculum is limited. Banding mycorrhizal inoculum 5 cm deep and 5 cm from 2-month-old citrus seedlings was as efficient as placing the inoculum directly on the roots at transplanting and more effective than seed pelleting (25). Using a tractor-drawn fertilizer bander, Ferguson (7) successfully inoculated 3 1/2-month-old citrus by applying granular mycorrhizal inoculum in a band 15 cm deep and 15 cm from the seedling row. Banding mycorrhizal inoculum 15 cm deep along the edges of 1-m wide citrus beds, prior to and after seeding, resulted in adequate mycorrhizal infection. Inoculation by this method was superior to seed pelleting but not as good as mixing inoculum into the soil or applying inoculum directly in the root zone. Placement of inoculum in a band alongside of corn and soybean seed did not result in adequate infection (19). Probably band placement must be in the zone of root proliferation to be effective.

Mixing inoculum with soil.--Mixing mycorrhizal inoculum with soil (Fig. 3) may be the most natural method for inoculating plants, but it can be inefficient since it often requires large amounts of inoculum to obtain rapid infection. For use in greenhouses and nurseries, however, mixing inoculum in the soil mix or bedding soil may be desirable. Broadcasting and rototilling mycorrhizal inoculum into citrus seedbeds gave maximum infection of citrus by mycorrhizal fungi as well as maximum growth responses (7). Mixing mycorrhizal inoculum into soil gave adequate mycorrhizal infection with soybean in small plots (34) and with citrus in nursery beds (38).

Seed pelleting.--Seed pelleting with mycorrhizal inoculum, similar to that which is practiced with *Rhizobium*, would be the easiest and most efficient method of inoculating plants with VA mycorrhizal fungi if it would provide consistently good infection. Most seed pelleting techniques using mycorrhizal fungi appear to be variations of the method devised by Hattingh and Gerdemann (14). In this method, mycorrhizal inoculum was extracted from a pot culture of sudan grass by wet sieving and decanting (9). Sievings were collected on 233, 149, and 74- μm sieves. In addition, roots retained on a 1-mm sieve were fragmented in a Waring blender and added to the sievings. Five ml of a 1% aqueous solution (w/v) of 400-centipoise methyl-cellulose was mixed with 35 ml of mycorrhizal sievings. This material was poured over 420 citrus seeds which were thoroughly mixed with the inoculum. The seed was dried until it could be handled without loss of inoculum. Using this method, citrus nursery stock was successfully infected with mycorrhizal fungi. However, use of this technique to inoculate citrus seed has given variable results (7, 25). In some cases, only 20% of the seedlings became immediately infected with the mycorrhizal symbiont, and layering, banding, or mixing mycorrhizal inoculum into the soil were more effective.

Crush and Pattison (4) added finely divided calcium carbonate to the methylcellulose-mycorrhizal sievings mixture used to pellet the seed. White clover plants which developed from pelleted seeds, became infected, but sowing seed above inoculum was more effective. Gaunt (8) pelleted sievings from pot cultures by mixing the mycorrhizal inoculum with 1% methylcellulose (w/v) and finely ground perlite. Although the extent of mycorrhizal infection was not presented, Gaunt concluded that mycorrhizal establishment on onion and tomato roots could readily be attained by pelleting seed. Hall (12) added white clover seed to mycorrhizal soil pellets as described above. He noted that if pellets were too moist or if substances were included which altered the pH, seed germination could be adversely affected.

Several researchers (7, 19, 32) have added inoculum directly with the seed. Powell (32) hand rolled white clover and rye grass seed in moist mycorrhizal inoculum and obtained up to 60% infected seedlings. In other experiments, 42-78% of the roots were mycorrhizal after inoculation in this manner, and it was concluded that mycorrhizal fungi could successfully be introduced by infesting seeds. Jackson et al (19), found that placement of inoculum with the seed was not an effective means of inoculating corn with mycorrhizal fungi. Ferguson (7) mixed mycorrhizal inoculum with citrus seed in a mechanical, tractor-drawn seed planter and sowed seed and inoculum together. Only about 48% infection was attained, but growth responses were observed.

Many authors (4, 7, 19, 26) have expressed the opinion that mycorrhizal inoculum must be placed in the root zone to be effective. Seed pelleting with mycorrhizal fungi can be effective but susceptible roots may pass through the zone of inoculum so rapidly that infection may not take place. Since many environmental factors affect the rate of root elongation as well as susceptibility to infection, it is not surprising that seed inoculations have met with variable success. Inoculum viability may be reduced during seed pelleting and fungicides used to coat seeds may have a harmful effect on mycorrhizal inoculum in the vicinity of seeds (7).

Pre-inoculation of transplanted seedlings.--For crops that will be transplanted, pre-inoculation in the seedbeds is probably the most effective method for inoculation. Pre-inoculated plants are frequently used in mycorrhizal field experiments (15, 21, 22, 31, 33, 35, 36). However, growth responses obtained in such experiments are subject to question unless it can be shown that mycorrhizal plants were of similar size and had similar nutrient concentrations at the time of transplanting.

In all experiments with mycorrhizal inoculation techniques, it is essential to have appropriate controls. In experiments where mixed inocula or spores are used, washings of the inocula containing associated microorganisms is the most adequate control. Autoclaved or irradiated inocula are unlikely to contain plant pathogens or beneficial microorganisms that might affect the outcome of the experiment. Where root fragments are present in inocula, non-mycorrhizal roots from plants grown under similar conditions may be added to controls.

During the course of the rapid developments in mycorrhizal research in the last few years, many techniques for inoculating plants have been developed. Obviously, many more will be devised as the need for special methods arises. However, the greatest need in this area is for techniques to rapidly assess the viability of inoculum so that plant responses to various mycorrhizal species and isolates can be meaningfully evaluated. Such methods should be extremely useful to assess the quality of batches of commercially prepared inoculum.

LITERATURE CITED

1. Azcon-G. de Aguilar, C., Azcon, R., and Barea, J. M. 1979. Endomycorrhizal fungi and Rhizobium as biological fertilizers for Medicago sativa in normal cultivation. Nature 279:325-327.

2. Black, R. L. B. and Tinker, P. B. 1977. Interaction between effects of vesicular-arbuscular mycorrhiza and fertilizer phosphorus on yields of potatoes in the field. Nature 267:510-511.
3. Clarke, C. and Mosse, B. 1981. Plant growth response to vesicular-arbuscular mycorrhiza. XII. Field inoculation responses of barley at two soil P Levels. New Phytol. 87:695-703.
4. Crush, J. R. and Pattison, A. C. 1975. Preliminary results on the production of vesicular-arbuscular mycorrhizal inoculum by freeze-drying, pages 485-493. in: F. E. Sanders, B. Mosse and P. B. Tinker, eds. Endomycorrhizas. Academic Press, London. 626 pp.
5. Daniels, B. A. and Graham, S. O. 1976. Effects of nutrition and soil extracts on germination of _Glomus mosseae_ spores. Mycologia 68:108-116.
6. Dommergues, Y. R., Diem, H. G., and Divies, C. 1979. Polyacrylamide-entrapped _Rhizobium_ as an inoculant for legumes. Appl. Environ. Microbiol. 37:779-781.
7. Ferguson, J. J. 1981. Inoculum production and field application of vesicular-arbuscular mycorrhizal fungi. Ph.D. Dissertation, University of California, Riverside, 117 pp.
8. Gaunt, R. E. 1978. Inoculation of vesicular-arbuscular mycorrhizal fungi on onion and tomato seeds. New Zealand J. Bot. 16:69-77.
9. Gerdemann, J. W. and Nicolson, T. H. 1963. Spores of mycorrhizal _Endogone_ species extracted from soil by wet sieving and decanting. Trans. Brit. Mycol. Soc. 46:235-244.
10. Godfrey, R. M. 1957. Studies on British species of _Endogone_. III. Germination of spores. Trans. Brit. Mycol. Soc. 40:203-210.
11. Hall, I. R. 1976. Response of _Coprosma robusta_ to different forms of endomycorrhizal inoculum. Trans. Brit. Mycol. Soc. 67:409-411.
12. Hall, I. R. 1979. Soil pellets to introduce vesicular-arbuscular mycorrhizal fungi into soil. Soil Biol. Biochem. 11:85-86.
13. Hall, I. R. 1980. Growth of _Lotus pedunculatus_ Cav. in an eroded soil containing soil pellets infested with endomycorrhizal fungi. New Zealand J. Agric. Res. 23:103-105.
14. Hattingh, M. J. and Gerdemann, J. W. 1975. Inoculation of Brazilian sour orange seed with an endomycorrhizal fungus. Phytopathology 65:1013-1016.
15. Hayman, D. S. and Mosse, B. 1979. Improved growth of white clover in hill grasslands by mycorrhizal inoculation. Ann. Appl. Biol. 93:141-148.
16. Hepper, C. M. 1979. Germination and growth of _Glomus caledonius_ spores: the effects of inhibitors and nutrients. Soil Biol. Biochem. 11:269-277.
17. Hepper, C. M. 1981. Techniques for studying the infection of plants by vesicular-arbuscular mycorrhizal fungi under axenic conditions. New Phytol. 88:641-647.
18. Hepper, C. M. and Smith, G. A. 1976. Observatons on the germination of _Endogone_ spores. Trans. Brit. Mycol. Soc. 66:189-194.
19. Jackson, N. E., Franklin, R. E., and Miller, R. H. 1972. Effects of vesicular-arbuscular mycorrhizae on growth and phosphorus content of three agronomic crops. Soil Sci. Amer. Proc. 36:64-67.
20. Jung, G., Mungier, G., Diem Hoang, D., and Dommergues, Y. R. 1981. Polymer-entrapped symbiotic microorganisms as inoculants for legumes and non-legumes. (Abstr.) Fifth North American Conference on Mycorrhizae, Quebec, Canada. p. 54.
21. Khan, A. G. 1972. The effect of vesicular-arbuscular mycorrhizal associations on growth of cereals. I. Effects on maize growth. New Phytol. 71:613-619.
22. Khan, A. G. 1975. The effect of vesicular-arbuscular mycorrhizal associations on growth of cereals. II. Effects on wheat growth. Ann. Appl. Biol. 80:27-36.

23. Kleinschmidt, G. D. and Gerdemann, J. W. 1972. Stunting of citrus seedlings in fumigated nursery soils related to the absence of endomycorrhizae. Phytopathology 62:1447-1453.

24. LaRue, J. H., McClellan, W. D., and Peacock, W. L. 1975. Mycorrhizal fungi and peach nursery nutrition. Calif. Agric. 29(5):6-7.

25. Menge, J. A., Lembright, H., and Johnson, E. L. V. 1977. Utilization of mycorrhizal fungi in citrus nurseries. Proc. Int. Soc. Citriculture 1:129-132.

26. Mosse, B. and Haymann, D. S. 1980. Mycorrhiza in agricultural plants, pages 213-230. in: P. Mikola, ed. Tropical Mycorrhiza Research, Clarendon Press. 270 pp.

27. Manjunath, A. and Bagyaraj, B. J. 1981. Components of VA mycorrhizal inoculum and their effects on growth of onion. New Phytol. 87:355-361.

28. Owusu-Bennoah, E. and Mosse, B. 1979. Plant growth responses to vesicular-arbuscular mycorrhiza. XI. Field inoculation responses in barley, lucerne and onion. New Phytol. 83:671-680.

29. Powell, C. L. 1976. Development of mycorrhizal infections from *Endogone* spores and infected root segments. Trans. Brit. Mycol. Soc. 66:439-445.

30. Powell, C. L. 1976. Mycorrhizal fungi stimulate clover growth in New Zealand hill country soils. Nature 264:436-438.

31. Powell, C. L. 1977. Mycorrhizas in hill country soils. III. Effect of inoculation on clover growth in unsterile soils. New Zealand J. Agric. Res. 20:343-348.

32. Powell, C. L. 1979. Inoculation of white clover and rye grass seed with mycorrhizal fungi. New Phytol. 83:81-86.

33. Powell, C. L. and Daniel, J. 1978. Growth of white clover in undisturbed soils after inoculation with efficient mycorrhizal fungi. New Zealand J. Agric. Res. 21:675-681.

34. Ross, J. P. and Harper, J. P. 1970. Effect of *Endogone* mycorrhiza on soybean yields. Phytopathology 60:1552-1556.

35. Saif, S. R. and Khan, A. G. 1977. The effect of vesicular-arbuscular mycorrhizal associations on growth of cereals. III. Effects on barley growth. Plant and Soil 47:17-26.

36. Schenck, N. C. and Hinson, K. 1973. Responses of nodulating and non-nodulating soybeans to a species of *Endogone* mycorrhiza. Agron. J. 65:849-850.

37. Smith, F. A. and Smith, S. E. 1981. Mycorrhizal infection and growth of *Trifolium subterraneum*: comparison of natural and artificial inoculum. New Phytol. 88:311-325.

38. Timmer, L. W. and Leyden, R. F. 1978. Stunting of citrus seedlings in fumigated soils in Texas and its correction by phosphorus fertilization and inoculation with mycorrhizal fungi. J. Amer. Soc. Hort. Sci. 103:533-537.

EVALUATION OF PLANT RESPONSE TO COLONIZATION BY VESICULAR-ARBUSCULAR MYCORRHIZAL FUNGI

A. Host Variables

R. G. Linderman and J. W. Hendrix

Introduction

Vesicular-arbuscular (VA) mycorrhizal fungi often have a positive quantitative effect on growth and vigor of plants. Indeed, the purpose of much mycorrhiza research is to utilize these fungi to accelerate growth or to produce plants of acceptable size using less resources, such as fertilizer. Therefore, evaluation of response by plants that have formed VA mycorrhizae (VAM) usually involves size determinations. Frequently the differences between inoculated and non-inoculated plants are obvious and can be recorded photographically (Plate 4D), but for experimental purposes, one must also quantify and document such differences.

Our intent in this chapter is to discuss the different parameters that can be measured to evaluate plant responses to VAM, and the experimental factors affecting these responses and their interpretation, under controlled greenhouse as well as field conditions.

Physical Parameters

There are many parameters that can be measured to describe relative growth differences between plants inoculated with VA mycorrhizal fungi and those not inoculated. Certainly not all of these parameters would be used for any one experiment; their selection will depend on the type of plant and its physiology as well as the specific objectives and duration of the experiment.

Tissue Weight.--Weights of shoots and/or roots can be taken as fresh weights, but more accurately as dry weights (after 72 hr at 70°C). Taking these two measurements separately will allow the calculation of a root/shoot ratio which is of particular importance in evaluating the quality of such plants as forest trees or woody ornamental plants. It is sometimes appropriate and desirable to periodically harvest just the shoots of plants that regenerate new shoot growth, such as grasses and forage legumes. Multiple harvests would then give responses in terms of total above-ground seasonal yield. When harvesting whole plants, it is desirable to mark the soil line to facilitate an accurate separation of roots and shoots.

Plant Height.--With many plants, linear extension of the main shoot upward or laterally is a valuable measure of response to VAM. Measurements should be from the base of the plant, at the soil line, to the tip of the terminal shoot bud. A type of measurement within this category would be the number and lengths of internodes.

Stem Diameter.--Stem diameter (stem caliper or root collar diameter) is useful for evaluating responses on many single stem plants. This is often the most important parameter for evaluation of forest tree seedlings and certain kinds of woody nursery stock, and also transplants of certain annuals, such as tomatoes and tobacco. Vernier or dial calipers should be used in preference to a ruler. The best place on the plant to measure must be determined by experience. Some plants vary in the amount of swelling at the soil line in a manner unrelated to true size. A consistent routine should be established wherein measurements are taken at a certain distance from the soil line, at the same internode, or between the same internodes. It may be necessary to measure at several places to determine the least variable point of measurement which correlates well with other more meaningful parameters requiring plant sacrifice.

Shoot Volume.--This parameter is useful as a nondestructive estimate of shoot size or weight. The formula height multiplied by the square of stem diameter has been used as an estimate of shoot size of Southern pines outplanted for long-term growth evaluation (15). Measurements based on the shoot diameter and heights of plants having multiple crowns or spherical geometry may also be useful, but experiments should be conducted to determine their relevance to other growth parameters.

Leaf Area and Number.--Determining the number of leaves and leaf area on a plant can be done non-destructively, but with many plants this method would be extremely time consuming or impossible. If leaves can be removed, however, their numbers and total area can be measured readily, the latter preferably with an electronic leaf area meter. Another method is to produce a photocopy of the leaves on a sheet of paper of known area and weight. The leaf images can be cut out and weighed; their weight will be proportional to their area.

Root Responses.--Roots colonized by VA mycorrhizal fungi exhibit little or no morphological change compared to non-mycorrhizal roots, other than reduced root hair development. Some plants, such as onion, produce a yellow pigment when they form VAM, but most plants do not. The pigment can be extracted and analyzed colorimetrically as a measure of colonization (3) (see Chapter 4). This pigmentation effect may be host-specific (13) or fungus-specific (1).

In the case of experiments with unrooted cuttings, one can count the number or volume of newly-formed roots. A reasonably accurate way to measure root volume is to wash the roots free of soil and determine the amount of water displaced when the whole root system is submerged in a vessel of water.

Root length can be measured directly, but the process is laborious and very difficult for fibrous root systems. Root lengths for plants with tap roots can be estimated by spreading the roots out on white paper and photocopying the root image, which serves as a record. The root image on the photocopy can then be traced with a map distance instrument to determine total length. Giovanetti and Mosse's method (10) for estimating root colonization can also provide an estimate of total root length in the sample.

Wilt Response.--Plants with VAM may be less likely to wilt under drought conditions. The magnitude of the wilt response can be measured by determining the water potential of the leaf using a pressure bomb or other apparatus (see Chapter 15B). Other plants with a distinct leaf petiole, such as cotton, exhibit a physical change in the angle of the petiole in relation to the stem, the angle increasing with loss of turgor leading to the wilt response (8).

Yield/Maturity.--Crop yield is usually the most important parameter to be measured. However, mycorrhizal fungi may also affect the time at which plants reach definitive physiological stages. These differences may be quantified in terms of time of flowering and the number and size of flowers; the number and size of fruits, bulbs, tubers, corms, etc., or, for some crops, amount of seed produced (20). For some plants, maturity may be indicated by the time the plant shoot dies back (bulb crops) or by the time of leaf drop in the fall. For perennials, a related phenomenon to be considered is the time of bud break and the beginning of active growth in the spring.

Transplant Survival.--Some plants normally are grown as seedling or cutting transplants. VAM may affect survival of transplants. This response may be recorded as either a plus or a minus to enable calculation of percentage of survival. In some cases, however, the response may be a difference in the time for the transplant to resume growth after overcoming the physical shock, and can be measured as number of days before signs of new growth (such as bud break or next new set of leaves) appear.

Disease Reaction.--There are reported cases (23) where plants with VAM exhibit a different response to plant pathogens than non-mycorrhizal plants. The response may be manifested as changes in plant growth, measured as discussed above. In other cases, the disease itself may be quantified by counting lesions or cankers, pathogen multiplication (as in a virus titer) (7), or plant survival.

Visual Changes.--Frequently, visible differences between plants with and without VAM are evident, sometimes when measurable physical parameters do not differ. Such differences, which may be subtle color or shape differences, may be recorded photographically. If the differences are due to deficiency of mineral elements such as P, N, Cu, or Zn or loss of chlorophyll, one could document those differences by analysis of mineral elements or extractable pigments (i.e., anthocyanin, chlorophyll).

Physiological or Chemical Parameters

The morphological changes that plants may exhibit in response to colonization by VA mycorrhizal fungi are the result of altered plant physiology. Many plant physiological processes are influenced by P-nutrition, which often is greatly affected by the plant having VAM. Page limitations for this chapter preclude description of the methods used to measure these processes, but one should be aware that they can be measured and that physiological responses can often be detected much earlier than morphological ones. For example, altered *photosynthetic rates*, as measured by CO_2 assimilation, ultimately would be reflected as a change in plant size, but only after considerable time lapse. Changes in *plant respiration*, *transpiration*, *hormones*, and *amino acid and protein concentrations*, all may take place well in advance of physical changes. Patterns of mineral nutrient uptake (i.e., N, P, K) can be measured by isotopic tracer methods, and many examples of such studies are reported in the literature. Workers frequently subject their experimental plants to tissue analysis to determine relative differences in major and minor element uptake, and/or soluble carbohydrates, between plants with and without VAM.

Some physiological responses have secondary and tertiary responses that also can be measured. For example, increased P uptake by plants with VAM may in turn affect membrane permeability, the pattern of root exudation, and the degree of root pathogen ingress which ultimately may influence plant size or survival. Root exudation rates or patterns can be measured by monitoring efflux of isotopes, i.e., ^{86}Rb as a tracer for K^+ (21), as well as efflux of other exuded organic substances or metabolites.

Another measurable process is nitrogen fixation which is influenced by P nutrition. Thus, a plant with VAM and *Rhizobium* nodules presumably will fix more nitrogen than a plant without VAM if P is limiting. Fixation of N_2 by the nodules can be measured by the acetylene reduction method (12).

Interpreting Plant Responses to VAM under Controlled Conditions

As in other research, experiments intended to determine how plants might respond to inoculation with VA mycorrhizal fungi are only as good as the experimental plan. This section discusses factors that need to be considered in the planning, design, and interpretation of such experiments. Some experiments that fail to demonstrate any differences in response to inoculation with VA mycorrhizal fungi or that have been misinterpreted can be traced to neglected crucial factors.

Growth Medium.--It is extremely important to know pertinent characteristics of the soil mix being used, or to choose a growth medium that will not be detrimental to the success of an experiment (5, 16). A positive host response to VAM is favored by a medium that is well-drained, not excessively organic, and with moderate P deficiency or P-fixing capacity. The medium may need to be steam pasteurized, gas fumigated, irradiated or microwaved to suppress or eliminate pathogens or indigenous VA mycorrhizal fungi that may interfere with the response to added VA mycorrhizal fungi. Most advances in exploiting mycorrhizal fungi in crop production still are empirically developed through evaluation of many media, fertilizers, watering regimes, and fungal isolates before choosing the superior combinations.

Fertility.--Nearly all plants can acquire mineral nutrients without having VAM if the nutrients are supplied in a readily available form and adequate amount. However, high levels of fertility, especially P, generally have a detrimental effect on the establishment of VA mycorrhizal fungi and on their potential to enhance plant growth by increasing P uptake. Conversely, having so little P available that even VAM can't acquire it may even suppress growth below the control level. Therefore, it is essential to conduct growth response studies at more than one fertility level, preferably using several levels of soluble P. Another way to establish different levels of available P is to fertilize with relatively insoluble forms of P or slow-release fertilizers. Slow-release fertilizers have great potential because they limit the amount of soluble nutrients at any one time but provide sufficient nutrients over a period of months for optimum plant growth and yet allow mycorrhizal development.

A desired outcome of varying available P is to establish how VAM-responsive or VAM-dependent the host is under the test conditions used. A VAM-responsive host will exhibit enhanced growth at low P levels, but progressively reduced or no growth enhancement at high levels. A VAM-dependent host would exhibit enhanced growth by VAM up to abnormally high P levels indicating that, for practical purposes, it requires VAM to acquire adequate P. The degree of dependence may be influenced by the availability of P as governed by the P-fixing capacity of the soil or medium; hence it is important to monitor the level of available P during the experiment to confirm whether sufficient soluble P is available.

Appropriate Controls.--Appropriate control treatments are a must in VA mycorrhiza research. Unless surface disinfested spores are used, the inoculum generally contains many organisms, some of which may be capable of enhancing or suppressing growth. Therefore, it is essential to have control treatments that include all organisms contained in the pot culture inoculum except the VA mycorrhizal fungus. Generally, these controls can be established by inoculating all plants with a solution prepared by passing pot culture inoculum through a 20 μm sieve that will remove spores and most hyphal fragments of VA mycorrhizal fungi, but allow other organisms to pass through. If roots with VAM or soil are used as the source of inoculum, then roots or soil from non-mycorrhizal plants should be added to control treatments. If VA mycorrhizal fungus spores are used as the source of inoculum, spore washings should be added to control treatments.

Influence of Root Pathogens.--Growth response experiments may at some point become contaminated with root pathogens that could mask growth effects due to the VAM. Thus, it is essential to examine roots at the time of harvest to verify their freedom from disease. Examination for root pathogens should involve little extra work, because roots are washed anyway in order to clear, stain, and examine for colonization by VA mycorrhizal fungi. Roots could also be plated on culture media to assay for root pathogens.

One major source of contamination may be the inoculum itself. In such cases, the inoculum should be decontaminated, perhaps by chemical treatment, to suppress or eliminate the pathogen. If the inoculum is to be spores only, these can be decontaminated by hypochlorite or antibiotic treatments (17). To minimize pathogen problems, one should avoid producing pot-culture inoculum on the same host to be studied in the experiment.

Multiple Harvests.--The formation of VAM may take several weeks. During this period the fungus is parasitic, receiving organic nutrients from its plant host, presumably benefiting its host little or not at all until it has developed an effective nutrient/water absorbing external mycelium in the soil. Thus, during the parasitic phase, mycorrhizal plants may actually exhibit a growth depression because the fungus is a nutrient sink. To avoid being misled by this response, it is desirable to conduct multiple harvests over a time period long enough to allow establishment of both the parasitic as well as the external mycelium phase. The latter may not form, or may be slow to form, because of inhibitory effects of the growth medium, and thus growth enhancement due to increased P uptake would be delayed or may not occur at all. Conversely, caution also is required in interpreting growth experiments conducted over a long period of time, especially where soil volume is restricted. Sometimes growth of plants of one treatment appear to catch up with growth of plants in another. Although this may well be a valid result, especially in the field, for container experiments it could simply be caused by the faster-growing plants becoming pot-bound or depleting nutrients. Such problems may be avoided by observing root development during the course of the experiment, perhaps terminating the experiment as soon as roots begin to crowd the containers, or by using larger containers.

Host Species.--It is important to recognize that there are certain plant families that typically cannot and/or do not form VAM (9): ectomycorrhizal families (Pinaceae, Betulaceae, Fagaceae), septate endomycorrhizal families (Orchidaceae, Ericaceae), and the typically non-mycorrhizal families (Brassicaceae, Chenopodiaceae, Fumariaceae, Cyperaceae, Commelineaceae, Urticaceae, Polygonaceae, and Proteaceae). There are also some plants that typically form ectomycorrhizae in nature that may form VAM if suitably inoculated (members of the families Myrtaceae, Salicaceae, some species of *Quercus* and *Alnus*).

Host Responsiveness vs. Dependency.--As mentioned earlier, if inoculation studies with VA mycorrhizal fungi are conducted at several P levels, it is possible to characterize the host as responsive or dependent; in the latter case, the host apparently does not respond even to higher levels of P (9). It is essential to determine the levels of available P during the experiment, not just the amount of soluble P added. Different media "fix" P to varying degrees so that lack of growth response could mean that there was insufficient available P for the host to acquire adequate amounts on its own. The same host tested in another soil with less P-fixing capacity could respond to soluble P even without VAM. Thus, a host may only depend on VAM in soils where P is sufficiently bound or is inaccessible as to be unavailable except to the absorbing hyphae of the VAM.

Colonization Level in Relation to External Hyphae.--It is necessary to determine the level of colonization by VA mycorrhizal fungi, preferably during and at the final harvest of the experiment (4). Several studies have shown that plant growth enhancement by VAM generally is related to colonization level and to the extent of external mycelium, the latter being necessary for P uptake (22). However, Graham et al (11) demonstrated that in some cases in which plant growth was not enhanced, external mycelium failed to develop although roots were heavily colonized. They used the method of Sutton and Sheppard (25) wherein the amount of external hyphae was related to the amount of sand those hyphae aggregated or bound together. The absence of external mycelium was presumably due to some inhibitory effect of the test soil. These findings may offer a plausible explanation for the lack of growth response in some inoculation experiments. Where significant colonization has occurred, Menge et al (16) have shown that soil factors such as pH, organic matter content, levels of P, Mn, Zn, and Cu, were correlated to lack of growth enhancement in the soils they examined.

Miscellaneous Considerations.--During the planning stage, other variables should be taken into account. Light and moisture levels are especially important. Too little light may seriously reduce the production of carbohydrate needed to establish a mutualistic relationship. The statistical design of the experiment should take into account the inherent genetic variability in seeds compared with clonally propagated plants and number of replications should be adjusted accordingly. The magnitude of differences one wants to detect will determine the number of replications required. Detection of true differences of small size will require more replications than those required to detect true differences of large size; experience with the experimental system may be required to make this judgment.

The form and quantity of inoculum may influence the growth response, in terms of both magnitude and time. Additionally, VA mycorrhizal fungi undoubtedly exist as biotypes adapted to particular environmental conditions. Choosing the wrong biotype for the test system or vice versa will likely nullify any meaningful growth response. It may be desirable, therefore, to test more than one isolate, species, or biotype in any given experimental system.

Single-spore techniques have rarely been used with VA mycorrhizal fungus species. Development of such techniques is needed, however, because meaningful research on biotypes is hampered by multispore-root fragment inoculation procedures usually employed.

Interpreting Plant Responses to VAM Under Field Conditions.

The number of papers dealing with field aspects of endomycorrhizal fungi is small compared to those done in containers in greenhouses. The reason is that field situations are enormously complex compared with more controllable experiments with "pure" cultures in containers, the latter certainly being complex enough. There is, however, considerable need to demonstrate beneficial growth responses to VAM in actual field situations. Some of the problems related to field research were recently discussed by Maronek et al (14), and will only be briefly reviewed here.

Assessment of Inoculum Potential.--Endomycorrhizal fungi are present in nearly all agricultural soils, but the task of evaluating their roles in crop production lies ahead and will depend largely on methods yet to be developed. With present methods, however, it is possible to accumulate more basic ecological data needed for planning of later field-scale experiments. Those with a knowledge of taxonomy can identify fungi present in particular field situations by planting hosts in field-collected soil samples and evaluating the cultures for colonization and sporulation after a suitable growth period (24). It is also possible to identify directly those spores present in field soils; though laborious and difficult, this procedure has the advantage of permitting quantification of spore numbers. Parallel with an assessment of fungal activity based on spore numbers should be an assessment of colonization potential of soils (6, 19). Colonization potentials are necessary because some soils contain fungi that readily colonize but are apparently non-sporulating species; in addition, spore populations may reflect dormant species, while spores of currently active fungi may not be present at the time soil samples are collected. Spore populations and colonization assays can be related to a number of age-old and new crop production techniques known to affect crop yields: crop rotation, winter cover crops, green manure crops, fertilization practices, no-till as opposed to conventional tillage production, and any number of other factors. These sampling studies should be conducted throughout the production or cropping season.

Soil Fumigation.--While field research in the absence of endomycorrhizal fungi is generally impractical, fungicides and soil fumigants affect mycorrhizal fungi and can be used in experiments to relate fungal and host relationships (for references, see 14). Soil fumigation offers great potential, but much work remains to be done in its development as an experimental tool in mycorrhiza research. While fumigants are lethal to endomycorrhizal fungi, there is evidence that under field conditions they seldom completely eliminate all propagules of VA mycorrhizal fungi (14). Information is needed on mycorrhizal colonization in relation to factors such as fumigant rates, different fumigants (there is evidence that chloropicrin is more lethal to mycorrhizal fungi than methyl bromide), soil temperature and soil type.

Introduction of Exotic Species or Isolates.--Efforts to achieve growth benefits with exotic fungal isolates introduced to natural soils harboring endemic populations has met with mixed success. Some methods for interpreting growth results are available (18, 26). More ecological research is needed to determine which may be a more fruitful approach, managing endemic fungal populations or introducing exotic species. If the latter, methods to prevent endemic species from competing with the introduced ones may need to be developed, although there is some evidence that the indigenous fungi do not interfere with colonization by inoculated fungi (2).

LITERATURE CITED

1. Abbott, L. K. and Robson, A. D. 1978. Growth of subterranean clover in relation to the formation of endomycorrhizas by introduced and indigenous fungi in a field soil. New Phytol. 81:575-585.
2. Abbott, L. K. and Robson, A. D. 1981. Infectivity and effectiveness of five endomycorrhizal fungi: Competition with endigenous fungi in field soils. Aust. J. Agric. Res. 32:621-630.
3. Becker, W. N. and Gerdemann, J. W. 1977. Colorimetric quantification of vesicular-arbuscular mycorrhizal infection in onion. New Phytol. 78:289-295.
4. Biermann, B. and Linderman, R. G. 1981. Quantifying vesicular-arbuscular mycorrhizae: A proposed method toward standardization. New Phytol. 87:63-67.
5. Biermann, B. and Linderman, R. G. 1982. Effect of plant growth medium and fertilizer P level on establishment of VA mycorrhizae and host growth response. J. Amer. Soc. Hort. Sci. (In press)
6. Bowen, G. D. 1980. Misconceptions, concepts and approaches in rhizosphere biology. Pages 283-304 _in_: Contemporary Microbial Ecology. D. C. Elwood, J. N. Hedger, M. J. Latham, J. M. Lynch, and J. H. Slater, eds. Academic Press. 438 pp.
7. Daft, M. J. and Okusanya, B. O. 1973. Effect of _Endogone_ mycorrhiza on plant growth. V. Influence of infection on the multiplication of viruses in tomato, petunia, and strawberry. New Phytol. 72:975-983.
8. Erwin, D. E., Moje, W. and Malca, I. 1965. An assay of the severity of Verticillium wilt on cotton plants inoculated by stem puncture. Phytopathology 55:663-665.
9. Gerdemann, J. W. 1975. Vesicular-arbuscular mycorrhizae. Pages 575-591 _in_: The development and function of roots. J. G. Torrey and D. T. Clarkson, eds. Academic Press Inc., London. 618 pp.
10. Giovannetti, M. and Mosse, B. 1980. An evaluation of techniques for measuring vesicular-arbuscular mycorrhizal infection in roots. New Phytol. 64:489-500.
11. Graham, J. H., Linderman, R. G. and Menge, J. A. 1982. Development of external hyphae by different isolates of mycorrhizal _Glomus_ spp. in relation to root colonization and growth of Troyer citrange. New Phytol. 90: (in press).
12. Hardy, R. W. F., Holsten, R. D., Jackson, E. K. and Burns, R. C. 1968. The acetylene-ethylene assay for N_2 fixation: laboratory and field evaluation. Plant Physiol. 43:1185-1207.
13. Khan, A. G. 1978. Vesicular-arbuscular mycorrhizas in plants colonizing black wastes from bituminous coal mining in the Illawarra region of New South Wales. New Phytol. 81:53-63.
14. Maronek, D. M., Hendrix, J. W. and Kiernan, J. 1981. Mycorrhizal fungi and their importance in horticultural crop production. Horticultural Rev. 3:172-213.
15. Marx, D. H. and Artman, J. D. 1979. The significance of _Pisolithus tinctorius_ ectomycorrhizae to survival and growth of pine seedlings on coal spoils in Kentucky and Virginia. Reclam. Rev. 2:23-31.
16. Menge, J. A., Jarrell, W. M., Labanauskus, C. K., Huzar, C., Johnson, E. L. V. and Sibert, D. 1982. Predicting mycorrhizal dependency of Troyer citrange on _Glomus fasciculatus_ in California citrus soils. Soil Sci. Soc. Am. J. 44: (in press).

17. Mertz, S. M., Jr., Heithaus, J. J., III and Bush, R. L. 1981. Mass production of axenic spores of the endomycorrhizal fungus *Gigaspora margarita*. Trans. Br. Mycol. Soc. 72:167-169.
18. Powell, C. L. 1976. Development of mycorrhizal infections from *Endogone* spores and infected roots. Trans. Br. Mycol. Soc. 66:439-445.
19. Powell, C. L. 1980. Mycorrhizal infectivity of eroded soils. Soil Biol. Biochem. 12:247-250.
20. Powell, C. L. L. 1981. Inoculation of barley with efficient mycorrhizal fungi stimulates seed yield. Plant Soil 59:487-489.
21. Ratnayake, M., Leonard, R. T. and Menge, J. A. 1978. Root exudation in relation to supply of phosphorus and its possible relevance to mycorrhizal formation. New Phytol. 81:543-552.
22. Sanders, F. E., Tinker, P. B., Black, R. L. B. and Palmerley, S. M. 1977. The development of endomycorrhizal root systems. I. Spread of infection and growth-promoting effects with four species of vesicular-arbuscular endophyte. New Phytol. 78:553-559.
23. Schenck, N. C. and Kellam, M. K. 1978. The influence of vesicular arbuscular mycorrhizae on disease development. Florida Agric. Expt. Sta. Bull. 798. 16 pp.
24. Schenck, N. C. and Smith, G. S. 1981. Distribution and occurrence of vesicular-arbuscular mycorrhizal fungi on Florida agricultural crops. Proc. Soil Crop Sci. Soc. Florida 40:171-175.
25. Sutton, J. C. and Sheppard, B. R. 1976. Aggregation of sand-dune soil by endomycorrhizal fungi. Can. J. Bot. 54:326-333.
26. Warner, A. and Mosse, B. 1980. Independent spread of vesicular-arbuscular mycorrhizal fungi in soil. Trans. Br. Mycol. Soc. 74:407-410.

EVALUATION OF PLANT RESPONSE TO COLONIZATION BY VESICULAR-ARBUSCULAR MYCORRHIZAL FUNGI

B. Environmental Variables

G. R. Safir and J. M. Duniway

INTRODUCTION

Considerable information is available in the literature on the effects of environmental variables on host plant responses to vesicular-arbuscular (VA) mycorrhizal infection and several reviews on the subject are available (1,5,10). Most of the literature on plant responses to mycorrhizae, however, is descriptive and much more research will have to be done on the physiology of mycorrhizal plants before we will know their responses to many of the environmental variables that influence plant and fungal growth in nature. It is impossible to discuss all the environmental variables that are likely to influence plant responses to VA mycorrhizae in this brief review and only some of the more important variables that can be manipulated under controlled conditions are discussed.

Before the impacts of environmental variables, are discussed, it should be noted that one cannot make generalizations about the influence of environmental variables on mycorrhizal plants for each host-endophyte-environment system may behave differently and should be characterized for the desired plant and or fungal responses before any experiments are conducted to answer specific questions about the physiological impacts of VA mycorrhizae. The potentials for VA mycorrhizae to promote plant growth are generally the greatest under conditions where the fungi can alleviate some physical-chemical constraints on root acquisition of nutrients from the soil. Therefore, optimum or nonlimiting soil conditions for the growth of nonmycorrhizal plants are rarely the conditions under which mycorrhizae promote growth the most. Furthermore, the fungus and host members of a mycorrhizal relationship may each respond differently to environmental factors. For example, spore germination, hyphal spread within the soil, infection levels, fungal survival and growth promotion may all respond differently to any one set of environmental conditions. The choice of conditions, therefore, depends on ones goals in research and, in some cases, on the extent to which one wishes to reproduce natural conditions.

Soil Conditions

With few exceptions, VA mycorrhizae are world-wide in distribution. North temperate podzols, very wet soils, and highly disturbed soils, such as coal spoils, may support relatively few natural infections by mycorrhizae (5). Other soil types present in grasslands, muck farms, rain forests, sand dunes and arid regions support variable levels of VA mycorrhizae. Each soil type used for mycorrhizal studies should be analyzed for the variables discussed below. Furthermore, soil strength and texture may affect plant responses to mycorrhizae. Soil strength or penetration resistance influences the rate at which roots can grow to reach water and nutrients in soil and soil texture influences the rates at which water and nutrients will flow or diffuse to root surfaces. Because the mycelium of mycorrhizal fungi may translocate nutrients and perhaps some limited amount of water from the bulk soil to roots they may promote growth more when soil strength and/or texture limit root function than when there are few such soil constraints on roots. Other general characteristics that ought to be considered include the presence and responses of microorganisms other than VA mycorrhizal fungi, inoculum distribution, the root length density that the host will achieve, and the total volume and nutrient content of the soil available to mycorrhizal and nonmycorrhizal root systems.

Phosphorus.-- The level of available phosphorus (P) in the soil has been shown to have dramatic effects on mycorrhizal infection levels and on the stimulation of plant growth by mycorrhizae (7). Different forms of P added to the soil such as rock phosphate, super phosphate, organic phosphate or solution P will have different effects on a mycorrhizal system, probably because of differences in phosphate solubility (5). In most instances high levels of soil phosphorus will inhibit mycorrhizal infection and growth promotion to some degree. These P effects will be strongly affected by other factors such as soil type, soil pH, soil nitrogen levels and the host-endophyte combination used. It has been hypothesized that the effects of soil P on infection levels are more directly expressed via their impact on host P nutrition. There is evidence, however, that under some conditions soil or host P levels may be unrelated to infection levels depending on the fungus strain and the host involved. It is advised that both soil and host P levels should be determined when characterizing a plant-fungus combination. There are several different methods than can be employed for soil P determinations and their efficiency is largely dependent on pH (9).

pH.-- Their is increasing evidence that various VA mycorrhizal fungi differ in their preference for certain pH ranges (7). The mechanisms for this are unknown, but it is known that soil nutrient solubility and plant nutrition can be altered by pH changes. The large effect of pH on P availability in soil has already been mentioned and soil pH influences the solubility and availability of other elements to roots in soil, including iron, manganese, copper, zinc, and toxic amounts of aluminum. The effects of soil pH on plant responses to mycorrhizae may, therefore, be very complex and are highly influenced by many physical-chemical features of the soil.

Nitrogen.--High levels of soil nitrogen (N) will generally have negative effects on mycorrhizal development and growth stimulation (1,5). This may not always be so, however, since similar levels of N fertilizer have both increased and decreased mycorrhizal infection. These effects may be related to initial soil N levels available, amount of soluble N within a given soil type, or possibly to soil microbial activities. The effects of nitrogen on mycorrhizae can also be strongly influenced by soil P availability (5). For example, at low and high soil P levels, N fertilization decreased mycorrhizal infection, whereas at moderate soil P levels, increased N levels increased infection. The actual soil P-N levels supporting increased or decreased infection levels are variable and depend on host utilization rates and availability of these nutrients in any given soil type. In any event, it is wise to determine soil nitrogen levels before beginning mycorrhizal experiments.

Water.--Soil water status probably has many direct and indirect effects on mycorrhizal infection and growth promotion. Excessively high soil water potentials reduce growth and infection by mycorrhizal fungi (10). Excessive soil moisture probably affects the fungus or plant indirectly via anaerobiosis. Thus, soils that are poorly drained and remain saturated for long periods will decrease the efficiency of mycorrhizal infection. There is also limited evidence suggesting that low soil water potentials will decrease mycorrhizal infection, growth promotion (8,10) and spore production. For example, in one study mycorrhizal infection levels decreased dramatically with each decrease of soil water potential from -2 to -14 bars (10). Also, exposure of mycorrhizal plants to several cycles of moderate drought decreased spore production and plant growth stimulation (8). The mycorrhizal plants, however, were able to maintain adequate P levels upon exposure to drought cycles whereas nonmycorrhizal plants were P limited under similar conditions.

VA mycorrhizae are reported to reduce plant resistance to water uptake and the effects of mycorrhizae on plant water status and, therefore, growth may interact strongly with both the ambient conditions and soil water status. For example, a decrease in resistance to water uptake may have the greatest impact under conditions of high transpirational demand and/or limited water availability to roots in soil. A further complication can occur because transpirational demand increases with plant size and plants that are larger because of mycorrhizal infection will deplete soil moisture supplies faster than will smaller nonmycorrhizal plants.

It is impossible to separate soil water availability from nutrient availability in that decreasing soil water status will usually decrease plant nutrient uptake. This decreased nutrient uptake becomes greater in the case of highly immobile elements such as phosphorus and reductions of plant phosphorus uptake of 50-70% are not uncommon in soils drier than -2 bars (11). Water potential influences both the resistance of soils to root penetration and the geometry of the diffusion paths for nutrient movement to root surfaces. Furthermore, salinity may influence mycorrhizae and their impacts on plant growth. Unfortunately, soil water potential changes with time and is an inherently difficult parameter to effectively manipulate, measure, and isolate as a variable in experiments on mycorrhizae. Water availability or the water potential at a given water content will be different for each soil type, therefore, soil characteristic curves should be established and at least some descriptive estimates of the soil water status in experiments should be given.

Temperature.--Evidence is accumulating which indicates that mycorrhizal plants behave differently than do nonmycorrhizal plants at high soil temperatures (6) such as those found in tropical soils. There are also indications that various VA mycorrhizal fungi respond differently to temperature changes (1,2,3,4). When dealing with the effects of temperature of mycorrhizal infection it should be kept in mind that plant species differ dramatically in their response to temperatures in growth, translocation, root extension, photosynthesis, etc and these differences would be expected to have an impact on mycorrhizal establishment and function.

Ambient Conditions

Mycorrhizal fungi consume photosynthetically fixed carbon from the host and conditions that are optimum for photosynthesis and carbohydrate translocation to roots are likely to stimulate mycorrhizal growth promotion. For example, those light intensities which favor maximum plant growth generally favor mycorrhizal infection (4,5) and it can also be assumed other factors that favor photosynthesis such as moderate to high atmospheric humidities and high CO_2 levels will also favor infection. The converse of this may also be true in that environmental factors that impair photosynthesis and plant growth may impair a growth response to mycorrhizae and may in some instances cause mycorrhizal plants to grow less than their nonmycorrhizal counterparts. On the other hand, as was suggested earlier, if mycorrhizae decrease plant resistance to water uptake the benefit of that decrease may be most apparent under conditions of high transpirational demand, i.e., conditions that are not optimum for photosynthesis. More sound research on the physiological ecology of mycorrhizal plants is needed before one can really suggest an optimum set of ambient conditions for any given piece of mycorrhizal research.

General Recommendations

Since most of the environmental parameters that have been tested affect the mycorrhizal fungus, the infection process and mycorrhizal growth promotion, it is important that the magnitude of these effects be examined singly or in combination. For example, if at a given fertility level, mycorrhizal infection occurs but does not stimulate plant growth, the physiological differences between infected and noninfected plants may be different than if growth promotion by mycorrhizae was evident. Most researchers now recognize that adequate controls such as similar sized, nonmycorrhizal controls (achieved by added soil nutrients, etc.) must be included along with the usually smaller nonmycorrhizal plants that occur at the low soil fertility levels in which mycorrhizal growth stimulation is usually studied. Furthermore, cultural practices such as mixed cropping and pest management practices are known to alter mycorrhizal function. These practices must also be considered if mycorrhizal experiments are to realistically represent agricultural conditions.

Note: The authors would like to thank D. S. Hayman and G. D. Bowen for helpful discussions.

LITERATURE CITED

1. Bowen, G. D. 1980. Mycorrhizal roles in tropical plants and ecosystems. Pages 166-189. in: Tropical Mycorrhiza Research. P. Mikola, ed. Oxford Univ. Press. 277 pp.
2. Cooper, K. M. and Tinker, P. G. 1981. Translocation and transfer of nutrients in vesicular-arbuscular mycorrhizae. IV. Effect of environmental variables on movement of phosphorus. New Phytologist 88:327-339.
3. Furlan, V. and Fortin, J. A. 1973. Formation of endomycorhizas by *Endogone calospora* on *Allium cepa* under three temperature regimes. Naturaliste Can. 100:467-477.
4. Hayman, D. S. 1974. Plant growth responses to vesicular-arbuscular mycorrhizae. VI. Effect of light and temperature. New Phytologist 73:71-80.
5. Hayman, D. S. 1982. Influence of soils and fertility on activity and survival of VA mycorrhizal fungi. Phytopathology 72:(In press).
6. Moawad, M. 1979. Ecophysiology of VA mycorrhiza in the tropics. Pages 197-209. in: The Soil-root Interface. J. L. Harley and R. S. Russell, eds. Academic Press, London.
7. Mosse, B. 1972. Influence of soil type and *Endogone* strain on the growth of mycorrhizal plants in phosphate deficient soil. Rev. Ecol. Biol. Soc. 9:529-537.
8. Nelsen, C. E. and Safir, G. R. 1982. Improved drought tolerance of mycorrhizal onion plants caused by improved phosphorus nutrition. Planta 154:(In Press).
9. Olsen, S. R., Cole, C. V., Watanabe, F. S. and Dean, L. A. 1954. Estimation of available phosphorus in soils by extraction with sodium bicarbonate. Circ. 939. USDA. Sup. of Documents US Gov. Printing Office, Washington 25, D.C. 19 pp.
10. Reid, C. P. P. and Bowen, G. D. 1979. Effects of soil moisture on VA mycorrhizae formation and root development in *Medicago*. Pages 211-219. in: The Soil-root Interface. J. L. Harley and R. S. Russel, eds. Academic Press, London.
11. Viets, F. G. 1972. Water deficits and nutrient availability. Pages 217-239. in: III. Water Deficits and Plant Growth. T. T. Kozlowski, ed. Academic Press, N.Y. 368 pp.

SPORE GERMINATION AND AXENIC CULTURE OF ENDOMYCORRHIZAE

A. Spore Germination of Vesicular-Arbuscular Mycorrhizal Fungi

Lidia S. Watrud

Introduction

Interest by researchers in many parts of the world in evaluating the potential of vesicular-arbuscular mycorrhizal fungi to biologically alleviate various stresses in plants (nutrient, disease, nematode, salt or drought), is one of long duration. However, availability of inoculum for large-scale testing has been a limiting factor. For the most part, greenhouse and field studies on the effects of inoculation of plants with vesicular-arbuscular mycorrhizae have employed non-aseptic "mixed" inocula (soil, roots, spores), obtained from natural or greenhouse pot culture soils. In studies dealing with the nutrition or physiology of the fungi themselves however, researchers have tended to favor the use of surface-disinfested spores. To date, successful axenic subculture of hyphae of vesicular-arbuscular mycorrhizae has yet to be reported. Some of the methods described below have been used to help define factors which enhance germination or growth of several vesicular-arbuscular mycorrhizal fungi, particularly those of the genera *Glomus* and *Gigaspora*.

Spore Surface-Disinfestation Methods

Antibiotic treatments with chloramine-T, streptomycin, and gentamicin, or chloramine-T and streptomycin, have been reported for surface-disinfestation of *Gigaspora* and *Glomus* respectively, (12, 14). Essentially, the procedures entail transferring isolated spores from one antibiotic solution to another (e.g., by use of cup sieves), followed by multiple rinsings with sterile water. Typical concentrations and exposure times for chloramine-T range from 2-5% (w/v), for 3-20 minutes, followed by 0.025% streptomycin sulfate for 20 minutes. Gentamicin sulfate (0.01%), may also be used as a separate treatment for 20 minutes, or used in combination with the streptomycin sulfate (12). Filter sterilization (0.22 micron), of freshly prepared antibiotic solutions is recommended. Application of vacuum during the initial part of the exposure to chloramine-T, and inclusion of a surfactant such as Tween 20 (0.05%, v/v), have also been reported to be helpful (12). Use of spores free of visible soil or root debris can help ensure the success of surface-disinfestation procedures. However, due to the common presence of mycoparasites and other microflora which may be associated with the spores (2, 16, 18), true surface sterility of treated spores cannot be counted upon. Apparent surface sterility of spores incubated in axenic laboratory systems can persist for many weeks, especially in non-liquid systems. An alternative, or supplement to the use of antibiotics is sodium hypochlorite. Household bleach (5% NaOCl), is diluted with nine parts of sterile water; spore exposure (for 2-3 minutes) is followed by several sterile water rinses (3, 10).

Effects of Media on Axenic Growth of Vesicular-Arbuscular Mycorrhizal Fungi

Most references cite the use of agar-solidified (0.75-2.0%, w/v), media to assess the effect of nutrients or physical factors on axenic growth and germination (6, 10, 21). Plating has been reported to have been done directly onto the agar surface, or onto cellulose membranes or filters placed on the agar media or soil. The degree of purity of the solidifying agent (bacteriological grade agar, purified agar, or agarose), may affect germination and/or growth of given isolates. Presumably, this may reflect inhibitory or stimulatory levels of metals or other contaminants in the agar sources (11). Plant parts, use of soil extract agar and activated charcoal have been noted to be stimulatory to some isolates (1, 6, 21). Sterilized soils may inhibit germination (3). Although nutrients such as yeast extract and peptone have been noted to be stimulatory to growth at concentrations (w/v), of 0.01-0.04%, and 1-5% respectively (6), a need for exogenous nutrients for germination has not been demonstrated.

Physical Factors

Temperature of incubation, pH, light, and salts, have each been noted to affect the germination and/or growth of vesicular-arbuscular mycorrhizal fungi (1, 4, 5, 6, 8, 13, 14, 15, 17, 19, 20). Tolerance of given isolates to extreme salt or pH conditions has been noted (5, 9). Generally, pH values in the range of 5-8 are tolerated. It has been noted that growth of some *Gigaspora* species may be favored by slightly acid conditions, and that growth of some *Glomus* species may be favored by slightly alkaline conditions (5). Incubation temperatures higher than 30°C are generally not well tolerated, except for isolates from warmer geographic areas (17). Common temperatures of incubation are 20-25°C. Incubation of cultures in the dark is recommended, since light has been noted to be inhibitory (15, 20).

Evaluating Axenic Growth

Plating of spores onto test media can be by individual spore placement, or by plating groups of spores. Tools which can be used include syringes, capillary tubes or Pasteur pipettes. Actual plating can be effected by expulsion of spore suspensions from those tools. Where drier spore inocula are desired, syringes having needles with bores smaller than the average spore diameters can be used to pick up groups of spores from aqueous suspensions contained in petri dishes. The spore groups can then be transferred to solid media with minimal water transfer.

Evaluation of the germination and growth from spores plated individually on solid or in liquid media is generally made by use of a dissecting microscope. Parameters which are used to evaluate the effects of media or physical conditions are percent germination, degree of elongation and branching of hyphae or germ tubes, and development of infection or reproductive structures (6, 7, 8, 13, 14, 20). Macroscopically, the "colony" diameter or radial hyphal growth ensuing from groups of spores plated centrally in a plate, has been used to compare test treatments (21).

LITERATURE CITED

1. Daniels, B. A. and Graham, S. O. 1976. Effects of nutrition and soil extracts on the germination of *Glomus mosseae* spores. Mycologia 68:108-116.
2. Daniels, B. A. and Menge, J. A. 1980. Hyperparasitization of vesicular-arbuscular mycorrhizal fungi. Phytopathology 70:584-588.
3. Daniels, B. A. and Menge, J. A. 1981. Evaluation of the commercial potential of the vesicular-arbuscular mycorrhizal fungus *Glomus epigaeus*. New Phytol. 87:345-354.
4. Daniels, B. A. and Trappe, J. M. 1980. Factors affecting spore germination of the vesicular-arbuscular mycorrhizal fungus *Glomus epigaeus*. Mycologia 72:457-471.
5. Green, N. E., Graham, S. O., and Schenck, N. C. 1976. The influence of pH on the germination of vesicular-arbuscular mycorrhizal spores. Mycologia 68:929-934.
6. Hepper, C. M. 1979. Germination and growth of *Glomus caledonius* spores: the effects of inhibitors and nutrients. Soil Biol. Biochem. 11:269-277.

7. Hepper, C. M. and Smith, G. A. 1976. Observations on the germination of Endogone spores. Trans. Br. Mycol. Soc. 66:189-194.
8. Hirrel, M. C. 1981. The effect of sodium and chloride salts on the germination of Gigaspora margarita. Mycologia 73: 610-617.
9. Hirrel, M. C. and Gerdemann, J. W. 1980. Improved growth of onion and bell pepper in saline soils by two vesicular-arbuscular mycorrhizal fungi. Proc.Soil Sci. Soc. Amer. 44:654-655.
10. Koske, R. E. 1981. Gigaspora gigantea: Observations on spore germination of a vesicular-arbuscular mycorrhizal fungus. Mycologia 73:288-300.
11. McIlveen, W. D. and Cole H., Jr., 1979. Influence of zinc on development of the endomycorrhizal fungus Glomus mosseae and its mediation of phosphorus uptake by Glycine max cultivar Amsoy-71. Agric. Environ. 4:245-256.
12. Mertz, S. M. Jr., Heithaus III, J. J., and Bush, R. L. 1979. Mass production of axenic spores of the endomycorrhizal fungus Gigaspora margarita. Trans. Br. Mycol. Soc. 72: 167-169.
13. Mosse, B. 1959. The regular germination of resting spores and some observations on the growth requirements of an Endogone sp. causing vesicular-arbuscular mycorrhiza. Trans. Br. Mycol. Soc. 42:273-286.
14. Mosse, B. 1962. The establishment of vesicular-arbuscular mycorrhiza under aseptic conditions. J. Gen. Microbiol. 27:509-520.
15. Schenck, N. C., Graham, S. O., and Green, N. E. 1975. Temperature and light effect on contamination and spore germination of vesicular-arbuscular mycorrhizal fungi. Mycologia 67:1189-1192.
16. Schenck, N. C. and Nicolson, T. H. 1977. A zoosporic fungus occurring on species of Gigaspora margarita and other vesicular-arbuscular mycorrhizal fungi. Mycologia 69:1049-1053.
17. Schenck, N. C. and Schroder, V. N. 1974. Temperature response of Endogone mycorrhiza on soybean roots. Mycologia 66:600-605.
18. Sward, R. J. 1981. The structure of the spores of Gigaspora margarita I. The dormant spore. New Phytol. 87:761-768.
19. Tommerup, I. C. and Kidby, D. K. 1980. Production of aseptic spores of vesicular-arbuscular endophytes and their viability after chemical and physical stress. Appl. and Environ. Microbiol. 39:1111-1119.
20. Watrud, L. S., Heithaus III, J. J., and Jaworski, E. G. 1978. Geotropism in the endomycorrhizal fungus Gigaspora margarita. Mycologia 70:449-452.
21. Watrud, L. S., Heithaus III, J. J., and Jaworski, E. G. 1978. Evidence for production of inhibitor by the vesicular-arbuscular mycorrhizal fungus Gigaspora margarita. Mycologia 70:821-828.

SPORE GERMINATION AND AXENIC CULTURE OF ENDOMYCORRHIZAE

B. Dual Culture of Endomycorrhizae

M. F. Allen and T. V. St. John

INTRODUCTION

Endomycorrhizae are widespread in both native and agricultural ecosystems and can alter productivity and stress tolerance of many plant species (10). However, the physiology of the two symbionts and the physiological, biochemical and ecological processes governing the interaction are still poorly understood. Gnotobiotic cultures grown under defined, reproducible conditions can be valuable tools for studying the association. Rhizosphere and root-inhabiting organisms in non-sterile systems may produce growth regulators (6) and alter nutrient status (7) which will influence mycorrhizal infection (4,17,19). The microorganism populations may, in turn, be altered in response to changing root exudation patterns with infection (5,20) which can further affect physiological reactions.

We will describe some techniques which can be used to maintain and study two-membered cultures of vesicular-arbuscular (VA) mycorrhizae. References to techniques for work with other endomycorrhizae can be found in this manual (Chapt. 1B) or other sources (16). We will assume that prior to transfer onto a described medium, axenic, germinated spores and seeds are available. Pregermination assures that the spores are viable and does not expose the sensitive germination step to possible inhibitors (12) in the assembled unit. These techniques can be found in this volume (Chapt. 8A) and additional references (2, 12,15,19,21). The choice of symbionts is important; host plants must be of appropriate size for the culture facility. Seeds and spores should be easy to sterilize and should germinate readily. Endophytes commonly used in two-membered cultures include *Gigaspora margarita*, *Glomus caledonius*, *G. mosseae*, and *G. fasciculatus*.

INFECTIONS IN AGAR OR HYDROPONICS

Several techniques and media are available which can be used to grow VA mycorrhizae in relatively defined conditions. Mosse (17) first established a VA mycorrhiza under gnotobiotic conditions by adding a bacterium (*Pseudomonas* sp.) or bacterial filtrates and by using only component impurities as the nitrogen (N) source. Mosse and Phillips (19) further improved the growth conditions by eliminating the bacterium and by manipulating the phosphate (P) amounts and sources. Ca-phytate in conjunction with plant macronutrients appeared to give the most intense infection and external hyphal growth. Mosse and Hepper (18) established root organ cultures using a modified White's growth medium and Ca-phytate as the P source. They noted that mycorrhizal development was enhanced by using a pressure cooker for medium sterilization as opposed to an autoclave. Allen et al (2) noted that filter sterilization of the Ca-phytate enhanced mycorrhizal establishment perhaps as a consequence of reduced mineralization of P. Subsequently, Allen et al (3) found that plants grown in the Ca-phytate medium could establish higher infection but had lower internal P concentrations than plants grown with inorganic P. Therefore, although Ca-phytate is known to contain contaminants of inorganic P (2,3) and other inositol phosphates (14), in a high nutrient regime the use of a complex organic P source such as phytates may be important to infection establishment.

More recent studies have indicated that other complex P sources (11) or a continuous flow of a dilute nutrient solution (15) can result in good mycorrhizal establishment. Hepper (11) found that intense mycorrhizal infection could be initiated and maintained on agar slants, washed 3mm Whatman filter papers, or

Fåhraeus slides. Her medium, pH 6.8, consisted of micro- and macronutrients but used bone meal as the P source. MacDonald (15) initiated mycorrhizae on agar coated slides and maintained them in a hydroponic culture chamber. He used a defined, low nutrient solution (e.g. P = 1 mg l^{-1}, N = 4.2 mg l^{-1}) and allowed the solution to drip over the mycorrhiza thereby improving aeration. He has maintained cultures for as long as 3 to 4 months by periodically replacing the nutrient solution.

Although each of these techniques has distinct advantages and disadvantages, we will describe only the procedure used by Allen et al (2) in detail; this technique or a similar procedure (i.e. 18,19) has been used successfully by several workers. Selection of a growth medium will depend on the objectives of the study but it should contain all essential elements for plant growth (9). The elements can be supplied either in a highly dilute medium with continuous replacement (15) or in higher concentrations along with a complex P source (2,3,11,18,19). Table 1 shows the medium used by Allen et al (2). Sterilization techniques can be critical. Allen et al (2) simultaneously autoclaved the inorganic portion of the medium containing the agar, and filter sterilized the portion containg the Ca-phytate. The two components were then mixed aseptically and distributed into deep dishes. Water purity, clean glassware, and controlled pH are critical to establishment and growth. We recommend, initially at least, that double distilled water be used as even distilled water running through tubing can contain heavy metals or organics which can be toxic to mycorrhizae (12). Known pH's should be tested and standardized, possibly for each plant-fungus combination. We recommend that the pH of the soil where the particular association occurs naturally or pH 5.5 to 6.8 (2,11) be attempted initially. Also, temperatures may be important and thus we suggest simulation of normal growing season conditions, at least during early tests. Placement of the sterile, germinated seeds and spores is critical. Transfers should be done aseptically, preferably in a laminar-flow hood, and the germinated spore or spores should be actually touching the host plant. Aeration is essential to establishment and maintenance of a mycorrhiza. If an agar support is used, consistency of the medium must not drastically reduce gaseous diffusion. A 1% agar medium appears optimal (Allen, personal observations). If a hydroponic system is used, aerial suspension of the roots and hyphae is necessary. Finally, sterility testing is a must. Often, in agar and hydroponic systems, contamination can be observed. Samples should be taken regularly and tested for the presence of slow-growing bacteria.

INFECTIONS IN SAND AND SOIL

There are some experimental contexts in which a defined growth medium is not critical, but vigorous plant and fungal growth are. Rapid plant growth appears to be inhibited in the closed culture vessels employed in agar methods, probably because of poor air circulation. In this section we describe some variations on the method of St. John et al (21) that allow good infections and vigorous plant growth in gnotobiotic conditions.

The initial experiments (21) were carried out with sand in a funnel. Nutrient solution was supplied by a cheesecloth or glass wool wick in the neck of the funnel. Subsequently, autoclaved clay pots with the drain hole plugged with silicon sealing compound have worked well. When the lower portion of the pot is kept wet, the walls act as a wick and keep the soil moist. Small pots have a somewhat better wick action than large pots. Drip systems and frequent hand watering are possible alternatives, but the last option may be quite inconvenient in some kinds of sterile facilities.

Inoculation of sterile seedlings with sterile, germinated spores should be carried out in a laminar-flow cabinet. Most leaves should be trimmed from the host plants to improve transplant survival. A small block of agar containing the pre-germinated spore and germ tube is sliced out with sterile forceps and inverted onto the surface of a lateral root. As few as one spore per plant has given good infection. Increasing the number of spores improves the probability

Table 1. Medium for growth of VA mycorrhizae in agar culture.

Component	mg/l in final solution
$Ca(NO_3)_2 \cdot 4H_2O$	341.
KNO_3	80.
KCl	65.
$MnCl_2 \cdot 4H_2O$	7.07
$ZnSO_4 \cdot 7H_2O$	2.68
KI	0.75
$Na_2SO_4 \cdot 10H_2O$	200.
$MgSO_4 \cdot 7H_2O$	738.
H_3BO_3	1.5
$Na_2MoO_4 \cdot 2H_2O$	0.00254
$CuSO_4 \cdot 5H_2O$	0.0201
$FeCl_3 \cdot 6H_2O$	4.16
*Ca-phytate	≈230.
Agar	10 g

*630 mg of Ca-phytate was suspended in 500 ml of twice distilled water and filter sterilized using a 0.22µm membrane filter. The filter retained ≈400 mg of the Ca-phytate (2). The rest of the components were dissolved in twice distilled water and autoclaved for sterilization, pH 5.5.

of infection but also increases the probability of introducing contaminants and extends the handling time, during which spores and roots are subject to desiccation. Petri dishes containing spores should be kept closed and plants awaiting inoculation should be kept in sterile water until needed. Roots of inoculated plants are inserted into a hole and the sand or soil firmed around the roots. Assembled units should be covered while in the laminar-flow cabinet to prevent desiccation and should be given a 24 h period without light to ease the stresses associated with transplanting.

Sand is a suitable rooting medium in some cases and can be sterilized by autoclaving. Unfortunately, pH is difficult to control in sand. A sandy, nutrient-poor soil maintained a stable pH and gave more vigorous infections than sand. Sterilization of soil is a difficult problem and the least objectionable method, gamma irradiation, is not always available. Autoclaving is suitable if a soil is selected that results in little autoclave toxicity.

Once the sterile host, spores, soil, and culture vessels have been assembled, the open-pot cultures must be kept sterile by maintaining a sterile environment. Kreutzer and Baker (13) described a plastic isolator or "sterile tent" and several other ways to keep growing plants sterile. Most experiments with this technique have been carried out in a commercially available sterile tent with a bank of plant growth lights above the tent. The system proposed by Emge et al (8) may also be useful.

The composition of the nutrient solution appears not to be critical, and any convenient formulation may probably be used provided phosphorus is kept at a low level. In most early experiments 17 µg of P per gram of soil was supplied as hydroxyapatite and soluble P was omitted from the nutrient solution. Other sparsely-soluble forms of P probably are equally suitable, but substantial amounts of soluble P inhibit infection (2,3). A nutrient solution such as that given by Epstein (9; p. 39) diluted to one quarter strength and with P as described above is recommended.

DISCUSSION

A controlled plant-endophyte growth system is essential for definitive physiological and biochemical experimentation and provides a useful tool for ecologists and taxonomists as well. However, establishment may not always be easy. We have made some tentative observations which need further testing but which anyone newly attempting these procedures should consider. First, light quality and intensity are critical for plant growth and mycorrhizal establishment, although optimal levels will vary depending on the host plant. High intensity lamps are highly recommended, preferably at a minimum of 500 to 700 $\mu Em^{-2}s^{-1}$, but good infection also has been reported at lower light levels (21). Also, water quality is important; if little or no infection results from repeated trials, the researcher may wish to try water from a different source since low levels of heavy metals or organic compounds from the water, glassware, or agar can inhibit fungal growth (11,18) and thus development of mycorrhizae. Plant and fungal densities may affect establishment and should be carefully defined (1,18). Aeration is a significant factor both due to oxygen requirements and, possibly, build up of ethylene or other volatiles. Sterility checks should be a continuing process.

Dual culture provides a valuable tool for mycorrhizal research (2,11,15, 18,21). Ecologists may wish to expand on existing systems to create microcosms containing additional organisms. Axenic systems, however, are essential for physiological and biochemical studies because of the hormonal, nutritional and physical alterations of the host plant caused by rhizosphere and internal microorganisms.

ACKNOWLEDGMENTS

We gratefully acknowledge the critical reviews by Drs. Martha Christensen, David C. Coleman and Russ Ingham, and thank Linda Finchum for typing and proofing. The methods described were developed, in part, under support from National Science Foundation Grants DEB 78-25222 and DEB 78-11201.

LITERATURE CITED

1. Allen, E. B. and Allen, M. F. 1981. Natural re-establishment of vesicular-arbuscular mycorrhizae following stripmine reclamation in Wyoming. J. Appl. Ecol. 17:139-147.
2. Allen, M. F., Moore, T. S., Jr., Christensen, M., and Stanton, N. 1979. Growth of vesicular-arbuscular-mycorrhizal and nonmycorrhizal *Bouteloua gracilis* in a defined medium. Mycologia 71:666-669.
3. Allen, M. F., Sexton, J. C., Moore, T. S., Jr., and Christensen, M. 1981. Influence of phosphate source on vesicular-arbuscular mycorrhizae of *Bouteloua gracilis*. New Phytol. 87:678-694.
4. Azcon, R., Azon-G. De Aguilar, C., and Barea, J. M. 1978. Effects of plant hormones present in bacterial cultures on the formation and responses to VA endomycorrhiza. New Phytol. 80:359-364.
5. Baker, K. F. and Cook, R. J. 1974. Biological Control of Plant Pathogens. W. H. Freeman and Co., San Francisco. 433 pp.
6. Brown, M. E. 1975. Rhizosphere microorganisms: opportunists, bandits, or benefactors? Pages 21-38 in: N. Walker, ed. Soil Microbiology. John Wiley, London. 262 pp.
7. Coleman, D. C., Cole, C. V., Anderson, R. V., Blaha, M., Campion, M. K., Clarholm, M., Elliot, E. T., Hunt, H. W., Shaefer, B., and Sinclair, J. 1977. An analysis of rhizosphere-saprophage interactions in terrestrial ecosystems. Ecol. Bull. (Stockholm) 25:299-309.
8. Emge, R. G., Melching, J. S., and Kingsolver, C. H. 1970. A portable chamber for the growth and isolation of infected plants. Plant Dis. Rep. 54:130-131.
9. Epstein, E. 1972. Mineral Nutrition of Plants: Principles and Perspectives. John Wiley, New York. 412 pp.
10. Hayman, D. C. 1980. Mycorrhiza and crop production. Nature 287:487-488.

11. Hepper, C. M. 1981. Techniques for studying the infection of plants by vesicular-arbuscular mycorrhizal fungi under axenic conditions. New Phytol. 88:641-647.
12. Hepper, C. M. and Smith, G. A. 1976. Observations on the germination of Endogone spores. Trans. Br. Mycol. Soc. 66:189-194.
13. Kreutzer, W. A. and Baker, R. 1975. Gnotobiotic assessment of plant health. Pages 11-21 in: G. W. Bruehl, ed. Biology and Control of Soil-borne Plant Pathogens. Am. Phytopathol. Soc., St. Paul, MN. 216 pp.
14. L'annunziata, M. F. and Gonzalez, J. 1977. Soil metabolic transformations of Carbon-14-myo-inositol, Carbon-14-phytic acid and Carbon-14-iron (III) phytate. Soil Organic Matter Studies I:239-253.
15. MacDonald, R. M. 1981. Routine production of axenic vesicular-arbuscular mycorrhizas. New Phytol. 89:87-94.
16. Moore-Parkhurst, S. and Englander, L. 1981. A method for the synthesis of a mycorrhizal association between Pezizella ericae and Rhododendron maximum seedlings growing in a defined medium. Mycologia 73:994-997.
17. Mosse, B. 1962. The establishment of vesicular-arbuscular mycorrhiza under aseptic conditions. J. Gen. Microbiol. 27:509-520.
18. Mosse, B. and Hepper, C. M. 1975. Vesicular-arbuscular mycorrhizal infections in root organ culture. Physiol. Plant Path. 5:215-223.
19. Mosse, B. and Phillips, J. M. 1971. The influence of phosphate and other nutrients on the development of vesicular-arbuscular mycorrhiza in culture. J. Gen. Micro. 69:157-166.
20. Ratnayake, M., Leonard, R. T., and Menge, J. A. 1978. Root exudation in relation to supply of phosphorus and its possible relevance to mycorrhizal formation. New Phytol. 81:543-552.
21. St. John, T. V., Hayes, R. I., and Reid, C. P. P. 1981. A new method for producing pure VAM-host cultures without specialized media. New Phytol. 89:81-86.

TAXONOMY OF ECTO- AND ECTENDOMYCORRHIZAL FUNGI

Orson K. Miller, Jr.

INTRODUCTION

This chapter provides a summary of the known or putative ecto- and ectendomycorrhizal fungi and a guide to the literature where complete descriptions of the species may be found. Illustrations and taxonomic tables are provided to enable the user to more accurately place the fungus in a family or genus. It is not meant, however, as a definitive reference source in itself to the accurate identification of ectomycorrhizal fungi. The use of the term ectomycorrhizae refers to both ecto- and ectendomycorrhizae unless otherwise stated. Many genera included here are considered to be putative mycorrhizal fungi even though the synthesis work has not as yet been accomplished.

Fungi which are now known to form ectomycorrhizae with higher plants (48) are in three classes within the Eumycota (Basidiomycotina, Ascomycotina and Zygomycotina). The largest number by far are Basidiomycotina (Tables 1-6), members of a few genera of Ascomycotina (Tables 1, 7) are known but only one genus and a few species have been identified in the Zygomycotina (Table 1). It appears that entire families within the Basidiomycotina have evolved as mycorrhizal symbionts of higher plant families (30). In other cases, only certain genera within a family are mycorrhizal or an entire family may be nonmycorrhizal (30).

The picture within the Ascomycotina is not clearly elucidated. The Tuberales are apparently an order of hypogeous ectomycorrhizal fungi. Ecological evidence of the occurrence of many Ascomycotina tentatively implicates more species as mycorrhizal associates (50). At this time there is too little evidence to speculate on any pattern of mutualism. In the Zygomycotina, the Endogonaceae in the sense of Gerdemann and Trappe (13) contain only two ectomycorrhizal species, both in the genus *Endogone*, (*E. lactiflua* Bk. & Br., Plate 4D and *E. flammicorona* Trappe & Gerd.). All other species are endomycorrhizal and in other genera. It is unlikely that there will be more than a few additional ectomycorrhizal species found within the Zygomycotina.

Perhaps the most important general reference works to genera and species of ectomycorrhizal fungi include Singer (39), Ainsworth et al (1, 2), Watling and Watling (53), Smith et al (42), Smith and Smith (43), and Miller and Farr (31). The placement of a species in a genus may be determined by consulting the above works. The tables presented here enable one to relate the genus to a family and indicate something about its host and distribution, as well as its general features. Important references to these taxa are indicated in the text. General guides by Pomerleau (36), Smith (41), Miller (29), Miller and Miller (32), Dahncke and Dahncke (8), and Cetto (3, 4, 5) provide many quality color illustrations and species descriptions. Detailed explanations of the microscopic features are covered in Largent et al (23).

SYSTEMATICS

The Russulales and their allies covered in Table 2 include two worldwide genera, *Lactarius* and *Russula*, which have lamellae and forcible spore discharge. *Lactarius* has recently been monographed in North America by Hesler and Smith (18) but no comparable North American monograph exists for *Russula*. Many species are covered by Romagnesi (37), Hennig (19), and Schaeffer (38). The distinctive amyloid ornamented, globose to ovoid spores and the heteromerous pileus trama separate these taxa from all other lamellate agarics. *Macowanites* (Plate 5A), *Arcangeliella* and *Cystangium*, all with a labyrinthoid gleba are the other three genera in the Russulaceae (Table 2). They do not forcibly discharge their spores but have basidia which are ballistosporic in structure (35). Four genera, placed in the Elasmomycetaceae by Pegler and Young (35), are statismosporic and the gleba is chambered as in *Zelleromyces* (Plate 5B) **but** have the typical spores and heteromerous trama of the Russulales. They are included with this latter family in the Russulales. Singer

I. BASIDIOMYCOTINA
- 1. Agaricales
  - Amanitaceae
  - Hygrophoraceae
  - Tricholomataceae
  - Entolomataceae
  - Cortinariaceae
  - Paxillaceae
  - Gomphidiaceae
  - Boletaceae
  - Strobilomycetaceae
- 2. Russulales
  - Russulaceae
  - Elasmomycetaceae
- 3. Gautieriales
  - Gautieriaceae
- 4. Hymenogastrales
  - Octavianinaceae
  - Hymenogastraceae
  - Rhizopogonaceae
  - Hydnangiaceae
- 5. Phallales
  - Hysterangiaceae
- 6. Lycoperdales
  - Mesophelliaceae
- 7. Melanogastrales
  - Melanogastraceae
  - Leucogastraceae
- 8. Sclerodermatales
  - Sclerodermataceae
  - Astraceae
- 9. Aphyllophorales
  - Corticiaceae
  - Cantharellaceae
  - Clavariaceae
  - Thelephoraceae

II. ASCOMYCOTINA
- 1. Eurotiales
  - Elaphomycetaceae
- 2. Pezizales
  - Humariaceae
- 3. Tuberales
  - Pseudotuberaceae
  - Hydnotryaceae
  - Geneaceae
  - Eutuberaceae
  - Terfeziaceae

III. ZYGOMYCOTINA
- 1. Mucorales
  - Endogonaceae

Table 1. Synopsis of Orders and Families containing ectomycorrhizal fungi.

and Smith (40) have fully described most of the gastroid genera and species in this complex. The family, Octavianinaceae, contains genera and species which have similar spore morphology (35) but are nonamyloid and not included, therefore, in the Russulales. Secotioid families of uncertain position are arbitrarily placed in the Hymenogastrales (Table 1).

Many species of Russula and Lactarius are known to form mycorrhizae (48). Recent works by Molina and Trappe (pers. comm.), Steven Miller and O. K. Miller (pers. comm.), and Malajczuk in Australia (pers. comm.) indicate that three species of Zelleromyces are ectomycorrhizal with higher plants. Field observations of species implicate the other genera listed in Table 2 as ectomycorrhizal.

Perhaps the largest number of ectomycorrhizal fungi are found among the eleven families and the 27 genera listed in Table 3. There are, no doubt, even more genera than those listed among these families which will be found to contain mycorrhizal species as widespread mycorrhizal investigations take place. Species in the Gomphidiaceae, many in the Amanitaceae and some in the Tricholomataceae and Cortinariaceae will not grow in pure culture. Field observations have been utilized to establish the connection of the fungus to its host in these cases.

Many of the families are cosmopolitan but the Gomphidiaceae (27, 28) and the Rhizopogonaceae are originally known only from the Northern Hemisphere where they are mycorrhizal associates of conifers (34). However, with the extensive planting of conifers, especially pine, in the Southern Hemisphere many of the Northern Hemisphere fungal associates are now beginning to be found there (22). Rhizopogon roseolus (Plate 4E) is especially common in Australian pine plantations but is known only in exotic plantings. Molina and Trappe (33, 34) report ectomycorrhizal synthesis with

Table 2. Ectomycorrhizal genera in the Russulales and their allies.

Genus	Family	Forcible spore discharge	Amyloid ornamented spores	Statismo-sporic	Latex	Stipe or stipe-columella	Host	Distribution
Russula	Russulaceae	+	+	-	-	+	BHR[a]	Cosmop.
Lactarius	Russulaceae	+	+	-	+	+	BHR	Cosmop.
Macowanites	Russulaceae	-	+	-	-	+	BHR	Cosmop.[b]
Arcangeliella	Russulaceae	-	+	-	+	+	BHR	Cosmop.
Cystangium	Russulaceae	-	+	-	-	+	Hd's[c]	S. Hem.
Elasmomyces	Elasmomycetaceae	-	+	+	-	+	BHR	Cosmop.[b]
Zelleromyces	Elasmomycetaceae	-	+	+	+	±	BHR	Cosmop.
Martellia	Elasmomycetaceae	-	+	+	-	-	BHR	Cosmop.[b]
Gymnomyces	Elasmomycetaceae	-	+	+	-	-	BHR	Cosmop.
Octavianina	Octavianinaceae	-	-	+	-	-	BHR	Cosmop.[b]
Sclerogaster	Octavianinaceae	-	-	+	-	-	BHR	N. Hem.
Stephanospora	Octavianinaceae	-	-	+	-	-	BHR	Cosmop.
Wakefieldia	Octavianinaceae	-	-	+	-	-	BHR	Cosmop.

a BHR = Broad host range

b Infrequently found in Southern Hemisphere

c Hd's = Hardwoods

Table 3. Ectomycorrhizal Agaricales and their secotioid allies.

Genus	Family	Forcible spore discharge	Amyloid pileus trama	Stipe or stipe-columella	Host	Distribution
Gomphidius	Gomphidiaceae	+	-	+	Conifers	N. Hem.
Chroogomphus	Gomphidiaceae	+	+	+	Conifers	N. Hem.
Cystogomphus	Gomphidiaceae	+	-	+	Conifers	Europe
Gomphogaster	Gomphidiaceae	-	-	+	Conifers	N. America
Brauniellula	Gomphidiaceae	-	+	+	Conifers	N. America
Rozites	Cortinariaceae	+	-	+	BHR[a]	Cosmop.
Descolea	Cortinariaceae	+	-	+	BHR	S. Hem.
Cortinarius	Cortinariaceae	+	-	+	BHR	Cosmop.
Inocybe	Cortinariaceae	+	-	+	BHR	Cosmop.
Hebeloma	Cortinariaceae	+	-	+	BHR	Cosmop.
Truncocolumella	Rhizopogonaceae	-	-	+	Conifers	N. Hem.
Rhizopogon	Rhizopogonaceae	-	-	-	BHR	Cosmop.
Hymenogaster	Hymenogastraceae	-	-	-	BHR	Cosmop.
Gautieria	Gautieriaceae	-	-	+	BHR	Cosmop.
Paxillus	Paxillaceae	+	-	+	BHR	Cosmop.
Neopaxillus	Paxillaceae	+	-	+	Cicads	S. Hem.
Amanita	Amanitaceae	+	-	+	BHR	Cosmop.
Limacella	Amanitaceae	+	-	+	Conifers	N. Hem.
Entoloma	Entolomataceae	+	-	+	BHR	Cosmop.
Tricholoma	Tricholomataceae	+	-	+	BHR	Cosmop.
Catathelasma	Tricholomataceae	+	-	+	BHR	Cosmop.
Armillaria	Tricholomataceae	+	-	+	BHR	Cosmop.
Leucopaxillus	Tricholomataceae	+	-	+	BHR	Cosmop.
Melanoleuca	Tricholomataceae	+	-	+	BHR	Cosmop.
Laccaria	Tricholomataceae	+	-	+	BHR	Cosmop.
Hydnangium	Hydnangiaceae	-	-	±	BHR	Cosmop.
Hygrophorus	Hygrophoraceae	+	-	+	BHR	Cosmop.

a BHR = Broad host range

Table 4. Ectomycorrhizal genera in the Boletaceae and Strobilomycetaceae, Agaricales.

Genus	Family	Forcible spore discharge	Ornamented spores	Spore color	Poroid	Boletinoid	Host	Distribution
Boletus	Boletaceae	+	-	olive-brown	+	-	BHR[a]	Cosmop.
Leccinum	Boletaceae	+	-	olive-brown	+	-	BHR	Cosmop.
Tylopilus	Boletaceae	+	-	pinkish brown	+	-	BHR	Cosmop.
Pulveroboletus	Boletaceae	+	-	olive-brown	+	-	Hd's[b]	N. America
Aureoboletus	Boletaceae	+	-	ochraceous buff	+	-	Hd's	N. America
Gyroporus	Boletaceae	+	-	yellow	+	-	BHR	Cosmop.
Boletopsis	Boletaceae	+	-	white	+	-	Conifers	N. Hem.
Austeroboletus	Boletaceae	+	-	olive-brown	+	-	Hd's	Aust. & New Zea
Boletinellus	Boletaceae	+	-	brown	+	+	Fraxinus	N. Hem.
Suillus	Boletaceae	+	-	olive-brown	+	+	Conifers	N. Hem.
Fuscoboletinus	Boletaceae	+	-	red brown	+	+	Conifers	N. Hem.
Phylloporus	Boletaceae	+	-	brown	-	-	BHR	Cosmop.
Gastroboletus	Boletaceae	-	-	brown	+	-	Conifers	Cosmop.
Strobilomyces	Strobilomycetaceae	+	+	brown	+	-	Hd's	Cosmop.
Porphyrellus	Strobilomycetaceae	+	+	brown	+	-	Hd's	Cosmop.
Boletellus	Strobilomycetaceae	+	+	brown	-	-	Hd's	Cosmop.
Chamonixia	uncertain position	-	-	buff to brown	+	-	BHR	N. America

a BHR = Broad host range
b Hd's = Hardwoods

Table 5. Ectomycorrhizal genera in four orders of non-secotioid Gasteromycetes.

Genus	Family	Mature Gleba	Host	Distribution
Mesophellia	Mesophelliaceae[a]	hard, dull white	Hd's	Australia & New Zealand
Astraeus	Astraceae[b]	powdery, brown	BHR	Cosmop.
Scleroderma	Sclerodermataceae[b]	powdery, purple-brown	BHR	Cosmop.
Pisolithus	Sclerodermataceae[b]	powdery, brown	BHR	Cosmop.
Melanogaster	Melanogastraceae[c]	gel-filled, black	BHR	Cosmop.
Alpova	Melanogastraceae[c]	gel-filled, orange	*Alnus*	N. Hem.
Leucogaster	Leucogastraceae[c]	gel-filled, white	BHR	Cosmop.
Leucophleps	Leucogastraceae[c]	gel-filled, white	Conifers	N. Hem.
Hysterangium	Hysterangiaceae[d]	viscous, green	BHR	Cosmop.

a Lycoperdales; b Sclerodermatales; c Melanogastrales; d Phallales.

Hd's = Hardwoods; BHR = Broad Host range.

Table 6. Ectomycorrhizal genera in the Aphyllophorales.

Genus	Family	Fruiting Body Type	Tissue	Host	Distribution
Piloderma	Corticiaceae	resupinate	membraneous	BHR[a]	Cosmop.
Byssoporia	Corticiaceae	resupinate	membraneous	BHR	Cosmop.
Amphinema	Corticiaceae	resupinate	membraneous	Conifers	N. Hem.
Thelephora	Thelephoraceae	effused reflexed to stipitate-pileate	flexible,tough	BHR	Cosmop.
Tomentella	Thelephoraceae	resupinate	cottony	BHR	Cosmop.
Cantharellus	Cantharellaceae	stipitate-pileate, ridged	fleshy	BHR	Cosmop.
Craterellus	Cantharellaceae	stipitate-pileate, smooth	fleshy	BHR	Cosmop.
Polyozellus	Cantharellaceae	stipitate-pileate, wrinkled	fleshy	Conifers	N. Hem.

a BHR = Broad host range

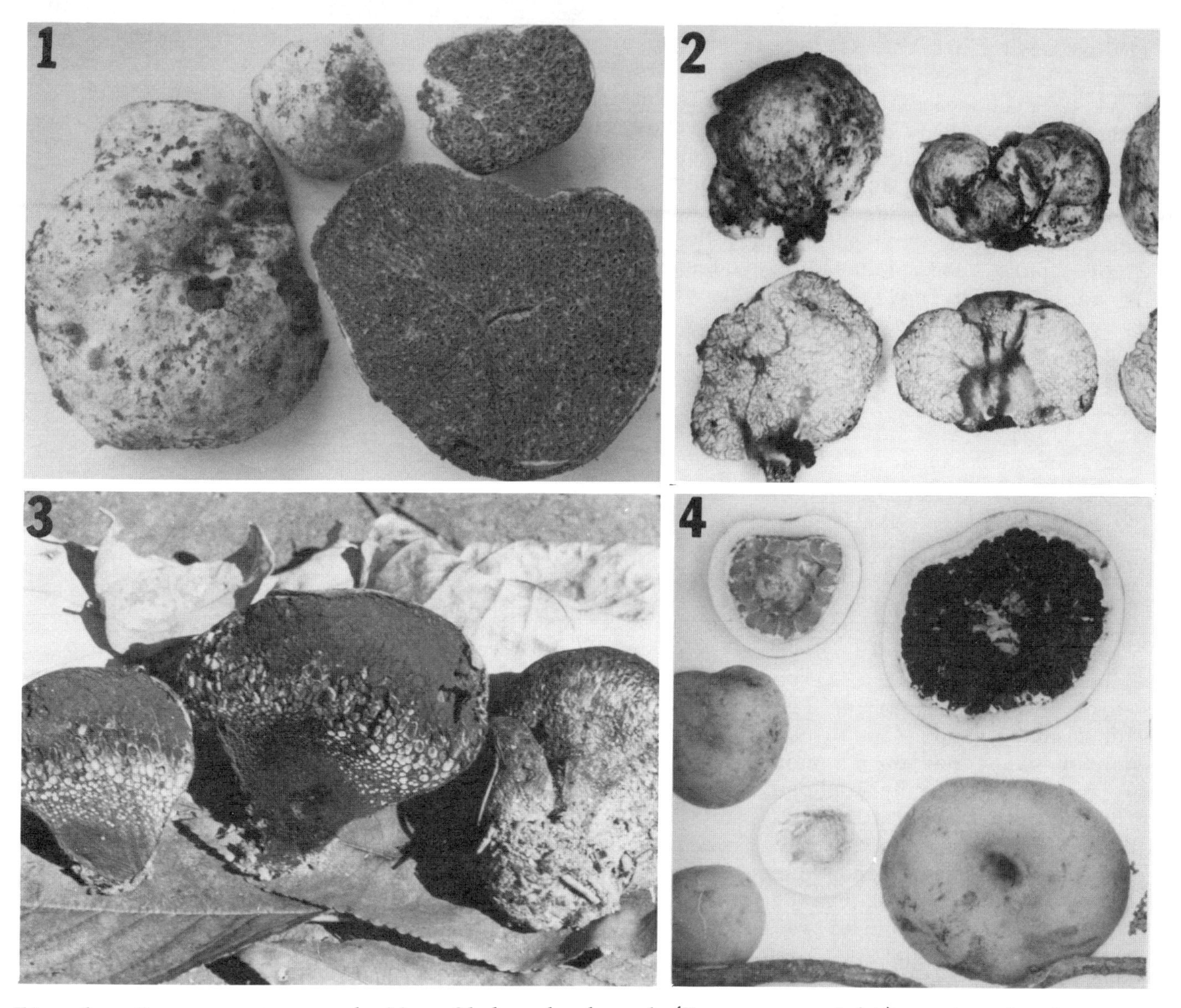

Fig. 1. Hymenogaster parksii. Gleba chambered (X sec. on right), not gelatinous, 1.0-2.5 cm broad. Photo by J. Trappe.
Fig. 2. Hydnangium carneum. Gleba pink, chambered (X sec. at bottom), 1.0-2.0 cm broad. Photo by J. Trappe.
Fig. 3. Pisolithus tinctorius. Gleba brown with peridioles (X sec. on left), 4.0-25 cm broad.
Fig. 4. Elaphomyces granulatus. Gleba not chambered, black to blackish-brown, powdery at maturity, 1.5-4.5 cm broad. Photo by J. Trappe.

conifers by Truncocolumella citrina (Plate 4F), Laccaria and numerous species of Rhizopogon (Table 3). Malajczuk and Trappe (pers. comm.) have achieved synthesis with Eucalyptus and species of Hymenogaster (Fig. 1) and Hydnangium (Fig. 2 and Table 3).

The boletes have been separated into two families based upon the presence or absence of spore ornamentation (Table 4). They are further divided into genera by differences in spore morphology, spore deposit color, the arrangement of the pores, staining reactions and a variety of other microscopic characters. Species of Suillus, Fuscoboletinus and Boletopsis are primarily associated with conifers in the Pinaceae (Table 4) native to the Northern Hemisphere. However, species in the other bolete genera are found as ectomycorrhizal associates with a wide variety of perennial angiosperms throughout the world. Smith and Thiers (44), Snell and Dick (45), and Thiers (46) have all produced comprehensive treatments of the boletes in North America. Thiers and Trappe (47) have published the most comprehensive work on Gastroboletus. Many other bolete studies are listed by Watling and Watling (53).

The non-secotioid Gasteromycetes listed in Table 5 are mostly hypogeous with the exception of the Astraceae and Sclerodermataceae in the Sclerodermatales. Astraeus has an earth-star appearance as shown by Miller in Fig. 33 (29) while the oval to club-shaped species of Scleroderma have a deep purple gleba when young (Plate 351, (29)). Lastly, Pisolithus tinctorius (Pers.) Coker & Couch is typified by a series of chambers (peridioles) which mature apically within an oval to club-shaped, foul smelling fruiting body (Fig. 3). Its cosmopolitan distribution and broad host range make it one of the most versatile mycorrhizal fungi known. The Mesophelliaceae with one genus, Mesophellia, is known only from Australia where Malajczuk (pers. comm.) has achieved synthesis with Eucalyptus diversicolor F. Muell.. It is the only family in the Lycoperdales reported so far with known mycorrhizal taxa. Hysterangium is typified by its viscous, green gleba (Plate 4C), disagreeable smell, and small dull white fruiting bodies which often stain pink when bruised. The Hysterangiaceae is considered to be a primitive member of the Phallales (stinkhorns) which are also typified by a green, viscous gleba which has a foetid odor at maturity(20,29). The order Melanogastrales is composed of two families, Melanogastraceae and Leucogastraceae. The gleba soon develops a gel and becomes rubbery and tough. The white gel-filled chambers of Leucophleps and Leucogaster are distinctive (Plate 4H) (54), while the orange, gel-filled gleba, typical of Alpova (Plate 4G) becomes dull red after a few minutes of exposure to the weather. Melanogaster is widely distributed (30) and typified by a black, gel-filled gleba (4, 5).

Molina and Trappe (33) have reported mycorrhizal synthesis of Pacific Northwest trees with Astraeus, Scleroderma, Pisolithus, Melanogaster, Alpova and Hysterangium (Table 5). Molina (pers. comm.) has also achieved synthesis with two species of Leucogaster. Fogel (12) reports the discovery of basidia in the genus Leucophleps which closely resembles Leucogaster and is undoubtably also ectomycorrhizal. Malajczuk and Trappe (pers. comm.) have also reported synthesis of Astraeus and Melanogaster with species of Australian Eucalyptus.

Among the Aphyllophorales listed in Table 6, Piloderma bicolor (Pk.) Julich (Syn. Corticium bicolor Pk.) and Thelephora terrestris Ehrl. per Fr. are perhaps the best known ectomycorrhizal species. Thelephora has been monographed by Corner (7) and the Athelieae monographed by Julich (21). Byssoporia terrestris (DC per Fr.) Larsen & Zak (Syn. Poria terrestris (DC per Fr.) Sacc.) forms mycorrhizae with conifers (25). Amphinema is also an ectomycorrhizal fungus associated with pines. Cantharellus cibarius Fr. forms ectomycorrhizae with conifers and in all likelihood the whole Cantharellaceae is a mycorrhizal family. There seems to be little doubt that many genera in the Clavariaceae also form mycorrhizae (30) as do the stipitate-pileate hydnums (Hydnaceae). However, no direct evidence of ectomycorrhizal formation has been presented for either family. Tomentella, according to Larsen (24), is a genus of 72 species distributed worldwide and it is quite likely that some or all of the species form ectomycorrhizae. However, the vast majority of the fungi in the Aphyllophorales (10) are decomposers of lignicolous substrates or plant parts (30).

There will undoubtably be more Ascomycotina found to form ectomycorrhizae with higher plants than those shown in Table 7. In fact, species in many of the genera are implicated from field observations as mycorrhizal by Trappe (50), Hawker (17), Mattirolo (26) and many others. Recently Chevalier and Frochot (6) have carried out successful field trials using Tuber melanosporum Vitt., the Perigord truffle,to form ectomycorrhizae with oak and hazel seedlings. Danielson (9) has achieved synthesis with a number of conifers and Sphaerosporella brunnea (Alb. & Schw. ex Fr.) Svercek & Kub., a small cup fungus. My students and I have recently observed the direct connection of Tuber shearii Pk. with ectomycorrhizae in pure Pinus strobus L. plantations. The discovery of new ectomycorrhizal Ascomycetes will certainly accelerate during the next decade. Major taxonomic works in the Tuberales are by Fischer (11), Gilkey (14, 15, 16), Hawker (17), and Trappe (51, 52).

Cenococcum geophilum Fr. (Syn. C. graniforme (Sow.) Ferde. & Winge), a probable anamorph of Elaphomyces, is a brown pigmented ectomycorrhizal fungus. Trappe (48, 49) has compiled a long list of conifer and hardwood hosts throughout the world. The mycorrhizal mantle of C. geophilum possesses a characteristic morphology, well illustrated by Trappe (50). The thick-walled, radially arranged tissue is strikingly

Table 7. Ectomycorrhizal genera in the Ascomycotina and one anamorph.

Genus	Family	Habit	Host	Distribution
Hydnotrya	Helvellaceae	hypogeous	BHR[a]	Eu. & N. America
Geopora	Pyronemataceae	hypogeous	Conifers	Eu. & N. America
Stephensia	Pyronemataceae	hypogeous	Conifers	Eu.
Sphaerosporella	Humariaceae	epigeous	Conifers	N. America
Tirmania	Pezizaceae	hypogeous	Composites	Africa
Genea[b]	Geneaceae	hypogeous	BHR	Cosmop.
Barssia	Balsamiaceae	hypogeous	BHR	West. N. America
Balsamea	Balsamiaceae	hypogeous	Hd's[c]	N. Hem.
Picoa	Balsamiaceae	hypogeous	Conifers	N. America
Pachyphloeus	Terfeziaceae	hypogeous	BHR	N. Hem.
Terfezia	Terfeziaceae	hypogeous	BHR	Cosmop.
Hydnobolites	Terfeziaceae	hypogeous	BHR	N. Hem.
Choiromyces	Terfeziaceae	hypogeous	Hd's	N. Hem.
Tuber	Tuberaceae	hypogeous	BHR	Cosmop.
Mukagomyces	Tuberaceae	hypogeous	Hd's	N. Hem.
Elaphomyces	Elaphomycetaceae	hypogeous	BHR	Cosmop.
Cenococcum[d]	Elaphomycetaceae	hypogeous	BHR	Cosmop.

a BHR = Broad host range
b Includes species previously placed in Genabea and Myrmecocystis (1)
c Hd's = Hardwoods
d Anamorph

similar to the tissue of the peridium in the genus Elaphomyces (50; Miller, unpub.). At this time, no one has induced C. geophilum to fruit in pure culture but it is highly probable that if it did it would be a species of Elaphomyces.

SUMMARY

We now recognize mycorrhizal genera within at least twenty-five families of Basidiomycotina and seven families of Ascomycotina (Table 1). Many of our discoveries of ectomycorrhizal fungi have been very recent (6, 9, 32, 34). The additional knowledge which we have gained concerning the taxonomic distribution of mycorrhizal fungi indicates specific patterns within the classes mentioned above. These patterns may have evolutionary significance at the family or generic levels in the Eumycota. The relationships between specific genera of conifers and sections of the genus Rhizopogon reported by Molina and Trappe (34) is almost certainly a forerunner of many yet undiscovered examples of the coevolution of both fungus and host. Conversely there are clear examples of families in the Basidiomycotina composed entirely of genera which have evolved as decomposers of plant and animal remains (30). The elucidation of these patterns of evolution and coevolution will greatly enhance our understanding of the role which the Eumycota play in each ecosystem.

LITERATURE CITED

1. Ainsworth, G. C., Sparrow, F. K. and Sussman, A. S. 1973. The Fungi. Vol. IV A Academic Press, New York, N. Y. 621 pp.

2. Ainsworth, G. C., Sparrow, F. K. and Sussman, A. S. 1973. The Fungi. Vol. IV B. Academic Press, New York, N. Y. 504 pp.
3. Cetto, B. 1978. Der Grosse Pilzefuhrer. Band 1. Arti Grafiche Saturnia, Trento, Italy. 669 pp.
4. Cetto, B. 1979. Der Grosse Pilzefuhrer. Band 2. Arti Grafiche Saturnia, Trento, Italy. 729 pp.
5. Cetto, B. 1979. Der Grosse Pilzefuhrer. Band 3. Arti Grafiche Saturnia, Trento, Italy. 635 pp.
6. Chevalier, G. and Frochot, H. 1981. Truffle production from artificially mycorrhizal plants, first results. (Abs.) Fifth North American Conf. on Mycorrhizae, Laval Univ., Quebec, Canada. p. 35.
7. Corner, E. J. H. 1968. A monograph of Thelephora (Basidiomycetes). Bief. Nov. Hed. 27:1-110.
8. Dahncke, R. M. and Dahncke, S. M. 1979. 700 Pilze in Farbfotos. AT verlag Aarau, Stuttgart. 686 pp.
9. Danielson, R. M. 1981. Ectomycorrhiza formation by an operculate Discomycete. (Abs.) Fifth North American Conf. on Mycorrhizae, Laval Univ., Quebec, Canada. p. 34.
10. Eriksson, J. and Ryvarden, L. 1973. The Corticiaceae of North Europe. Fungi Flora, Oslo II:60-286.
11. Fischer, E. 1938. Tuberineae. Naturl. Pflanzenfamilien. 5b:1-42.
12. Fogel, R. 1979. The genus Leucophleps (Basidiomycotina, Leucogastrales). Can. J. Bot. 57:1718-1728.
13. Gerdemann, J. W. and Trappe, J. M. 1974. The Endogonaceae in the Pacific Northwest. Mycol. Mem. No. 5, 76 pp.
14. Gilkey, H. M. 1939. Tuberales of North America. Oregon State Univ., Corvallis, Oregon. 63 p.
15. Gilkey, H. M. 1947. New or otherwise noteworthy species of Tuberales. Mycologia 39:441-452.
16. Gilkey, H. M. 1962. New species and revisions in the order Tuberales. Mycologia 53:215-220.
17. Hawker, L. E. 1954. British hypogeous fungi. Phil. Trans., Roy. Sci. Lon. 237:429-546.
18. Hesler, L. R. and Smith, A. H. 1979. North American species of Lactarius. Univ. of Michigan Press, Ann Arbor, Mich. 841 p
19. Hennig, B. 1970. Handbuch fur Pilzfreude. Band 5. Gustav Fischer Verlag, Jena, DDR. 391 pp.
20. Horak, E. 1963. Fungi austroamericani V Beitrag zur Kenntnis der Gattungen Hysterangium Vitt., Hymenogaster Vitt., Hydnangium Wallr. and Melanogaster Cda. in Sudamerika (Argentinien, Uruguay). Syd. Ann. Mycol. 17:197-205.
21. Julich, W. 1972. Monographie der Athelieae (Corticiaceae, Basidiomycetes). Willdenowia 7:1-283.
22. Lamb, R. J. 1979. Factors responsible for the distribution of mycorrhizal fungi of Pinus in Eastern Australia. Aust. For. Res. 9:25-34.
23. Largent, D., Johnson, D. and Watling, R. 1977. How to Identify Mushrooms to Genus III: Microscopic Features. Mad River Press, Inc., Eureka, CA. 148 pp.
24. Larsen, M. J. 1974. A contribution to the taxonomy of the genus Tomentella. Mycol. Mem. No. 4, 145 pp.
25. Larsen, M. J. and Zak, B. 1977. Byssoporia gen. nov.: Taxonomy of the mycorrhizal fungus Poria terrestris. Can. J. Bot. 56:1122-1129.
26. Mattirolo, O. 1934. Rapporti simbiotici sviluppatisitra il tartufo 'branchelto' (Tuber borchii Vittadini) et i proppi americana detti canadesi. Ann. R. Accad. Agric. Torino. 70:3-10.
27. Miller, O. K., Jr. 1971. The genus Gomphidius with a revised description of the Gomphidiaceae and a key to the genera. Mycologia 63:1129-1163.
28. Miller, O. K., Jr. 1973. A new gastroid genus related to Gomphidius. Mycologia 65:226-229.
29. Miller, O. K., Jr. 1980. Mushrooms of North America. 4th printing. E. P. Dutton, Inc., New York, N. Y. 368 pp.

30. Miller, O. K., Jr. 1982 . Ectomycorrhizae in the Agaricales and Gasteromycetes. Can. J. Bot.60: (in press).
31. Miller, O. K., Jr. and Farr, D. F. 1975. An Index of the Common Fungi of North America. J. Cramer, Vaduz, Germany. 206 pp.
32. Miller, O. K., Jr. and Miller, H. H. 1980. Mushrooms in Color. E. P. Dutton, Inc., New York, N. Y. 286 pp.
33. Molina, R. and Trappe, J. M. 1982 . Patterns of ectomycorrhizal host specificity and potential among Pacific Northwest conifers and fungi. For. Sci. 28: (in press).
34. Molina, R. and Trappe, J. M. 1981. Ectomycorrhiza formation, host specificity and taxonomic relationships in the genus Rhizopogon. (Abs.) Fifth North American Conf. on Mycorrhizae, Laval Univ., Quebec, Canada. p. 30.
35. Pegler, D. N. and Young, T. W. K. 1979. The Gasteroid Russulales. Trans. Br. Mycol. Soc. 72:353-388.
36. Pomerleau, R. 1980. Flore des Champignons au Quebec. Les Editions La Presse, Ltee, Montreal, Canada. 652 pp.
37. Romagnesi, H. 1967. Les Russules d'Europe et d'Afrique du Nord. Bordas, Paris. 998 pp.
38. Schaeffer, J. 1933. Russula--Monographie. Ann. Mycol. 31:305-516.
39. Singer, R. 1975. The Agaricales in Modern Taxonomy. J. Cramer, Vaduz, Germany. 912 pp.
40. Singer, R. and Smith, A. H. 1960. Studies on Secotiaceous Fungi. IX. The Astrogastraceous Series. Bull. Torrey Bot. club. 21:1-112.
41. Smith, A. H. 1975. A field guide to Western Mushrooms. Univ. of Michigan Press, Ann Arbor, Mich. 280 pp.
42. Smith, A. H., Smith, H. V. and Weber, N. S. 1979. How to know the gilled mushrooms. Wm. C. Brown Co., Dubuque, Iowa. 334 pp.
43. Smith, H. V. and Smith, A. H. 1973. How to know the non-gilled fleshy fungi. Wm. C. Brown Co., Dubuque, Iowa. 402 pp.
44. Smith, A. H. and Thiers, H. 1971. The Boletes of Michigan. Univ. of Michigan Press, Ann Arbor, Mich. 417 pp.
45. Snell, W. and Dick, E. 1971. The Boleti of Northeastern North America. J. Cramer, Weinheim. 115 pp.
46. Thiers, H. D. 1975. California mushrooms. A field guide to the boletes. McMillan Inc., New York, N. Y. 261 pp.
47. Thiers, H. D. and Trappe, J. M. 1969. Studies in the genus Gastroboletus. Brittonia 21:244-254.
48. Trappe, J. M. 1962. Fungus associates of ectomycorrhizae. Bot. Rev. 28:538-606.
49. Trappe, J. M. 1964. Mycorrhizal hosts and distribution of Cenococcum graniforme. Lloydia 27:100-106.
50. Trappe, J. M. 1971. Mycorrhiza-Forming Ascomycetes. Pages 19-37 in: Mycorrhizae. E. Hacskaylo ed. USDA Misc. Publ. 1189. 255 pp.
51. Trappe, J. M. 1971. A synopsis of the Carbomycetaceae and Terfeziaceae (Tuberales). Trans. Br. Mycol. Soc. 57:85-92.
52. Trappe, J. M. 1979. The orders, families and genera of hypogeous Ascomycotina (Truffles and their relatives). Mycotaxon 9:297-340.
53. Watling, R. and Watling, E. 1980. A literature guide for identifying mushrooms. Mad River Press Inc., Eureka, CA. 121 pp.
54. Zeller, S. M. and Dodge, C. W. 1924. Leucogaster and Leucophleps in North America. Missouri Botan. Gard. II:389-410.

MORPHOLOGY AND DEVELOPMENT OF ECTO- AND ECTENDOMYCORRHIZAE

Hugh E. Wilcox

INTRODUCTION

An ectomycorrhiza is characterized as a root/fungus association in which the fungus grows as a mantle on the surface of the root and penetrates the cortex intercellularly to produce a network known as the Hartig net. In some otherwise ectomycorrhizal infections, persistent intracellular infections occur in the cortical cells and the composite organ is then characterized as an ectendomycorrhiza. Despite the relative simplicity of their basic structure, ectomycorrhizae display a wide array of shapes, colors, and sizes depending on host and fungal species. Ectendomycorrhizae are somewhat less diverse in morphology.

Purposes of morphological investigations most often fall into one of the four following categories: (1) to provide quantitative data on the ecological variability of the mycorrhizal condition; (2) to characterize definitely recognizable mycorrhizal types and to relate these to different species of fungal symbionts for taxonomic purposes; (3) to obtain physiological information on the morphogenetic effects of each symbiont on the other; and (4) to assess the mycorrhizal status of nursery stock and its effect on outplanting.

Methodology in morphological investigations of ecto- and ectendomycorrhizae can be divided into two separate but interrelated aspects: the siting and features of mycorrhizae in relation to the architecture and growth characteristics of the root system; and the detailed description of distinctive mycorrhizae sampled for anatomical study. Whether the emphasis is on one or the other of these approaches or on both depends upon the objectives of the investigator.

RELATIONSHIPS OF MYCORRHIZAE TO THE ROOT SYSTEM

Recognition of Modifications in Root Morphology Induced by Mycorrhizae

Ecto- and ectendomycorrhizae develop from the mutualistic interaction between an active fungus and a developing feeder root of a woody plant. The characteristics of the resultant composite organ result from the morphogenetic effects of each partner on the other. Environmental factors affect the quantitative expression of the relationship, but the unique characteristics of the mycorrhizal organ itself result from the interaction of the two organisms. Since mycorrhizal activity involves terminal rootlets, a study of mycorrhizal development must focus on the characteristics of meristem behavior in these regions. However, recognition of the modifications induced by mycorrhizae requires knowledge of non-mycorrhizal root systems under favorable conditions or under common environmental stresses.

Ectomycorrhizae begin to appear in many seedlings from one to three months after germination. The primary root or radicle can be invaded by some ectomycorrhizal fungi, but fewer mycorrhizae develop from it than subsequently from first, second, or higher order laterals. The initial root system, in form and habit of growth, is characteristic for each species, but is later altered by environmental conditions (29). Different species vary in the relative flexibility of their initial root habit, and also in their ability to form appreciable numbers of mycorrhizae during the first year.

It can be generalized that most mycorrhizal root systems show two classes of roots distinguishable at a glance as short or small diameter "feeder" roots which are ephemeral, and long and more prominent skeletal roots which form the framework of the root system (30, 31). Mycorrhizal fungi may colonize either type of root, converting short roots to recognizable forms of mycorrhizae but making less apparent modifications of long roots.

Ectomycorrhizal Seedlings in Synthesis Cultures

Objectives. -- Aseptically germinated seedlings of many woody plant species have been used in pure culture synthesis of mycorrhizae. Common objectives have been to determine the ability of a fungus to form ectomycorrhizae; to check the mycorrhizal specificity of a host species by testing various putative ectomycorrhizal fungi; and to select the optimal ectomycorrhizal fungal species or strains for their ability to form large numbers of mycorrhizae or to promote seedling growth. The conditions necessary for successful formation of mycorrhizae in mycorrhizal synthesis experiments are discussed elsewhere (Chap. 11).

Limitations. -- Although the morphological characteristics of synthesized ectomycorrhizae are frequently described, some objections have been raised regarding their validity. Mantle characteristics are most often questioned. Mantle color may be influenced by substrate pH and its thickness may be increased by excessive glucose concentration in the medium. The lengths of primary and first and second order long roots may also be affected by nutrient levels in the substrate and possibly by the accumulation of deleterious excretions. The characteristics of root colonization and mycorrhizal formation in first year seedlings may be different from that of older seedlings because ectomycorrhizal fungi ultimately adapt their colonization behavior to hierarchical relationships arising between the various branch orders. Since seedling growth is more or less continuous during the first year, it is not possible to study the relationships between patterns of ectomycorrhizal colonization and seasonal axial increments of root growth. It is rarely possible to prolong synthesis experiments for studies during a second or third growing season. Finally, ectomycorrhizal fungi which in nature invade older seedlings in a later successional sequence may either not produce ectomycorrhizae in aseptic synthesis during the first year or may develop mycorrhizae so late in the experiment as to lead to the unwarranted conclusion that they are not mycorrhizal or are of little consequence.

Justifiable Uses. -- Mycorrhizal synthesis cultures are valuable for screening fungi for their ability to form ectomycorrhizae (Chap. 11), and with suitable modifications can be used for morphogenetic studies. A modification which permits direct observation of the development of mycorrhizae and the patterns of rhizomorph and mantle formation has been used in the author's laboratory for several years. The procedure is described below.

Special Method for Studying Patterns of Rhizomorph and Mantle Development. -- In this procedure the seedling root system is on filter paper in a large-diameter culture tube with a liquid nutrient medium. The seedling can either be completely within the culture tube, or the top can be freed to the air if there is a tight seal around the stem. A tube 32-35 mm in diameter and 300 mm in length, half filled with solution, is sufficient to permit 12-16 weeks of growth of a pine or spruce seedling placed in the tube at 2 weeks of age. Longer periods of observation are possible if the solution is replenished aseptically. Studies with *Pinus resinosa* Ait. have shown that many mycorrhizal fungi invade the primary root, and extensive mycorrhizae have been obtained with fungal species such as *Pisolithus tinctorius* (Pers.) Coker & Couch and *Suillus subluteus* (Pk.) Snell in Slipp & Snell in the relatively short interval of 12-16 weeks after inoculation. The pattern of developing mycorrhizae and the attendant rhizomorphs can be seen in Plate 6A.

The fungi anchor the root system to the filter paper, and at the conclusion of the study the filter paper can be removed and flattened in a pan of shallow water without disturbing the mantles, rhizomorphs, or spatial arrangement of the root system. In the case of *Suillus*, the use of vital stains in the water reveals a gossamer mycelial surface shrouding the long roots and arching outward over the emergent mycorrhizae. The immobilization of the root system and the enhancement of its two-dimensional aspect are advantageous for macrophotography. Samples of mycorrhizae can be removed along with portions of the filter paper and embedded for studies of undisturbed mantles using optical or electron microscopy. This tube system offers numerous possibilities for morphological studies of ectomycorrhizal development, particularly in relation to the colonizing behavior of particular fungi within the whole root system.

Suitability of Tube Technique for Histochemical Studies of Mycorrhizal Organization. -- Histochemical techniques now exist for examining a variety of biochemical systems and their occurrence in tissues and cells, and it can be expected that these methods will be used to follow metabolic activities during the development of various ectomycorrhizal structures. The clean and relatively undisturbed ectomycorrhizal structures of the filter paper system are ideally suited for the various staining procedures, techniques of enzyme localization, and autoradiographic procedures which comprise the field of histochemistry (9).

Histochemical studies of ectomycorrhizae have been made for starch and other polysaccharides (15, 27), phenolic compounds (14, 27), and polyphosphate granules (1, 13). These studies need to be extended to a greater number of species and related to the developmental features of the individual root systems. Additional types of histochemical studies doubtless merit investigation.

Morphological Studies of Nursery Seedlings

Limited Diversity of Ectomycorrhizae. -- In most coniferous nurseries, mycorrhizae develop spontaneously before the end of the first year following seeding. Initial delays in mycorrhizal formation, however, may cause a considerable percentage of short roots to escape infection the first season. During the second year mycorrhizal development proceeds rapidly and practically all short roots become mycorrhizal (12). Isolations of mycorrhizal fungi from coniferous nursery seedlings have shown that the ectomycorrhizae are formed by relatively few fungal species compared to the diversity found under natural forest conditions. Often sporophores of one or two Hymenomycetes are found in the second year nursery beds. A few other ectomycorrhizal fungi are present, but either do not fruit or produce inconspicuous fructifications.

Ectendomycorrhizal Associations. -- In many forest nursery seedlings of *Pinus* and *Larix* a predominating mycorrhizal association exists which is distinguishable by a coarse intercellular net of large bulbous hyphae, a weak mantle, and persistent coarse intracellular hyphae which also occupy the cortical cells of the long roots (11, 24, 34). This so-called E-strain ectendomycorrhizal association of Mikola also occurs on other members of the Pinaceae, most notably *Picea*, but in these it forms only ectomycorrhizae and is therefore less readily detectable. Other ectendomycorrhizal associations also appear to be present.

Anomalous Infections. -- Often neither the pioneering mycorrhizal species nor E-strain ectendomycorrhizal fungi are adequate to produce mycorrhizae on all the short roots of the seedlings. Uninvaded short roots may either undergo cortical collapse and early deterioration or be invaded intracellularly by non-mycorrhizal fungi. The root responds by the formation of tannins which fill the cortical cells or suberins which incrust their walls, destroying their functioning ability (23).

Colonizing Behavior of Different Mycorrhizal Fungi. -- Morphological studies of nursery seedlings have provided valuable information on the differences in colonizing behavior of different mycorrhizal species as well as details of their seasonal and successional behavior. Additional studies are needed to separate the effects on root development caused solely by fungi from the range of normal root behavior. It is increasingly evident that many anatomical features characteristically ascribed to fungal invasion are also found in uninfected root systems. *The presence or absence of the Hartig net remains the most reliable index of mycorrhizal development; the presence of a fungal mantle is less reliable and seems to vary with root diameter and the status of fungal activity.* Ectendomycorrhizae generally produce sparse mantles, but may form voluminous mantles on actively-growing coralloid mycorrhizal clusters. Much work remains to be done on the factors controlling short root formation and on factors controlling their conversion to mycorrhizae.

Mycorrhizal Development in Relation to Seasonal and Intra-Seasonal Periods of Root Growth. -- Cycles of lateral root initiation and mycorrhizal development correspond to seasonal and environmentally induced periods of root growth activity. Particular attention should be directed to features of the long roots as well as those of short roots, remembering that often a continuum of root sizes exists and mycorrhizal fungi frequently adapt to the root branches with intermediate characteristics. Morphological study of mycorrhizal development must consider meristematic activity in both the mother root and in the short root which is undergoing transformation. This dictates the sampling of major hierarchical units such as major entire first order laterals with all their attendant long and short root branches.

Methods of Sampling Nursery Seedlings. -- Ordinarily little difficulty is encountered in excavating intact root systems from nursery beds during the first year. However, two- and three-year-old seedlings are difficult to sample without considerable damage to the root system as well as the nursery bed. One expedient is to sample intact first order laterals from between nursery beds. A principal first order lateral is located by carefully removing the soil at the base of a border seedling to expose its attachment to the primary root. After severing the root at the point of attachment, a small pointed tool can be used to progressively uncover the distal portions of the lateral and all its ramifications. Such a sample preserves the important morphological features associated with branch order, seasonal axial increments of growth, zones of meristematic activity, patterns of mycorrhizal development, etc. This sampling procedure should be repeated at other positions along the nursery bed to determine variability.

Time of Sampling. -- A detailed study of the development of mycorrhizal infection in seedlings requires sampling at frequent intervals. Laiho and Mikola (12) sampled nursery beds in Finland at weekly intervals during the first growing season and at two-week intervals during the second. Similar intervals are recommended for any initial study, but subsequent adjustments will probably need to be made, depending on the rhythms of seedling growth and the nature of the problems being investigated. If the investigator is interested only in the ultimate characteristics of fully-developed mycorrhizae and not their morphogenesis, the best period for sampling is in the autumn after the main period of root growth is over and mycorrhizal development has progressed to the most terminal regions of the mother roots.

Recording and Measurement of Morphological Features. -- Portions of root systems for study can either be photographed or photocopied (using the type of copier in which the specimen can be placed on a stationary sheet of glass rather than revolved around a drum.) This provides a visual record of branching patterns and mycorrhizal features upon which can be recorded the portions of roots sampled for fungal isolation or anatomical study.

The lengths and distribution patterns of second and higher order long root branches from an individual first order lateral frequently differ between ectomycorrhizal and ectendomycorrhizal infections. Extensive development of ectomycorrhizae may inhibit the elongation of long roots and transform probable long root primordia into mycorrhizae, whereas ectendomycorrhizal development appears to have little effect on lengths and numbers of long roots. Because of these differing morphogenetic effects of mycorrhizal fungi, measurements of long root numbers and their individual or accumulative lengths may be germane to an assessment of the mycorrhizal condition, provided the infection is predominately of one type. Also essential to this morphological assessment are observations of the general appearance of short roots and mycorrhizae, their distributions, and their relationship to meristematic regions (33). The distinction between ecto- and ectendomycorrhizae cannot be made without anatomical observations. In *Pinus* both types may produce similar appearing mycorrhizal clusters (Plate 5F,H).

Clearing Roots for Studying Patterns of Fungal Colonization. -- Major root branches can be stored in formalin-acetic acid-alcohol (FAA) (10) and subsequently cleared for determining patterns of fungal colonization in the various branch orders, the presence of ecto- and ectendomycorrhizae (Plate 6G), and the position of the metacutized junction between successive axial increments of growth. Ectomycorrhizal root systems can be cleared with the procedures of Phillips and Hayman (26) for pigmented endomycorrhizal roots. For roots which do not bleach adequately in hydrogen peroxide the procedures can be modified by moving the roots back and forth between the hydrogen peroxide and potassium hydroxide until bleached.

Ectomycorrhizae with thick mantles and Hartig nets are difficult to examine after clearing, but they can be teased apart with micro dissecting needles and individual macerated cells observed for details of shape, patterns of Hartig net, and presence of any intracellular fungi (Plate 6H).

Freehand and Microtome Study of Mycorrhizal Anatomy. -- Freehand transverse and longisections can be quickly prepared and mounted on a microscope slide in lactic acid or lactophenol with a trace of cotton blue to screen roots for ecto- or ectendomycorrhizal infections (Plate 6E,F). Terminal root portions of interest to the investigator can be processed by ordinary microtechnique procedures (10) for anatomical study. Serial transverse and longitudinal sections can be stained for mycorrhizal details by the Conant quadruple stain (10) (Plate 6C,D) or the chlorazol black E and Pianese 111-B stain (35) (Plate 6B). The Conant stain is advantageous for handling large numbers of slides because once the timing has been determined it is unnecessary to examine individual slides during the processing. Unfortunately, clove oil used as the solvent throughout the schedule is costly. The combination of safranin, crystal violet, fast green, and gold orange in the Conant stain yields slides which are excellent for study purposes but do not photograph well with black and white panchromatic film. Chlorazol black E and Pianese 111-B requires more careful monitoring of slides during the staining operation but is less costly. The combination of chlorazol, malachite green, acid fuchsin, and Martius yellow yields slides which photograph well in black and white as well as color.

Multiformity of Ectomycorrhizae. -- The structural variations of ectomycorrhizae in nature have not been fully explored, yet it is clear that the morphological diversity and the number of fungal species involved are both vastly greater than in forest nurseries. Initially, morphological studies were concentrated on classifying ectomycorrhizae into recognizable types for studies of seasonal development, distribution, and functional life of terminal root apices in different forest stands of various forest species. More recently, morphological study has turned to the search for criteria for identifying the species of fungi involved in the various mycorrhizal associations.

Techniques Used for Sampling Forest Ectomycorrhizae. -- The remarkable concentration of short root apices near the upper soil surface in northern coniferous forests has dictated the use of small sampling volumes to estimate mycorrhizal frequency. Mikola and Laiho (25) used a sampling tube of 2.5 cm diameter and sampled to a depth of 7 cm to study mycorrhizal relations in a northern spruce forest in Finland. Root density was highest in the middle of the humus layer with an average of 31 root tips per cm^3. The mycorrhizae were classified into a few recognizable types based on a scheme originally proposed by Melin (22) and modified by Mikola and Laiho (25).

The value of the soil tube method for sampling mycorrhizae rests on the fact that there are vastly more small-diameter root tips than large-diameter root tips in a volume of soil, and a very high proportion of small diameter tips are mycorrhizal. Improvements in the core sampling technique (16, 17, 18, 20) have gone hand-in-hand with the development of more sophisticated methods of classifying ectomycorrhizae. These later investigations have utilized the detailed classification scheme devised by Dominik (6, 21) in central Europe. Dominik's classification proposed artificial categories, or "genera," based on macroscopic and microscopic morphological features such as color, structure, surface characteristics of the mantle, etc. More will be said about this scheme in the final section of this paper.

Developmental Studies of Ectomycorrhizae in Relation to Root Morphology. -- There are many morphological problems which require more than the characterization of mycorrhizal apices, and for such problems core samples may be inadequate. There are large numbers of fine roots which bear these mycorrhizae. Therefore the root sample must be long enough to distinguish the fine long roots from short roots (4, 7, 32). As mentioned in the discussion of nursery seedlings, emphasis on the qualitative changes which occur in the mycorrhizal apex may serve a classification purpose but does not reveal the effects of mycorrhizal colonization on the morphology of long roots nor the interrelationship between long and short root development. Also interrelationships between mycorrhizal development and the health of a forest stand are often revealed by study of the morphological features of the last several orders of root branching. The mortality of thin long roots and the accompanying death of young mycorrhizae are often features of unhealthy root systems (32).

Knowledge of seasonal patterns of root growth activity and patterns of fungal colonization contributes to the understanding of mycorrhizal development, responses of mycorrhizae to various environmental influences, and competitive relationships between ectomycorrhizal fungal species. As important as the need for healthy mycorrhizae is the concomitant need for an adequate number of vigorous mother roots and subordinate mother roots.

Sampling of Major Hierarchical Units for Developmental Studies. -- Sampling procedures similar to those given for studying the infection in the root systems of nursery seedlings also apply to natural root systems. The sampling should consider major hierarchical units, but taken in the reverse direction, starting from the ultimate rootlets and working basipetally through successively earlier root branches. It is essential to obtain as many of the fine ultimate branches as possible in their natural positions in the last few orders of branching. The size and appearance of these functional units or "root boughs" (2) is determined by host species and environmental conditions. For example, in *Pinus resinosa* the three classes of long roots (pioneer, mother, and subordinate mother) may each show different features of mycorrhizal development (31).

It is obvious that single roots and small root branch systems of mature trees should be excavated carefully by hand, using small-diameter tools. In some cases an old-fashioned ice pick may be the best tool. In mor soils with a definite layer of unincorporated humus, the developing mycorrhizae can often be seen in profusion merely by using fingers to scratch off the unconsolidated 01 layer. A block of matted roots and mycorrhizae can be obtained by cutting four sides of a block with a shovel through the interlacing roots into the underlying soil layers. The block can be lifted out after severing a few small-diameter roots. The underlying decomposed humus layer often has a fine granular structure like fine sawdust, and practically all of it can be shaken out of the root mat, leaving the fine ramifications of the abundant network of roots practically clean (28). Where advance reproduction is present (volunteer seedlings in the forest understory), the uppermost first order laterals can often be pulled up *in toto* and shaken free of the adhering detritus.

The above sampling procedure must be somewhat modified in sites where a mull has developed, or where (as is frequently the case) the humus relationship is intermediate, having the structure of crumb mull and an abnormally high humus content. Here superficial excavations should be made into the upper layers of the mineral soil (A1 and A2) to uncover the small woody roots, which should be followed to their ultimate branchings. In this manner it is often possible to obtain smaller ascending laterals in their entirety.

Preparation of Samples for Morphological Studies. -- These samples of aggregated groupings of ultimate root branches provide information on growth characteristics of the root system, including seasonal and intraseasonal growth increments, patterns of root replacement, unique characteristics of mycorrhizal patterns, etc. The roots should be placed in plastic bags with moist sphagnum for transport to the laboratory. After washing out, the roots should be examined under a dissecting microscope in a shallow dish of water and portions selected for study. Selected material can be photographed or photocopied and processed as discussed previously for the examination of nursery seedlings.

Time of Sampling Forest Mycorrhizae. -- Similar considerations apply to the sampling of forest tree mycorrhizae as for nursery seedlings. All studies of the development and function of mycorrhizae should be coordinated with the seasonal growth activity of the root system. Most tree species have two peak periods of growth activity, one in the spring and the other in the late summer or early fall. Sampling should be frequent at these times and also in the autumn when root elongation slows and mycorrhizal development continues progressively closer to the meristematic regions of the larger diameter long roots.

Morphological studies concerned primarily with the classification of ectomycorrhizae as discussed in the following section may be best served by autumn sampling periods.

Reasons for Detailed Anatomical Examination of Individual Mycorrhizae. -- Most investigations of the detailed anatomical characteristics of various distinct ecto- and ectendomycorrhizae are intended to reveal information for classification and identification purposes. The desire to identify as many tree-fungus associations as possible is motivated by taxonomic and ecological considerations, but such a classification system would also benefit physiologists, pathologists, nursery managers, and others concerned with the benefits of mycorrhizae.

Distinctive vs. Non-Distinctive Ectomycorrhizae. -- Those ectomycorrhizae which have been selected for detailed morphological investigation are the ones most conspicuous to the unaided eye and uniform in appearance. These traits suggest that these mycorrhizae belong to a uniform type produced by one or a small group of predominating fungi; and that this type, once described, will continue to catch the attention of researchers. The features which have rendered these mycorrhizae most conspicuous include a wide-range of characteristic colors, a frequent thickening, and an assortment of shapes and branch patterns. In *Pinus* they may fork dichotomously, and repeated forking may lead to coralloid or nodular structures (Plate 5E,F). *In most other ectomycorrhizal host species they are pinnately or racemosely branched.* (Plate 5G).

In addition to distinctive ectomycorrhizae, there are many inconspicuous ectomycorrhizae with the requisite fungal mantle and Hartig net, but which possess no characteristic color or shape. There appear to be large numbers of these nondescript ectomycorrhizae, which may be as beneficial to their host as their more flamboyant counterparts. It is unfortunate that so little attention is being given to the development and role of non-distinctive mycorrhizae.

Qualitative Differences in Structure Between Mycorrhizae and Uninfected Roots. -- Caution must be exercised in the study of the anatomical features of presumptive mycorrhizal rootlets. Many of the mycorrhizal features reported earlier have been found to occur also in slow growing uninfected rootlets. A few *bona fide* qualitative structural differences exist between mycorrhizae and uninfected roots, and these are (1) a mantle of fungal tissue covering the root and penetrating centripetally between the cortical cells to form the Hartig net; and (2) radial enlargement of the cortical cells surrounded by the Hartig net. The lack of root hairs, limited root cap, and precocious vascular development found in ectomycorrhizae can also be found in uninfected short roots.

Methods for Characterization of the Fungus Mantle and Hartig Net. -- Because of the few qualitative differences between mycorrhizal and uninfected roots, there are relatively few potential morphological criteria for distinguishing mycorrhizal types. There is the appearance under the stereomicroscope, including general form or branching pattern, surface texture and color of the mantle, the presence or absence of rhizomophs, etc. Additional information on mantle, Hartig net, rhizomorphs, and mycelium characteristics is provided by microscopic examination of freehand and microtome sections.

The stain combinations used by the author to study the above features were discussed earlier. Other stain combinations which have often been used for prepared sections are orseillin BB and crystal violet (5, 22), safranin and fast green (10), and Cartwright's picro aniline blue (8, 19).

Dominik's Scheme for the Generalized Classification of Ectomycorrhizae. -- Mention has been made of the classification scheme of Dominik (6) which creates mycorrhizal form genera based primarily on color and structure of the fungus mantle. This scheme has been used extensively by Australian and European workers.

However, increasing dissatisfaction has arisen with its use because of difficulties in applying the color criteria, and because the artificial "genera" are not referrable to natural classifications of either the fungus or its host. A more complete critique of Dominik's key appears in an article by Zak (38).

Characterization of Distinctive Ectomycorrhizae for Particular Host Species. -- Several less inclusive classification schemes have been devised by various workers to characterize distinctive ectomycorrhizae for particular host species. These efforts, each restricted to one or two closely related host species, are more likely to designate mycorrhizal types that either lead to a single fungal species or to a group of closely related species. Zak also reviewed these restricted classification schemes and discussed some of the principles necessary for successful characterization and identification of ectomycorrhizae (38). He tabulates the features most likely to be useful for the construction of practical keys to identify ectomycorrhizae and their associated fungi, and notes that the first satisfactory keys will probably be constructed for individual tree species and for specific habitats.

The best examples of sound morphological descriptions of ectomycorrhizae are those of Chilvers (3) for *Eucalyptus* and Zak (36, 37) for *Pseudotsuga menziesii*. Chilvers surveyed mycorrhizal types over several years and restricted their selection to the eight most distinctive and common types, these all differing according to several easily recognizable criteria of low variability. The *Pseudotsuga* mycorrhizae characterized and classified by Zak were distinctive in appearance and the fungi were identified from associated sporocarps. Both authors have presented photographs or drawings to illustrate the important details described for each distinct ectomycorrhiza. The descriptions are based on stable macroscopic and microscopic characteristics and on color reactions to chemical treatments. Although the papers of Chilvers and Zak provide an important starting point for new investigations, new criteria will arise, some fortuitously and others through necessity.

Research Needs and Possibilities. -- Morphological investigations of distinctive ectomycorrhizae and identifications of fungal symbionts are valuable, but constitute only a partial attack on the problem of ectomycorrhizal development. For the very common but non-distinctive mycorrhizae it may be desirable to first isolate the mycorrhizal fungi and confirm their symbiotic nature in synthesis culture. Following recognition of the characteristics of these fungi, it might be possible to make a developmental and morphological study of their effects in nature. An understanding of mycorrhizal habit in non-distinctive fungi can only be obtained by a consideration of the root system as an entity. Specialized sampling procedures will have to be followed, similar to those presented earlier in this chapter.

LITERATURE CITED

1. Ashford, A. E., Ling-Lee, M. and Chilvers, G. A. 1975. Polyphosphate in eucalypt mycorrhizas: a cytochemical demonstration. New Phytol. 74:447-453.
2. Cannon, W. A. 1954. A note of the grouping of lateral roots. Ecology 35:293-295.
3. Chilvers, G. A. 1968. Some distinctive types of eucalypt mycorrhiza. Aust. J. Bot. 16:49-70.
4. Chilvers, G. A. and Pryor, L. D. 1965. The structure of eucalypt mycorrhizas. Aust. J. Bot. 13:245-259.
5. Cohen, I. and Doak, K. D. 1935. The fixing and staining of *Liriodendron tulipifera* root tips and their mycorrhizal fungus. Stain Tech. 10:25-32.

6. Dominik, T. 1959. Synopsis of a new classification of the ectotrophic mycorrhizae established on morphological and anatomical characteristics. Mycopathol. Mycol. Appl. 11:359-367.
7. Harley, J. L. 1940. A study of the root system of beech in woodland soils with especial reference to mycorrhizal infection. J. Ecol. 28:107-117.
8. Jackson, L. W. K. 1947. Method for differential staining of mycorrhizal roots. Science 105:291-292.
9. Jensen, W. A. 1962. Botanical Histochemistry. Principles and Practice. W. H. Freeman & Co., San Francisco. 408 pp.
10. Johansen, D. A. 1940. Plant Microtechnique. McGraw-Hill, New York. 523 pp.
11. Laiho, O. 1965. Further studies on the ectendotrophic mycorrhiza. Acta For. Fenn. 79:1-35.
12. Laiho, O. and Mikola, P. 1964. Studies on the effect of some eradicants on mycorrhizal development in forest nurseries. Acta For. Fenn. 77:1-34.
13. Ling-Lee, M., Chilvers, G. A. and Ashford, A. E. 1975. Polyphosphate granules in three different kinds of tree mycorrhiza. New Phytol. 75:551-554.
14. Ling-Lee, M., Chilvers, G. A. and Ashford, A. E. 1977. A histochemical study of phenolic materials in mycorrhizal and uninfected roots of *Eucalyptus fastigata* Deane and Maiden. New Phytol. 78:313-328.
15. Ling-Lee, M. Ashford, A. E. and Chilvers, G. A. 1977. A histochemical study of polysaccharide distribution in eucalypt mycorrhizas. New Phytol. 78:329-335.
16. Marks, G. C. 1965. The classification and distribution of the mycorrhizas of *Pinus radiata*. Aust. For. 29:238-251.
17. Marks, G. C., Ditchburne, N. and Foster, R. C. 1968. Quantitative estimates of mycorrhiza populations in radiata pine forests. Aust. For. 32:26-38.
18. Marks, G. C. and Foster, R. C. 1967. Succession of mycorrhizal associations on individual roots of radiata pine. Aust. For. 31:193-201.
19. Masui, K. 1926. A study of the ectotrophic mycorrhiza of *Alnus*. Mem. Coll. Sci., Kyoto Imperial Univ. Ser. B II:190-209.
20. Mejstrik, V. 1971. The classification and frequency of ectotrophic mycorrhizas on *Pinus radiata* D. Don. in New Zealand. Plant Soil 34:753-756.
21. Mejstrik, V. and Dominik, T. 1969. The ecological distribution of mycorrhiza of beech. New Phytol. 68:689-700.
22. Melin, E. 1923. Experimentelle Untersuchungen über die Konstitution und Ökologie der Mykorrhizen von *Pinus sylvestris* und *Picea abies*. Mykologische Untersuch. 2:73-331.
23. Meyer, F. H. 1974. Physiology of mycorrhiza. Annu. Rev. Plant Physiol. 25:567-586.
24. Mikola, P. 1965. Studies on the ectendotrophic mycorrhiza of pine. Acta For. Fenn. 79:1-56.
25. Mikola, P. and Laiho, O. 1962 Mycorrhizal relations in the raw humus layer of northern spruce forests. Comm. Inst. For. Fenn. 55:1-13.
26. Phillips, J. M. and Hayman, D. S. 1970. Improved procedures for clearing roots and staining parasitic and vesicular-arbuscular mycorrhizal fungi for rapid assessment of infection. Trans. Br. Mycol. Soc. 55:158-161.
27. Piché, Y., Fortin, J. A. and Lafontaine, J. G. 1981. Cytoplasmic phenols and polysaccharides in ectomycorrhizal and non-mycorrhizal short roots of pine. New Phytol. 88:695-703.
28. Romell, L. G. and Heiberg, S. O. 1931. Types of humus layer in the forests of northeastern United States. Ecology 12:567-608.
29. Toumey, J. W. 1929. Initial root habit in American trees and its bearing on regeneration. Int. Congr. Plant Sci. Proc. Vol. 1:713-727.

30. Wilcox, H. 1962. Growth studies of the root of incense cedar, *Libocedrus decurrens*. II. Morphological features of the root system and growth behavior. Am. J. Bot. 49:237-245.
31. Wilcox, H. 1964. Xylem in roots of *Pinus resinosa* Ait. in relation to heterorhizy and growth activity. Pages 459-478 *in* M. Zimmerman, ed. Formation of wood in forest trees. Academic Press, Inc., New York, N. Y. 562 pp.
32. Wilcox, H. E. 1968. Morphological studies of the root of red pine, *Pinus resinosa*. I. Growth characteristics and patterns of branching. Am. J. Bot. 55:247-254.
33. Wilcox, H. E. 1968. Morphological studies of the roots of red pine, *Pinus resinosa*. II. Fungal colonization of roots and the development of mycorrhizae. Am. J. Bot. 55:686-700.
34. Wilcox, H. E. 1971. Morphology of the ectendomycorrhizae in *Pinus resinosa*. Pages 54-68 *in* E. Hacskaylo, ed. Mycorrhizae. U. S. D. A. Forest Service Misc. Publ. 1189. Supt. Documents, Washington, D. C. 255 pp.
35. Wilcox, H. E. and Marsh, L. G. 1964. Staining plant tissues with chlorazol black E and Pianese III-B. Stain Tech. 39:81-86.
36. Zak, B. 1969. Characterization and classification of mycorrhizae of Douglas fir. I. *Pseudotsuga menziesii* + *Poria terrestris* (blue- and orange-staining strains.) Can. J. Bot. 47:1833-1840.
37. Zak, B. 1971. Characterization and classification of mycorrhizae of Douglas fir. II. *Pseudotsuga menziesii* + *Rhizopogon vinicolor*. Can. J. Bot. 49:1079-1084.
33. Zak, B. 1973. Classification of ectomycorrhizae. Pages 43-78 *in* G. C. Marks and T. T. Kozlowski, eds. Ectomycorrhizae: their ecology and physiology. Academic Press, New York, N. Y. 444 pp.

ISOLATION, MAINTENANCE, AND PURE CULTURE MANIPULATION OF ECTOMYCORRHIZAL FUNGI

Randy Molina and J. G. Palmer

INTRODUCTION

Isolation and experimental manipulation of ectomycorrhizal fungus cultures have been critical in developing all phases of ectomycorrhizal research. *In vitro* culture studies provide insight into the basic biology and processes of the fungal symbionts, e.g. growth responses to varying pH, temperature, and moisture regimes, mineral and carbohydrate physiology, and production of enzymes and hormones. Cellular interactions between fungus and host as well as with other microorganisms, including pathogens, can be critically examined *in vitro*. Isolation and comparative study of diverse fungal species and isolates within species provide a basis for selecting specific isolates for artificial inoculation of nursery stock (28, 71).

A tremendous body of literature exists on pure culture studies of ectomycorrhizal fungi; we will not attempt to review all aspects. Our purpose is to describe several techniques common in study of ectomycorrhizal fungus cultures. Isolation, maintenance, nutritional requirements, ectomycorrhiza synthesis, and basic precepts will be stressed. Variations of the methods described may work equally well to meet unique or exacting circumstances of individual researchers. Alternative techniques are also described in many of the references cited and in recent mycology texts (26, 72).

ISOLATION OF ECTOMYCORRHIZAL FUNGI

Ectomycorrhizal fungi are most commonly isolated from sporocarp tissue but can also be isolated from surface sterilized ectomycorrhizae, sclerotia, rhizomorphs, and, in some cases, from sexual spores. Sporocarp isolation is generally preferred because species can be identified and little pretreatment of fungal material is required. Ectomycorrhizal fungi can differ strongly in ease of isolation from sporocarps and growth in culture. However, consistent trends in isolatability are evident within certain taxonomic groupings. For example, most species of *Suillus* and *Rhizopogon* are easily isolated and grow well in culture. In the genus *Amanita*, species in subgenus *Amanita* are more easy to isolate and grow in culture than species of subgenus *Amanitopsis*. Only a few species have been successfully isolated in some genera, e.g. *Russula*, and in still others no successful isolations have been made, e.g. *Gomphidius*. Many species within the following genera can be routinely isolated from sporocarps: *Alpova*, *Amanita* *Astraeus*, *Boletus*, *Cortinarius*, *Fuscoboletinus*, *Hebeloma*, *Hymenogaster*, *Hysterangium*, *Laccaria*, *Lactarius*, *Leccinum*, *Melanogaster*, *Paxillus*, *Pisolithus*, *Rhizopogon*, *Scleroderma*, *Suillus*, and *Tricholoma*. The inability to isolate several species stems from ignorance on the exacting nutrient requirements which they likely derive from hosts. Further nutritional studies are needed to bring these recalcitrant species into culture.

Collection of sporocarps.--Collectors should familiarize themselves with the seasonal fruiting of fungi so that sporocarps in good condition and in various developmental stages can be collected. Young sporocarps are preferred for direct isolations. Fully matured specimens should also be collected, however, so that species can be reliably determined. Place sporocarps into waxed paper bags or wrap in waxed paper to retard drying (avoid plastic bags because sporocarps deteriorate rapidly in a "non-breathing" container). In the case of mushrooms, cut the stem from a mature specimen in good condition, place the cap over a white card, cover with waxed paper, and lay cap over card flat in the collection container to obtain a spore print. Record field notes on macroscopic sporocarp characters (pay attention to color changes when specimens are bruised) and all potential hosts in the vicinity immediately upon collection. It's best to familiarize oneself ahead of time with critical field characters needed to identify anticipated fungus collections as well as the ectomycorrhizal hosts in the area. Seek prior advice or, even better, participation of a taxonomic mycologist.

Isolations should be done as soon as possible after collection; best results are often obtained when done immediately in the field (53, 71). Properly stored sporocarps of many species, however, can yield successful isolations after several days. While in the field, avoid freezing, heating (direct sunlight) or drying of sporocarps. If isolations are to be attempted one or more days after collection, the specimens should be refrigerated (3-5°C).

Accurate identification of fungi is critical to interpreting research results. If identifications are uncertain, request confirmation by a mycologist familiar with the group in question. We stress that properly dried voucher specimens must be prepared for all isolates and accessioned into an active mycological herbarium for future reference (1); include the specimens from which the isolates were actually obtained.

Collections of ectomycorrhizae, rhizomorphs and *Cenococcum* sclerotia.--Purpose and circumstance may dictate the selection and collection of ectomycorrrhizae for direct isolation. Select roots with as little adhering debris as possible. Ectomycorrhizae in some substrates, e.g. rotten wood, are cleaner than in others, e.g. mineral soil. Collect enough material for 100-200 isolation attempts for each desired ectomycorrhiza type. Isolate as soon as possible after collection. Fine roots can dehydrate rapidly but preserve well if placed together with a little moist soil, humus, or moss in a "non-breathing" container (tightly closed can, jar, or plastic bag). If necessary, refrigrate root samples (3-5°C) until used. Rhizomorphs should be collected and stored similarly.

Cenococcum geophilum Fr. is most easily isolated from its hard black sclerotia (69). The long-lived sclerotia abound in most stands of ectomycorrhizal hosts and can be extracted from soil samples any time of year. To collect sclerotia, rake away the upper humus and sample the organic-mineral soil interface to a depth of 10-15 cm; at least five subsamples (trowel fulls) should be collected around the immediate vicinity of the host plant to yield 1 to 2 liters of soil. Take the samples back to the laboratory and, if neccessary, store them in tightly closed, "non-breathing" containers under refrigeration until sclerotia are extracted. Isolation should be attempted as soon as possible, but samples can be stored for many weeks without measureable loss of sclerotium viability.

Preparation for isolation.--A clean work area with still air is important to minimize contamination and generally is as good as specially designed isolation hoods. In the field, we have had good success using a small portable isolation chamber constructed of white-painted plywood with a slanted, transparent, plexiglass shield on top (Fig. 1); a rectanglular 15 X 40 cm opening in front allows easy hand entry and maneuverability within the box for flaming metal tools, tissue extractions and transfers onto nutrient agar. This portable chamber is easily set up on picnic tables in campgrounds, on stumps or logs, or on tables in travel accomodations. Any bench space in a clean laboratory is ideal for isolating. However, field material often harbors pests such as mites which can enter and consequently contaminate culture vessels. Thus, field collections should not be brought into areas where stock and experimental cultures are kept.

Necessary tools and materials include fine tipped, heat-sterilizable scalples, transfer needles, forceps, alcohol or gas lamps, and a general disinfectant such as 5% sodium hypochlorite or 95% ethanol. Abundant test tubes or petri plates with nutrient agar should be available; tubes are less subject to contamination than plates. At the Corvallis Forestry Sciences Laboratory, we routinely use modified Melin-Norkrans nutrient agar (MMN) (Table 1) with dextrose as the carbon source and potato dextrose agar. Addition of antibiotics can prove useful for reducing contamination, especially for isolations from roots.

Isolation from sporocarp tissue.--If available, select young sporocarps free of rot and insect (larvae) damage. Brush off adhering debris, especially from the stipe base; if the base is difficult to clean, cut it off. For mushrooms, cut a shallow (1-2 mm) slit across the middle of the cap suface and along the length of one side of the stipe. On tuber-like hypogeous sporocarps, the initial shallow cut should circumscribe roughly one-half of the fruit body. Do not cut through the entire specimen to expose interior tissue because this will drag surface contaminants into the cuts. To expose interior tissue, gently pull apart the sporocarp along the initial shallow cuts, using fingertip pressure; placing a fingernail into the initial cut will lend more leverage for breaking sporocarps composed of tough tissue. Quickly scan the exposed interior surface for areas of sound tissue free from contact with obvious contaminating sources. With a fine tipped scalpel (flame sterilized and cooled), cut and loosen small pieces (2-5 mm cubes) of sound tissue. Transfer the tissue explants with the scalpel or a sterile transfer needle directly onto nutrient agar in tubes or plates.

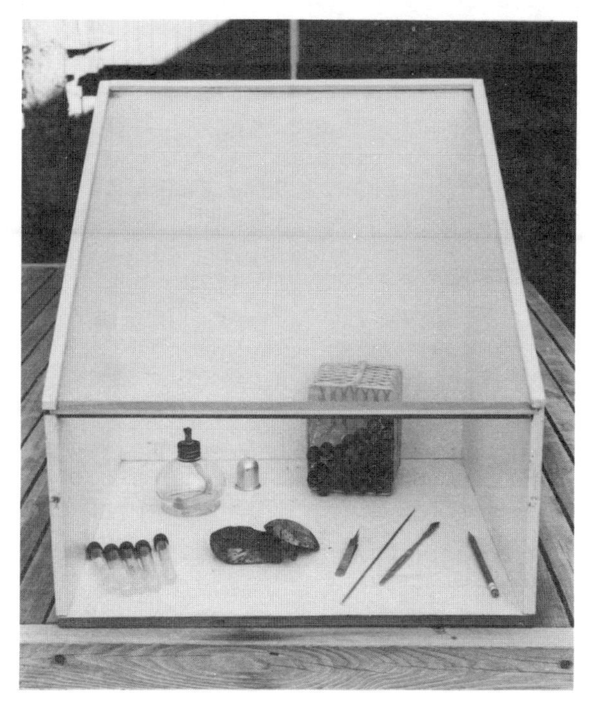

Fig. 1. Portable culture chamber for field isolations. Plywood frame and clear, plexiglass top reduce air movement; open front provides easy hand access.

Because sporocarp tissues are often strongly differentiated, attempt isolations from two to three locations on the sporocarps, e.g. cap and stipe (tissue from over the gills often seems to grow best). For most hypogeous fungi, such as *Rhizopogon* spp., the center of the gleba is best for isolation. With *Scleroderma* spp., the interior of the leathery peridial tissue is good. Isolations should be attempted from more than one specimen for greater assurance of success. The number of individual tissue transfers depends on the importance of obtaining an isolate and one's experience in isolating that particular species. For example, with fresh, larva-free sporocarps of *Rhizopogon* or *Suillus* spp., 6 to 10 tissue transfers are generally enough. If you are not familiar with the fungus species and an isolate is critically needed, 10 to 20 or more attempts should be made. Tissue transfers are routinely incubated at 20°C, but room temperature is often adequate.

After 3 to 4 days, begin observations under a stereomicroscope for initial fungus growth and contamination. Fungi which are easily isolated and grow well in culture produce visible mycelial growth 4 to 7 days after isolation; other fungi may take 2 to 6 weeks to show any sign of growth from the tissue. Most species "bush-out" in all directions from the tissue explant. Others, like *Laccaria* spp., grow only submerged in the agar or prostrate on the agar surface. Cultures should be frequently observed over the first few weeks and characteristics of emerging colonies and presence of contaminating microbes recorded. Check particularly for uniformity in culture characteristics of the mycelia emerging from the various tissue explants. Several uniform cultures emerging from different explants provide initial evidence of sucessful isolation. After the suspected ectomycorrhizal fungus is established on the agar substrate and no contaminants are visible, mycelium from the new colony edge should be aseptically transferred with a transfer needle onto fresh nutrient agar to set-up a stock culture. Ectomycorrhizal fungi that grow well in culture will be ready for initial transfer after 3 to 4 weeks. Slow growing fungi may take 2 to 4 months before they are ready for transfer and are easily lost in subsequent transfers. Tissue explants of recalcitrant fungi may show no contamination, but new hyphae never emerge from the tissue.

Bacteria and several "weed" fungi (Trichoderma, Aspergillus, Penicillium, and yeast spp.) are the most common contaminants and are easily detected by presence of a glistening bacterial slime or characteristic fungus conidia. Contaminants can swamp the tissue explant and agar substrate before the ectomycorrhizal fungus can begin to grow. Some fungi such as Laccaria spp. routinely yield bacterial associates upon isolation. However, the fungi often out-grow the bacteria and cultures are cleaned-up by aseptically removing a bit of mycelium from the colony edge and transfering it onto fresh substrate. This may have to be done repeatedly until bacteria are absent. Cultures with questionable bacterial contamination should be grown in a nutrient broth wherein proliferation of any bacteria present will cloud the liquid.

A mycorrhiza synthesis is needed to confirm that a species isolated for the first time is the intended species and if it forms ectomycorrhizae. This determination is even more difficult when two or more different-appearing fungi are isolated from one fungus collection. Familiarity with the cultural characteristics of the fungus species or comparison to known cultures of that fungus greatly aid culture confirmation. Clamp connections indicate that the fungus is a basidiomycete, but not all basidiomycetes form clamps. Culture descriptions of many ectomycorrhizal fungi are available in the literature, as are suggested ways to identify ectomycorrhizal fungi from cultural characteristics (59, 65, 72). As emphasized later, a pure culture mycorrhiza synthesis provides the strongest evidence of isolation of an ectomycorrhizal fungus.

Isolation from ectomycorrhizae.--Many ectomycorrhizal fungi can be isolated from their ectomycorrhizae and occasionally from large rhizomorphs, provided they can grow on agar media. Best results are obtained when fresh ectomycorrhizae, initially free of adhering debris, are treated immediately upon collection (3, 73, 76, 79, 80). Isolation success from individually treated ectomycorrhizae varies considerable between different fungi but is usually less than 20% and frequently less than 5%. Thus, isolation should be attempted from a large number of mycorrhizae.

After collection, ectomycorrhizae are sorted and the cleanest ones selected. The following basic treatments are then recommended by Zak (73, 76, 79, 80): 1) rinse the roots under tap water to remove loosely adhering debris; 2) place rootlets into a perforated plastic vial and shake vigorously for 3 min in a mild detergent solution (sonication in a water bath during or prior to step 2 further aids in loosening surface debris); 3) rinse off detergent with tap water; 4) soak for 4 min in 100 ppm mercuric chloride or 5-20 sec in 30% hydrogen peroxide; 5) rinse immediately with 2 liters of sterile distilled water and 6) aseptically transfer individual ectomycorrhizae (or rhizomorph pieces) onto nutrient agar in test tubes or plates. If results are poor, try varying the length of soaking in the surface sterilant. A variation on the soaking time is to put large numbers of root tips into the surface sterilant in a sterilized Petri plate and transfer some to sterile distilled water in another plate at intervals, starting at 2 and ending at 6 min, with the expectation that some point in that time span will prove to be just right. See Zak (73, 76, 79, 80) for greater detail on experimental procedures for this subject. Other methods of surface sterilization are described by Chu-Chou (3) and Slankis (63).

Surface sterilized ectomycorrhizae are incubated and growth is checked as described for isolation from sporocarps. Emergent growth of the mycorrhizal fungus usually occurs 2 to 4 weeks after treatment. If the identity of the fungus symbiont is known, e.g., reisolation of a fungus used in experimental inoculations, then it is relatively easy to compare culture characteristics to stock cultures and confirm reisolation. If the fungus symbiont is not known, comparisons with cultures taken from sporocarps found in the same vicinity, examination for clamps connections, and mycorrhizal syntheses must be done to confirm its ectomycorrhiza forming ability and indicate possible identity.

Isolation from Cenococcum sclerotia.--C. geophilum sclerotia are easily extracted from soil samples as described by Trappe (69). Small soil samples

(25-50 ml) are placed in an evaporating dish and wetted to make a slurry with free water on its surface. The dish is gently hand swirled so the low-density soil fraction containing sclerotia floats to the surface and can be poured off into another dish. Large volumes of soil can also be wet sieved through a 0.344 mm screen. Final fractions are placed in water under a stereomicroscope and the black sclerotia removed with forceps.

Sclerotia are treated similarly to ectomycorrhizae for isolation. Select large, clean sclerotia, free of obvious cracks or breakage. Remove small adhering debris by sonication in a water bath for 10 min. To surface sterilize, Trappe (69) recommends soaking for 10 to 20 min in 30% hydrogen peroxide. However, other surface sterilants work well and one may want to vary soaking time and sterilants to achieve maximum success. Rinse sclerotia in sterile water and aseptically transfer onto nutrient agar (MMN works well). As with ectomycorrhizae, attempt isolation from 30 or more sclerotia per soil sample. Incubate as previously described. Near-white to gray hyphal tips usually appear after 2 weeks. Established colonies are easily recognized as coarse hyphae which become jet black behind the tips (16, 69, 70).

Isolation from sexual spores.--Regardless of technique, ectomycorrhizal fungi are rarely isolated from basidio- or ascospores and so will not be covered in detail. Fries (8) has recently developed a promising technique involving germination of spores in the presence of activated charcoal to remove inhibitors plus the yeast *Rodotorula glutinus* (Fries) Harrison, which acts as a stimulant. Other Fries studies (9, 10) have also shown that mycelia of the same fungus and living roots of axenic seedlings can further stimulate spore germination. Readers are referred to Fries (8, 9, 10) for more details on experimental procedures and to Palmer (57) for information on spore drop techniques.

MAINTENANCE IN CULTURE

When ectomycorrhizal fungus cultures are needed over several years, stock cultures should be maintained separately from working cultures. Stock cultures are most commonly stored on nutrient agar slants in test tubes under refrigeration (3-5°C). At the Corvallis Forestry Sciences Laboratory we store four slants (13 X 100 mm screw-capped glass tubes with 3 ml agar) for each fungus isolate in a refrigerator at 3°C, most commonly on MMN nutrient agar but sometimes also on potato dextrose agar or both. Each isolate is transferred every 3 to 4 months. Transferred cultures are incubated at 20°C (usually for about 1 to 2 weeks or until the mycelium is firmly established on the agar) and then immediately placed in cold storage until the next transfer period. With this regime, the cultures grow slowly during most of their storage life and thus have less chance to modify their growth physiology as they sometimes do when cultures are kept in a constantly active growth state. Not all ectomycorrhizal fungi store well under such conditions; some prefer warmer incubation temperatures or more frequent transfers. Experiment with variations of the procedure to find the best storage conditions for troublesome species.

Marx and Daniel (31) emphasize that many ectomycorrhizal fungi change growth habits and may lose their ability to form ectomycorrhizae after long storage periods in a growing state. They found, however, that several ectomycorrhizal fungi can be stored for up to 3 years and probably longer in sterile cold water. Fungi are first grown on nutrient agar in plates from which 8 mm diam agar plugs are cut from the colony edge; 10 to 15 plugs are then aseptically transfered into 2.5 X 15 cm glass screw-capped test tubes containing 25 ml sterile distilled water and stored at 5°C in darkness. When needed, the fungus is retrieved by aseptically removing and transferring a mycelial disk onto fresh nutrient agar. This technique will not work with all fungi, so one must test each isolate for retrievability from this system before obligating stock cultures to this storage method.

Thorough records must be kept for each isolate maintained in stock cultures. Important data include identification number, isolation source (sporocarp, ectomycorrhiza, rhizomorph, etc.), date of isolation, species identifier and voucher (herbarium) number, associated hosts in the immediate vicinity of collection, location of collection and habitat type, experimentally determined hosts, and best storage medium and conditions. If isolates are used in experimental inoculations and then reisolated, the new isolate should be given a different identification number and the experimental conditions noted.

Mycorrhiza researchers regularly share fungus cultures for various research purposes. Firmly sealed and carefully packaged cultures can easily be sent long distances by air mail. Be sure to include complete background data for each isolate so that the receiver can cite this information in subsequent publications. Check for possible quarantine restrictions before mailing cultures abroad.

GROWTH IN PURE CULTURE

Nutrients.--The hexose, d-glucose, is used by almost all ectomycorrhizal fungi. Therfore this sugar is included as the sole or principal carbon source in synthetic media and as the principal or supplementary source in semisynthetic substrates. The two most commonly used concentrations are 5 g/l and 20 g/l. Other 6-carbon compounds are not utilized by as many fungi. Sorbose is mostly nonutilizable. Sugars with 3-, 4-, and 5-carbons are not used, or growth is poor. Among disaccharides, cellobiose, maltose, trehalose, and sucrose are satisfactory for many isolates. However, sucrose may dissociate to glucose and fructose at high temperatures, low acidity, or both. A few isolates grow well on trisaccharides or 3- to 6-carbon alcohols. However, phosphate ions at concentrations above those commonly used may reduce growth on some groups of carbohydrates. In general, soluble polysaccharides are more utilizable than the insolubles and polysaccharide derivatives. A few organic acids or their salts, e.g. citric, fumaric, malic, propionic, and succinic, support growth for a few isolates. Acetates are usually inhibitory. (See references 6, 11, 20, 21, 55, and 58).

Ammonium chloride (0.5 g/l) and ammonium nitrate (1.0 g/l) are most commonly used as supplemental nitrogen sources in semisynthetic nutrient solutions. In synthetic formulae, ammonium tartrate (0.5, 1, and 5 g/l) is most used. Other ammonium salts from 0.25 to 1.0 g/l and asparagine at 1.0 g/l are satisfactory. Nitrates are mostly nonutilizable. Consequently, ammonium nitrate is probably a good choice for initial attempts to isolate or culture a new species. In any case, a rapid rise or fall in pH can be expected with use of inorganics if the solution is unbuffered. That ammonium tartrate is a buffer, as is potassium monobasic or dibasic phosphate, probably explains its perferential use as the nitrogen source. However, experimental procedures that necessitate exclusion of any additional and possibly utilizable carbon usually precludes use of ammonium tartrate, ammonium salts or other organic acids, asparagine and other proteinaceous compounds. In addition, many of these substances are growth regulators, which make them undesirable for use in growth regulatory studies. (See references 20, 24, 37, 39, 40, 48, and 55).

Macronutrients other than nitrogen and carbon have not been investigated with specific intent to determine exact quantities or essentialities. The same applies to micronutrients. No doubt this reflects the complex and time-consuming procedures required. Experience indicates that phosphorus, potassium, sulfur, and magnesium must be present in relatively large amounts and copper, iron, manganese, molybdenum, and zinc in small amounts for growth of nearly all fungi. Each of the macronutrients is commonly added to both semisynthetic and synthetic formulae. Potassium and phosphorus are added as potassium di-hydrogen phosphate in either the 0.5 g/l or 1.0 g/l quantity. Magnesium and sulfur are added as magnesium sulfate ($MgSO_4 \cdot 7H_2O$) at 0.5 g/l in nearly every case.

At least one micronutrient is added to most semisynthetic solutions and several to most synthetic ones, more commonly as single chemicals rather than composite solutions. Iron in the form of ferric chloride near the level of 0.005 g/l is common although 1.0 - 0.3 ppm of iron is considered to meet the essential amount. Unfortunately, ferric chloride will diffuse into most types of glass, which makes storage in stock solution relatively short-term. Consequently ferric nitrate, ferric sulfate, or ferric citrate is often substituted, but each adds a small quantity of a macronutrient. Inorganic iron salts precipitate from solutions in increasing amounts as the pH rises above 4.0. Organic ferric salts or inorganics accompanied by an organic acid will stay in solution. At the Corvallis Forestry Sciences Laboratory we routinely incorporate the chelated iron product Geigy's Sequestrene (sodium ferric diethylenetriamine pentaacetate) into synthetic media at 0.02 g/l with good results. Zinc is added to nearly all semisynthetic and synthetic formulations as zinc sulfate commonly at levels between 1 and 5 mg/l. Conversely, inclusions of copper and manganese, which are probably essential for at least some isolates, are infrequent. Calcium as calcium chloride is more often added than is manganese. Quantities vary between 50 and 200 mg/l with 55.5 mg/l (5 ml of a 0.1 molar solution) most frequently used. Such anions as calcium, iron, and zinc become unavailable at a pH near 7.4, but an elevated pH is infrequent with these acidophiles. (See references 21, 27, 55, and 57).

In regard to vitamins and growth regulators, many ectomycorrhizal fungi are heterotrophic for thiamine; a few are heterotrophic for biotin. Neither vitamin is regularly added to semisynthetic media. Thiamine is added to most synthetic solutions usually at 50 μg/l but varies between 1 and 100 μg/l. Biotin is also added to some solutions before sterilization. The usual rate is 5 μg/l or 10 μg/l. Both should be added if the fungus has not previously been cultured or isolated. Stimulation of growth by other growth regulators is frequent; growth suppression is unlikely at the concentrations used, i.e. less than 1.0 g/l. Inoculation of single compounds and natural mixtures, e.g. casein hydrolysate, frequently increase growth but rarely to a significant level. Duplicable formulae of amino acids reported to stimulate casein hydrolysate (46) and vitamin complexes (18, 46) or both at 1.0 mg/l may initiate culture or increase growth. (See references 20, 39, 41, 45, 55, and 62).

Environmental Conditions.--In axenic culture, responses are isolate specific, but most fungi are acidophilic (pH optima 4.5-5.0, ranges 3.5-6.0), aerobic (require gaseous oxygen), mesothermal (optima between 18°C and 25°C), and photoinactive (no induction of zonation or pigmentation). In the usual closed-container culture air movement and relative humidity are uncontrollable. Osmotic pressure is initally a function of ionized and soluble compounds per unit water but subsequently varies with absorption, secretion, and lysis of mycelium. Outside of pH, our standard conditions (temperature 22°C, relative humidity 50%, constant light provided by equal mixtures of cool white and daylight fluorescent bulbs) indirectly affect culture chamber conditions. Stabilization of pH with buffers is possible but will complex some experimental procedures, especially if phosphates or organic acids or salts are involved. (See references 11, 14, 22, 27, 33, 37, 47, 49, and 55).

Media.--Semisynthetic formulations are most commonly used for propagation, inoculum preparation, storage on agar, fruiting, and other experimental procedures. The most common has been the Modess (49) modification of the original Hagem (15) or a variant of this formula (see Table 1). The MMN solution (Table 1) represents a Marx (27) enrichment of a Melin-Norkrans formula. The original solution (54) was a synthetic solution (no malt extract) with only 25 μg of thiamine and 2.5 g of glucose. Formulae used in eastern Europe often replace or augment malt extract with casein or yeast hydrolysate or both. (See references 4, 18, 23, and 47).

Synthetic nutrient media are used primarily to examine carbon and nitrogen nutrition and mycelial growth regulation by various compounds. Formulations are numerous; two are tabularized in Table 1. The Melin and Rama Das (46) formula included neither the synthetic casein hydrolysate nor the vitamin mixtures, and most users do not indicate inclusion or exclusion in experiments. With reference to "see Text" listed in Table 1, a micronutrient solution at 2 ml/l is always included in our solution: dissolve $Fe(NO_3)_3 \cdot 9H_2O$ at 1.45 mg, $ZnSO_4 \cdot 7H_2O$ at 0.88 mg, and $MnSO_4 \cdot H_2O$ at 0.41 mg in 700 ml of distilled water, acidify with H_2SO_4 to give a clear solution, and dilute to 1,000 ml with distilled water (57). (See references 21, 24, and 33).

TABLE 1. Semisynthetic and synthetic nutrient formulae commonly used for propagation, inoculum preparation, experimentation, and storage of ectomycorrhizal fungi in axenic culture

	SEMISYNTHETIC		SYNTHETIC	
NUTRIENTS	"Hagem" (Modess 1941)	MMN (Marx 1969)	Melin & Rama Das (1954)	Palmer & Hacskaylo (1970)
Malt Extract	5.0 g	3.0 g		
d-Glucose	5.0 g	10.0 g	20.0 g	5.0 g
NH_4Cl	0.5 g			0.5 g
$(NH_4)_2HPO_4$		0.25 g		
$(NH_4)_2C_4H_4O_6$			0.5 g	
KH_2PO_4	0.5 g	0.5 g	1.0 g	0.5 g
$MgSO_4 \cdot 7H_2O$	0.5 g	0.15 g	0.5 g	0.5 g
$CaCl_2$		0.05 g		
$FeC_6H_5O_6 \cdot 3H_2O$			0.5 ml (1% solution)	
$FeCl_3$	10 drops (1% solution)	1.2 ml[b] (1% solution)		
NaCl		0.025 g		
$ZnSO_4$			0.5 ml[a]	
Biotin				5.0 μg
Thiamine			1.0 μg	1.0 mg
Thiamine HCl		100 μg		
Micronutrients				(see Text)
Distilled H_2O	1,000 ml	to 1,000 ml	to 1,000 ml	to 1,000 ml
pH after autoclaving	4.6 - 4.8	5.5 - 5.7	Unlisted	4.5 - 4.7

[a] 1:500 Zinc in aqueous solution.
[b] Zak (unpublished) substitutes 0.02 g Sequestrene (Geigy).

Use of natural formulations is infrequent. However, Zak (74) regularly used the Lacy and Bridgmon (19) potato dextrose (PD) and Russian workers, the Voznyakovskaya and Ryzhkova cabbage dextrose (CD) as reported by Lobanow (23). The formula for the PD is 1) rehydrate 22 g of dehydrated potatoes in 178 g distilled water, 2) add 20 g d-glucose followed by agar if desired, and 3) stir well, 4) utilize all but potato sediment, and 5) autoclave for 20 min. Formula for the CD is: 1) weigh 100 g cabbage 2) cook for 10 min in 1 liter of tap water (we use distilled or deionized water), 3) filter, 4) adjust pH to 5.0-5.6, 5) add 3% glucose, 6) add a 0.5% solution of thiamine at the rate of 1.0 ml/10 liters of nutrient solution, 7) add agar if desired, and 8) autoclave the solution.

Techniques.--The common substrate is water that has been singly or doubly distilled or deionized or purified by some combination. Gelling with 2 ± 0.5% agar of various purities has advantages for routine propagation. Growth of some fungi can be initiated or accelerated in aerated particulate substrates such as mixtures of vermiculite ± peat. The pH of vermiculite tends to rise with time but can be stabilized at least in uninoculated mixtures (32, 61).

Containers made of the same material, even from the same lot of plastic, provide relatively homogenious conditions during prolonged experiments. At the Madison Forest Products Laboratory we prefer borosilicate glass test tubes (16 x 150 mm), Petri dishes (100 x 20 mm), and narrow mouth Erlenmeyer flasks (300 ml). Large Petri dishes enable inclusion of additional agar and delay drying. The flasks capped with 50 ml Griffin low-form beakers stablilize gas exchanges, do not rattle much on shakers, and provide good surface aeration. Unscratched borosilicate glass also provides uniform tranmission of quantity as well as wave-lengths of light irradiated from a constant source.

Inoculation of agar plugs or chunks to fresh nutrient agar usually gives commensurate growth if there is no film of liquid water at the implantation point. Any method used to inoculate a liquid substrate has limitations. Specific techniques often work well only with certain isolates. In nearly every case, the initial growth lag phase is shorter for floating than for submerged inoculum, even in those fungi that produce asexual reproduction structures such as conidia or oidia. We prefer the "fuzzing" technique if mycelial dry weight is to be measured: 1) cut plugs from agar at the edge of a colony (a 5 mm diameter cork borer is excellent), 2) transfer to fresh agar with the mycelium on the upper side of the plug, 3) keep plates at 21 ± 2°C, 4) observe at least once each day for horizontal extension of hyphae all around the elevated periphery, and 5) remove and float plug on the liquid substrate. Readers should be aware that at some time after "fuzzing," the hyphae will reflex. Thus, for isolates that project hyphae horizontally, a number of plugs much in excess of the number needed should be prepared. For nutrient studies in which the absence of key nutrients is essential, cover slips covered with a complete nutrient agar can be placed upon agar lacking the experimental nutrient, and the fungus is inoculated onto the cover slip's agar surface. Some hyphae will then grow out over the deficient agar from which new plugs can be removed and subsequently placed onto fresh deficient agar for growth response measurements. Inoculum of many fungi can be prepared by using a shake-culture in which mycelium will pelletize. (See reference 75).

Dry weights can be determined by several methods. If no reproductive forms develop and little fragmentation occurs in any experimental treatment, J. G. Palmer and E. Hacskaylo use the following method: 1) collect mycelium from a single replicate on a fine mesh screen placed on a funnel, 2) wash with distilled water, 3) remove mycelium when liquid no longer drips through, 4) mold into wad with the top of the mycelial mat to the outside, 5) place on a blotter with or without an underlying screen, 6) move frequently (every 30 seconds initially) as moisture diffuses through the blotter until uniform solidification occurs, 7) place in a beaker, 8) dry in a forced-air oven at 90°C for 34 hours, 9) cool to room temperature in a dessicator over dry dessicant, and 10) weigh. If there are many colonies or mycelial fragments, mycelium must be collected on pre-washed, pre-numbered, pre-dried, and pre-weighed spun-glass filter papers. We mount them in a Buchner funnel and apply a vacuum during washing and drying before placing in the oven.

PURE CULTURE ECTOMYCORRHIZA SYNTHESIS

Melin (34, 35, 36, 38) developed the pure culture synthesis technique to experimentally demonstrate the ability of known fungus isolates to form ectomycorrhizae with specific hosts under pure culture conditions. His pioneering studies were instrumental in establishing which major taxa of higher fungi were involved in these associations. This information provided a framework for predicting ectomycorrhizal host-fungus associations based on field observations of sporocarp-host associations (see Trappe, 67). Use of the pure culture synthesis technique has also led to discovery of important physiological aspects of the symbiosis, including uptake of nutrients and water by the fungus and translocation to the host (5, 42, 44), movement of photosynthate from host to fungus (43), interactions of growth regulating substances (12, 64), host to host transfer of carbohydrates via a shared fungal symbiont (60), protection against root pathogens (29), effects of temperature on mycorrhiza development (30), specificity and compatibilty between fungus and host (50, 51, 52), and several other processes. This basic technique will continue to provide new information on the development and function of ectomycorrhizal symbioses.

One must recognize the artificiality of the pure culture synthesis and limit extrapolation of results to natural situations. Positive synthesis results are conclusive and confirm the ability of that particular host-fungus combination to form ectomycorrhizae. Negative results, however, are not conclusive in themselves, but do suggest that the union of organisms in question seems unlikely.

Synthesis apparatus.--Size, arrangement, and complexity of the synthesis apparatus will depend upon the purpose of investigation. Melin (34, 35, 36, 38) primarily used flasks containing sterile sand moistened with a nutrient solution into which an aseptically germinated seedling and single fungus culture were introduced. Several investigators have modified Melin's techniques, often trying more complex arrangments to reduce the artificial nature of the enclosed system (17). With increased complexity also comes increased risk of contamination. Hacskaylo (13) greatly improved the system by using vermiculite instead of sand as the substrate; vermiculite provides better aeration and moisture holding capacity than sand. Marx and Zak (32) further improved the substrate by stabilizing the acidity with an addition of finely ground sphagnum peat moss. For a 2-liter flask culture, Zak (78) mixed 840 ml of vermiculite with 60 ml of finely ground peat moss in the flask, moistened this substrate with 550 ml of MMN nutrient solution, capped the flask with an inverted glass having a cheesecloth-wrapped cotton plug cemented to the bottom (Fig. 2), and autoclaved the assembly for 45-60 min at 120°C. Final pH is around 5.0. Molina (50, 51, 52) reports excellent seedling growth and ectomycorrhiza development in the following modified system: a 300 X 38 mm glass test tube is filled with 110 ml of vermiculite and 10 ml of peat moss, moistened with 70 ml of MMN nutrient solution, capped with an inverted 50 ml glass beaker, and autoclaved for 15 min at 120°C (Fig. 3). This arrangement allows for numerous syntheses to be run simutaneously in a relatively small area (Fig. 3). Pachlewski and Pachlewska (56) also reported good mycorrhiza syntheses in large test tubes but used a solid agar substrate rather than peat moss and vermiculite. Several alternative synthesis apparati and arrangements have been developed for specific experimentation but will not be described here (See references 2, 7, 17, 57, and 68).

Preparation of fungi.--Fungi are introduced into the synthesis vessel either as agar culture transfers or from liquid cultures. Agar cultures are first grown on nutrient agar in Petri plates for 3 to 4 weeks. Two to 4 agar plugs, 5-8 mm diam, are then aseptically removed and placed 1 to 2 cm deep around the germinant in the synthesis vessel. Liquid cultures are prepared by growing agar culture transfers in screw-capped glass prescription bottles containing 150 ml of sterile nutrient solution (MMN) with small bits of broken glass (about 1 cm deep) in the bottom. Liquid cultures are grown for 3 to 6 weeks at 20°C and hand shaken weekly to fragment the actively growing mycelium. Five to 10 ml of the liquid culture are then aseptically transferred into the synthesis vessel with a sterile, wide-bore pipette. Alternatively, the fungi can be grown in 20 X 150 mm screw capped glass test tubes filled with 10-15 ml of MMN solution and small bits of broken glass. After incubation and weekly shaking as described above, the glass lip of the tube is flamed and the entire liquid culture is aseptically poured into a single synthesis vessel. Because the liquid cultures are shaken just prior to inoculation, the numerous mycelial fragments will yield rapid and uniform colonization of the synthesis substrate.

Aseptic seed germination.--With minor modifications, the following procedure developed by Bratislav Zak (unpublished) for aseptic germination of Douglas-fir (*Pseudotsuga menziesii* (Mirb.) Franco) seed can be used for many ectomycorrhizal hosts. Select healthy seed, free of obvious defect, place them into a perforated plastic vial and treat as follows: 1) rinse in running tap water for 60 min; 2) wash in mild detergent solution (e.g., 2 to 3 drops of Tween 20 in 500 ml water) on a shaker for 90 min; 3) rinse in running tap water for 90 min; 4) soak in 30% hydrogen peroxide on a shaker for 60 min; 5) rinse aseptically with 2 liters of sterile distilled water; 6) shake excess water from vial and empty seed into a sterile Petri dish for planting.

Seeds are aseptically planted into nutrient agar in vials or plates. Planting should be done in a sterile transfer room or laminar flow hood when available. Incubate planted seed for 7-10 days at room temperature and check for contamination. Contaminated vials are discarded, or, if seeds are planted in plates, individually contaminated seed and the agar on which they rest can be cut out and removed, provided this is done before contaminants sporulate.

Figs. 2,3. Pure culture synthesis apparati. 2, Standard 2-liter flask culture with 4 mo. inoculated ponderosa pine seedling. 3, *In situ* arrangement of 38 x 300 mm glass synthesis tubes with bases submersed in cool tap water.

Noncontaminated seed are refrigerated (3-5°C) for 30 days (if stratification is needed) and then placed under artificial light at room temperature to germinate. The following soak periods in 30% hydrogen peroxide work well for other host species: *Pinus* spp., 30-60 min (66); *Tsuga heterophylla* (Raf.) Sarg. and *Picea sitchensis* (Bong.) Carr., 30 min (66); *Larix occidentalis* Nutt., 45 min; *Eucalyptus* spp., 10-30 min (25); *Arbutus menziesii* Pursh., 20 min (78); *Arctostaphylos* spp., 3 hr soak in concentrated sulfuric acid followed by 30 min in 30% hydrogen peroxide (77); *Alnus* species, 15 min (50, 51). Many host species will also germinate well without cold treatment. Surface sterilization with 30% hydrogen peroxide does not work well with hosts having resinous or pitchy seedcoats such as *Abies* spp.

Germinants are ready when radicles are approximately 1 to 2 cm long and cotyledons are still within the seed coat; germinants with longer radicles are difficult to plant. Use a hooked transfer needle with a long sterilizable handle for planting. Prepare a small planting hole by pushing aside some of the surface substrate. Aseptically transfer the germinant into the synthesis vessel so that the entire radicle is inserted in the substrate. Cover the seedcoat with a few mm of substrate. The fungus can be inoculated at this time for hosts that grow vigorously, such as *Pinus* or *Pseudotsuga* spp. Host species that grow slowly during the first month, e.g. *Alnus*, *Tsuga*, and *Picea* spp., are best inoculated 1 month after planting so that seedlings can develop adequately to cope with fungus competition. Because planting and inoculation necessitate exposing sterile synthesis vessels to the air for considerable periods, both procedures should be carried out in an air filtered transfer room or flow hood when available to reduce contamination risk.

Environmental set-up.--Several environmental set-ups can be used for pure culture syntheses, but two conditions are of prime concern. First, the area should be very clean and have a minimum of air turbulence to lessen the chance of contamination. Second, a build-up of heat within the synthesis vessel (greenhouse effect), particularly in the rooting substrate, should be avoided. At the Forestry Sciences Laboratory in Corvallis, Bratislav Zak designed a simple laboratory set-up for pure culture syntheses, consisting of a tank 15 cm deep filled 12 cm deep with constantly circulating unheated tap water and illuminated by an overhead combination of 35 watt fluorescent tubes and 25 watt incandescent bulbs (ca. 10500 lx) set for a 15-hr day (Fig. 3). The water bath cools the substrate. Standard growth chambers work well for mycorrhiza syntheses but special attention must be paid to contamination by rapidly circulating air and temperature control. Syntheses are also frequently carried out in greehouses, but the large fans commonly used for cooling present high contamination risk; added protection of the seal is needed under these conditions.

Synthesis evaluation.--Typical ectomycorrhiza syntheses are completed after 4 to 6 months, depending on the growth rates of fungus and host. If glass test tubes or agar substrates are used, ectomycorrhiza formation can often be seen directly through the glass walls. After a synthesis attempt is completed, a small bit of substrate should be aseptically removed from the vessel, transfered onto nutrient agar, incubated, and checked for contamination and reisolation of the original fungus. The seedling is removed intact from the synthesis vessel and its roots gently washed free of substrate with tap water. The entire root system is then placed under water in a Petri dish and observed with a stereomicroscope for ectomycorrhiza formation and characters. Specific observations should be recorded on the relative abundance of ectomycorrhizae present, mantle color and texture, presence of rhizomorphs and hyphal strands, length and branching pattern of ectomycorrhizae, and the overall size and health of the seedling. Suspected ectomycorrhizae are then sectioned by hand or microtome to characterize the mantle tissue and, most importantly, to confirm the presence and extent of Hartig net development. Any unusual or unique characters should also be noted. Root systems are easily stored in FAA for later sectioning and observations. Recording of other data may be important, depending on the investigator's purpose. Ectomycorrhizae are best photographed fresh.

LITERATURE CITED

1. Ammirati, J. 1979. Chemical studies of mushrooms: the need for voucher collections. Mycologia 71:437-441.
2. Bigg, W. L. and Alexander, I. J. 1981. A culture unit for the study of nutrient uptake by intact mycorrhizal plants under aseptic conditions. Soil Biol. Biochem 13:77-78.
3. Chu-Chou, M. 1979. Mycorrhizal fungi of Pinus radiata in New Zealand. Soil Biol. Biochem. 11:557-562.
4. Crafts, C. B. and Miller, C. O. 1974. Dectection and identification of cytokinins produced by mycorrhizal fungi. Plant Physiol. 54:586-588.
5. Duddridge, J. A., Malibari, A. and Read, D. J. 1980. Structure and function of mycorrhizal rhizomorphs with special reference to their role in water transport. Nature 287:834-836.
6. Ferry, B. W. and Das, N. 1968. Carbon nutrition of some mycorrhizal Boletus species. Trans. Br. Mycol. Soc. 51:795-798.
7. Fortin, J. A., Piche, Y. and Lalonde, M. 1980. Technique for the observation of early morphological changes during ectomycorrhiza formation. Can. J. Bot. 58:361-365.
8. Fries, N. 1978. Basidiospore germination in some mycorrhiza-forming hymenomycetes. Trans. Br. Mycol. Soc. 70:319-324.
9. Fries, N. 1981. Recognition reactions between basidiospores and hyphae in Leccinum. Trans. Br. Mycol. Soc. 77:9-14.
10. Fries, N. and Birraux, D. 1980. Spore germination in Hebeloma stimulated by living plant roots. Experientia 36:1056-1057.
11. Giltrap, N. J. and Lewis, D. H. 1981. Inhibition of growth of ectomycorrhizal fungi in culture by phosphate. New Phytol. 87:669-675.
12. Graham, J. H., and Linderman, R. G. 1980. Ethylene production by ectomycorrhizal fungi, Fusarium oxysporum f. sp. pini, and by aseptically synthesized ectomycorrhizae and Fusarium-infected Douglas-fir roots. Can. J. Microbiol. 26:1340-1347.
13. Hacskaylo, E. 1953. Pure culture synthesis of pine mycorrhizae in terra-lite. Mycologia 45:971-975.

14. Hackaylo, E., Palmer, J. G. and Vozzo, J. A. 1965. Effect of temperature on growth and respiration of ectotrophic mycorrhizal fungi. Mycologia 57:748-756.

15. Hagem, O. 1910. Untersuchungen über norwegische Mucorineen II. Skrift. Videnskabs-Selskabet I Christiana 1910. I. Math. Naturvid. Kl. No. 4:1-152.

16. Hatch, A. B. 1934. A jet-black mycelium forming ectotrophic mycorrhizae. Svensk Bot. Tidskr. 28:369-383.

17. Hatch, A. B. 1937. The physical basis of mycotrophy in Pinus. Black Rock. For. Bull No. 6, 168 pp.

18. Khudyakov, Y. P. and Voznyakovskava, Y. N. 1951. Chistye kul'tury mikoriznkh gribov. Mikrobiologiya 20:13-19.

19. Lacy, M. L. and Bridgmon, G. H. 1962. Potato-dextrose agar prepared from dehydrated mashed potatoes. Phytopathology 52:173.

20. Laiho, O. 1970. Paxillus involutus as a mycorrhizal symbiont of forest trees. Acta For. Fenn. 106:1-72.

21. Lamb, R. J. 1974. Effects of D-glucose on utilization of single carbon sources by ectomycorrhizal fungi. Tran. Br. Mycol. Soc. 63:295-306.

22. Lindeberg, G. 1944. Über die Physiologie ligninabbauender Bodenhymenomyzeten. Studien an Schwedischen Marasmius-Arten. Symb. Bot. Upsal. 8:1-183.

23. Lobanow, N. W. 1960. Mykotrophie der Holzpflanzen. VEB Deutscher Verlag der Wissenschaften. Berlin (DDR). 352 p.

24. Lundeberg, G. 1970. Utilization of various nitrogen sources, in particular bound soil nitrogen, by mycorrhizal fungi. Stud. For. Suec. 79:1-95.

25. Malajczuk, N., Molina, R. and Trappe, J. M. 1982. Ectomycorrhiza formation in Eucalyptus. I. Pure culture syntheses, host specificity and mycorrhizal compatibility with Pinus radiata. New Phytol. 91:(in press).

26. Malloch, D. 1981. Moulds: Their isolation, cultivation, and identification. Univ. of Toronto Press, Canada, 99 pp.

27. Marx, D. H. 1969. The influence of ectotrophic mycorrhizal fungi on the resistance of pine roots to pathogenic infections. I. Antagonism of mycorrhizal fungi to root pathogenic fungi and soil bacteria. Phytopathology 59:153-163.

28. Marx, D. H. 1980. Ectomycorrhiza fungus inoculations: a tool for improving forestation practices. Pages 13-71. in: P. Mikola, ed. Tropical Mycorrhiza Research, Oxford University Press, Oxford. 270 pp.

29. Marx, D. H. and Davey, C. B. 1969. The influence of ectotrophic mycorrhizal fungi on the resistence of pine roots to pathogenic infections. III. Resistence of aseptically formed mycorrhizae to infection by Phytophthora cinnamomi. Phytopathology 59:549-558.

30. Marx, D. H., Bryan, W. C., and Davey, C. B. 1970. Influence of temperature on aseptic synthesis of ectomycorrhizae by Thelephora terrestis and Pisolithus tinctorius on loblolly pine. For. Sci. 16:424-431.

31. Marx, D. H. and Daniel, W. J. 1976. Maintaining cultures of ectomycorrhizal and plant pathogenic fungi in sterile water cold storage. Can. J. Microbiol. 22:338-341.

32. Marx, D. H., and Zak, B. 1965. Effect of pH on mycorrhizal formation of slash pine in aseptic culture. For. Sci. 11:66-75.

33. McLaughlin, D. J. 1970. Environmental control of fruitbody development in Boletus rubinellus in axenic culture. Mycologia 62:307-331.

34. Melin, E. 1921. Über die Mykorrhizenpilze von Pinus silvestris L. und Picea abies (L.) Karst. Svensk Bot. Tidskr. 15:192-203.

35. Melin, E. 1922. Untersuchungen über die Larix Mycorrhiza. I. Synthese der Mykorrhiza in Reinkulture. Svensk Bot. Tidskr. 16:161-196.

36. Melin, E. 1923. Experimentelle Untersuchungen über die Birken und Espenmykorrhizen und ihre Pilzsymbionten. Svensk Bot. Tidskr. 17:479-520.

37. Melin, E. 1925. Untersuchungen über die Bedeutung der Baummykorrhiza. Eine ökologisch-physiologische Studie. Gustav Fischer. Jena. 152pp.

38. Melin, E. 1936. Methoden der experimentellen Untersung Mykotropher Pflanzen. Handb. Biol. Arbeitsmeth., Abt. 11:1015-1108.

39. Melin, E. 1954. Growth factor requirements of mycorrhizal fungi of forest trees. Svensk Bot. Tidskr. 48:86-94.
40. Melin, E. 1959. Mycorrhiza. Hand. Pflanzenphysiol. 11:605-638.
41. Melin, E. and Lindeberg, G. 1939. Über den Einfluss von Aneurin und Biotin auf das Wachstum einiger Mykorrhizenpilze. Vorläufige Mitteilung. Bot. Not. 1939:241-245.
42. Melin, E. and Nilsson, H. 1950. Transfer of radioactive phosphous to pine seedlings by means of mycorrhizal hyphae. Physiol. Plant. 3:88-92.
43. Melin, E. and Nilsson, H. 1957. Transport of C^{14} labelled photosynthate to the fungal associate of pine mycorrhiza. Svensk Bot. Tidskr. 51:166-186.
44. Melin, E. and Nilsson, H. 1958. Translocation of nutritive elements through mycorrhizal mycelia to pine seedlings. Bot. Notis. 111:251-256.
45. Melin, E. and Norkrans, B. 1942. Über den Einfluss der Pyrimidin- und der Thiazolkomponente des Aneurins auf das Wachstum von Wurzelpilzen. Svensk Bot. Tidskr. 36:271-286.
46. Melin, E. and Rama Das, V. S. 1954. Influence of root-metabolities on the growth of tree mycorrhizal fungi. Physiol. Plant. 7:851-858.
47. Mexal, J. and Reid, C. P. P. 1973. The growth of selected mycorrhizal fungi in response to induced water stress. Can. J. Bot. 51:1579-1588.
48. Mikola, P. 1948. On the physiology and ecology of Cenococcum graniforme especially as a mycorrhizal fungus of birch. Inst. Forest. Fenn. Commun. 36:1-104.
49. Modess, O. 1941. Zur Kenntnis der Mykorrhizabildner von Kiefer und Fichte. Symb. Bot. Upsal. 5:1-146.
50. Molina, R. 1979. Pure culture synthesis and host specificity of red alder mycorrhizae. Can. J. Bot. 57:1223-1228.
51. Molina, R. 1981. Ectomycorrhizal specificity in the genus Alnus. Can. J. Bot. 59:325-334.
52. Molina, R. and Trappe, J. M. 1982. Patterns of ectomycorrhzial specificity and potential among Pacific Northwest conifers and fungi. For. Sci. 28:(in press).
53. Moser, M. 1958. Die künstliche Mykorrhizaimpfung an Forstpflanzen. I. Erfahrungen bei der Reinkulture von Mykorrhizapilzen. Forstwiss. Centralbl. 77:32-40.
54. Norkrans, B. 1949. Some mycorrhiza-forming Tricholoma species. Svensk Bot. Tidskr. 43:485-490.
55. Norkrans, B. 1950. Studies in growth and cellulolytic enzymes of Tricholoma, with special reference to mycorrhiza formation. Symb. Bot. Upsal. 11:1-126.
56. Pachlewski, R. and Pachlewska, J. 1974. Studies on symbiotic properties of mycorrhizal fungi of pine (Pinus silvestris L.) with the aid of the method of mycorrhizal synthesis in pure cultures on agar. For. Res. Inst., Warsaw, Poland. 228 p.
57. Palmer, J. G. 1971. Techniques and procedures for culturing ectomycorrhizal fungi. Pages 132-144. in: E. Hacskaylo, ed. Mycorrhiza. USDA Misc. Publ. 1189. 255 pp.
58. Palmer, J. G. and Hacskaylo, E. 1970. Ectomycorrhizal fungi in pure culture I. Growth on single carbon sources. Physiol. Plant. 23:1187-1197.
59. Pantidou, M. E. and Groves, J. W. 1966. Culture studies of Boletaceae. Some species of Suillus and Fuscoboletinus. Can. J. Bot. 44:1371-1392.
60. Reid, C. P. P., and Woods, F. W. 1969. Translocation of C^{14}- labeled compounds in mycorrhizae and its implications in interplant nutrient cycling. Ecology 50:179-187.
61. Richard, C. and Fortin, J. A. 1974. Distribution géographique, écologie, physiologie, pathogénécite et sporulation du Mycelium radicis atrovirens. Phytoprotection 55:67-88.
62. Schisler, L. C. and Volkoff, O. 1977. The effect of safflower oil on mycelial growth of Boletaceae in submerged liquid cultures. Mycologia 69:118-125.

63. Slankis, V. 1958. An apparatus for surface sterilization of root tips. Can. J. Bot. 36:837-842.
64. Slankis, V. 1973. Hormonal relationships in mycorrhizal development. Pages 232-298. in: G. C. Marks and T. T. Kozlowski, eds. Ectomycorrhizae, Their Ecology and Physiology. Acad. Press, N.Y. 444 pp.
65. Thoen, D. 1977. Identification of ectomycorrhizal fungi by thin layer chromatography. Bull. Br. Mycol. Soc. 11:39-43.
66. Trappe, J. M. 1961. Strong hydrogen peroxide for sterilizing coats of tree seed and stimulating germination. J. Forestry 59:828-829.
67. Trappe, J. M. 1962. Fungus associates of ectotrophic mycorrhizae. Bot. Rev. 28:538-606.
68. Trappe, J. M. 1967. Pure culture synthesis of Douglas-fir mycorrhizae with species of Hebeloma, Suillus, Rhizopogon, and Astraeus. For. Sci. 13:121-130.
69. Trappe, J. M. 1969. Studies on Cenococcum graniforme. I. An efficient method for isolation from sclerotia. Can. J. Bot. 47:1389-1390.
70. Trappe, J. M. 1971. Mycorrhiza-forming Ascomycetes. Pages 19-37. in: E. Hacskaylo, ed. Mycorrhiza. USDA Misc. Publ. 1189. 255 pp.
71. Trappe, J. M. 1977. Selection of fungi for ectomycorrhizal inoculation in nurseries. Ann. Rev. Phytopathol. 15:203-222.
72. Watling, R. 1981. How to identify mushrooms to genus V. Cultural and developmental features. Mad River Press, Inc., Eureka, California. 169 pp.
73. Zak, B. 1971. Characterization and classfication of Douglas-fir mycorrhizae. Pages 38-53. in: E. Hacskaylo, ed. Mycorrhiza. USDA Misc. Publ. 1189. 255 pp.
74. Zak, B. 1971. Characterization and classification of mycorrhizae of Douglas-fir. II. Pseudotsuga menziesii + Rhizopogon vinicolor. Can. J. Bot. 49:1079-1084.
75. Zak, B. 1972. Flotation of excised root and fungal cultures on liquid media. Can. J. Microbiol. 18:536-538.
76. Zak, B. 1973. Classification of ectomycorrhizae. Pages 43-78. in: G. C. Marks and T. T. Kozlowski, eds. Ectomycorrhizae, Their Ecology and Physiology. Acad. Press, N.Y. 444 pp.
77. Zak, B. 1976. Pure culture synthesis of bearberry mycorrhizae. Can J. Bot. 54:1297-1305.
78. Zak, B. 1976. Pure culture synthesis of pacific madrone ectendomycorrhizae. Mycologia 68:362-369.
79. Zak, B. and W. C. Bryan. 1963. Isolation of fungal symbionts from pine mycorrhizae. For. Sci. 9:270-278.
80. Zak, B. and D. H. Marx. 1964. Isolation of mycorrhizal fungi from roots of individual slash pines. For. Sci. 10:214-222.

PRODUCTION OF ECTOMYCORRHIZAL FUNGUS INOCULUM

D. H. Marx and D. S. Kenney

INTRODUCTION

The need of many species of forest trees for ectomycorrhizae was initially observed when attempts to establish plantations of exotic pines in parts of the world deficient in the fungal partners routinely failed until the essential fungi were introduced. The need of pine and oak seedlings for ectomycorrhizae has also been demonstrated in the afforestation of former treeless areas, such as the grasslands of Russia and the Great Plains of the United States (51). Even on clearcut pine lands (64) or on amended adverse sites (62) nonmycorrhizal pine seedlings do not survive or grow well until indigenous fungi form ectomycorrhizae on their roots.

Ectomycorrhizae are formed by fungi belonging to the higher Basidiomycetes, Ascomycetes, and zygosporic Phycomycetes of the Endogonaceae (8,79,82). The host plants of these fungi are predominately trees belonging to the Pinaceae (pines, fir, larch, spruce, hemlock), Fagaceae (oak, chestnut, beech), Betulaceae (alder, birch), Salicaceae (poplar, willow), Juglandaceae (hickory, pecan), Myrataceae (eucalyptus), Ericaceae (*Arbutus*), and other families (49).

Ectomycorrhizal fungi have been introduced into deficient soils in various inocula to provide seedlings with adequate ectomycorrhizae to create man-made forests. Most research on inoculation with ectomycorrhizal fungi has been based on two premises. First, any ectomycorrhizal association on roots of tree seedlings is far better than none. Success in correcting deficiencies has contributed greatly to our understanding of the importance of ectomycorrhizae to trees. Second, some species of ectomycorrhizal fungi on certain sites are more beneficial to trees than other fungal species that naturally occur on such sites. Much work in recent years with a few fungal species has been aimed at selecting, propagating, manipulating, and managing the more desirable fungal species to improve tree survival and growth.

The majority of past work on inoculation with ectomycorrhizal fungi has been done in nurseries for the production of bare-root or container-grown tree seedlings. Most future work with inoculations will undoubtedly continue to concentrate on seedling production in nurseries. However, the inoculation of tree seeds with spores of ectomycorrhizal fungi, in a manner similar to inoculation of legume seeds with *Rhizobia*, to improve seedling establishment following direct seeding operations could become a very important alternative to planting nursery grown seedlings, especially on the more remote sites or those of very rough terrains.

No ectomycorrhizal fungus has been shown to complete its life cycle in the absence of mycorrhizal association in a natural environment (10). Also, there is no evidence that these fungi grow saprophytically in a natural forest soil (69). Thus, in inoculation programs, since hyphae cannot grow from the inoculum to roots, inoculum must be placed in the rooting zone of seedlings where roots can grow into the inoculum. Once initial root infection takes place, extramatrical mycelium originating from the initial infection spreads rapidly to infect other roots.

Most ectomycorrhizal fungi produce above-ground sporophores. The spores produced in these sporophores are disseminated great distances by wind, rain, insects, and small animals. The more dense the tree stands supporting ectomycorrhizal fungi and the closer they are to nursery areas, the greater the chances are for rapid natural ectomycorrhizal development on the nursery seedlings. In the southern United States, ectomycorrhizae appear on nursery-grown tree seedlings in the spring as early as 6 to 8 weeks after seed germination. In nurseries surrounded by pine and oak stands that produce abundant sporophores of ectomycorrhizal fungi, early ectomycorrhizal development occurs even in nursery soils fumigated just a few days prior to seeding. The most common ectomycorrhizal fungus naturally colonizing fumigated nursery soil in the United States is *Thelephora terrestris*, a symbiont that produces numerous sporophores starting in early spring (45).

The majority of reports on inoculation techniques with ectomycorrhizal fungi involve Basidiomycetes on pines, oaks, and eucalypts. Several types of natural and laboratory-produced inocula and several methods of application have been used through the years. Many of the techniques have proven successful, others have not. Frequently, conflicting results are encountered (33).

Soil Inoculum

The most widely used natural inoculum, especially in developing countries, is soil or humus collected from established pine plantations (51). This soil inoculum is either mixed into the rooting medium (usually a 5 to 10 percent volume), broadcast 0.5 to 1 cm deep onto soil and watered into the soil around young seedlings, or suspended in water (1 kg soil in 20 ℓ of water) and poured onto seedlings. The latter two procedures are usually done three to four times during the nursery season. Better results are obtained with freshly collected soil than with soil collected and stockpiled for several months. In Morocco, soil inoculum collected from oak and eucalyptus stands is mixed with clay and animal manure. This mixture is shaped into a hexagonal solid pot (about 10 x 10 cm) and allowed to dry until it is very hard. A small depression on the upper surface of the "Morrocan pot" is seeded and filled with more soil inoculum. Several months later these container-grown seedlings which have variable quantities of ectomycorrhizae are outplanted intact. A major drawback with soil or humus inoculum is that the species of ectomycorrhizal fungi in the inoculum cannot be controlled. There is no assurance that this inoculum contains the most desirable fungi for the tree species to be produced or the site to which the seedlings are to be outplanted. Transportation of large volumes of soil inoculum is difficult. Soil inoculum may also contain harmful microorganisms and noxious weeds in addition to the ectomycorrhizal fungi. Some of these microorganisms may be potentially harmful not only to the tree seedling crop (51), but also to nearby agricultural crops (28).

Tree seedlings with ectomycorrhizae or excised ectomycorrhizae have also been used as an inoculum source for new seedling crops. In France, a truffle-producing fungus, *Tuber melanosporum*, has been established in nursery beds of young seedlings by transplanting seedlings with *Tuber* ectomycorrhizae formed under laboratory conditions into the nursery bed of new seedlings. In Indonesia, seedlings of *Pinus merkusii* with abundant ectomycorrhizae are planted at 1- to 2-m intervals in new seedbeds. The extramatrical mycelium developing from the "nurse" seedlings infect roots on the younger seedlings (51). Other workers (7,23) have harvested ectomycorrhizae from established trees and used them as inoculum. This successful technique has been used only on a limited basis in research trials. A great deal of time and care is required to obtain a sufficient quantity of viable ectomycorrhizae. At least 1 kg of ectomycorrhizae should be mixed with 1 m^3 of soil.

In most instances, use of the above inocula assures that most seedlings will have some ectomycorrhizae. Frequently, however, the symbiotic fungi in these inocula are not identified and their degree of benefit to the seedlings is unknown. The use of natural inoculum, however, does satisfy the first premise mentioned earlier--that any ectomycorrhizae on seedlings used in forest regeneration are better than none.

Spore Inoculum

Sporophores and spores of various fungi have been used as inoculum to form specific ectomycorrhizae on tree seedlings. Whole or chopped sporophores are dried before use. They are essentially spore inoculum, since the vegetative matrix of the sporophore is killed by dessication during drying or by decomposition when added to soil. According to Trappe (83), the first attempts to use specific fungi to form ectomycorrhizae on seedlings date back to the 18th century. Sporophores of truffle fungi were added to planting holes of oak seedlings in new plantations in attempts to enhance truffle production. These inoculations were done nearly 75 years before the term "mycorrhiza" was coined and over 100 years before the true nature of ectomycorrhizal associations was demonstrated. There is no way of determining if these inoculations were successful.

Gastromycetes, such as the puffball-producing genera *Rhizopogon*, *Scleroderma*, and *Pisolithus*, produce numerous basidiospores that are easier to collect in large quantities than those of mushroom-produced ectomycorrhizal fungi. Various authors (5,20,21,60) have demonstrated the value of basidiospores as inoculum. By a variety of techniques, many scientists have successfully used basidiospore inoculum of *Pisolithus tinctorius* to form specific ectomycorrhizae on pine seedlings (6,29, 37,40,43,44,63). Spores of *P. tinctorius* are collected by crushing sporophores with ruptured peridia over a 25 to 30 mesh screen which allows the mature dry spores to pass through. Screened basidiospores are air-dried for several days at low humidity, then stored at 5°C. Spores collected and stored for several years in this fashion have been used to form *Pisolithus* ectomycorrhizae on pine. Since basidiospores of *P. tinctorius* and many other fungi will not germinate in the laboratory, spore viability can only be determined with ectomycorrhizal synthesis tests. A simple, effective inoculation procedure involves dusting dry spores of *P. tinctorius* onto soil around young seedlings and leaching them into the root zone with irrigation water. This inoculation has been successful on bare-root (43) and container-grown (63) pine seedlings. In most tests, 1 to 2 mg of spores per seedling have been applied. There are about 1.1×10^6 basidiospores of *P. tinctorius* per mg.

Inoculum composed of spores mixed with a moistened carrier, such as vermiculite, kaolin or sand, can be broadcast onto soil then mixed into the nursery soil (40,44) or mixed directly into the growing medium of containers (29,35,65) to form *P. tinctorius* ectomycorrhizae on pine seedlings.

Hydromulch is used to cover seeds in many tree nurseries in the United States. Hydromulch is a wood pulp or finely shredded and mildly pulped, recycled paper that is mixed with water in a special machine and blown onto seedbeds to a depth of 0.3 to 0.6 cm. Basidiospores of *P. tinctorius* suspended in water with a wetting agent have been mixed with hydromulch in the machine and used to cover seeds of *P. taeda*. Irrigation water and rain washed the spores from the hydromulch into the root zone and *Pisolithus* ectomycorrhizae were formed on about 75 percent of the seedlings after one growing season (40).

Several nurserymen in the United States have used another spore inoculation method. They collect sporophores of *P. tinctorius* in the fall and store them at 5°C. In the spring the sporophores with numerous dry spores are broken up into particles 2 to 4 mm in diameter, mixed at undetermined ratios with pine straw or sawdust, and placed as a mulch on the seedbeds. Unfortunately, this method yields highly variable quantities of *Pisolithus* ectomycorrhizae on seedlings at the end of the season.

Seed inoculation with basidiospores is another technique employed with some success. Theodorou (76) collected sporophores of *Rhizopogon luteolus* then air-dried and crushed them into a fine powder. Seeds of *P. radiata* were mixed with the finely divided sporophores in water. The coated seeds contained approximately 1.9×10^6 spores each. Seedlings which developed from spore-coated seeds formed abundant *Rhizopogon* ectomycorrhizae in sterile soil and in nonsterile soil that was deficient in ectomycorrhizal fungi. Theodorou and Bowen (78) found that inoculating seeds with spores from freeze-dried sporophores would work, but it required up to 100 times more spores to form the same quantity of ectomycorrhizae. It took 10 times more spores if the spores were air-dried for only 2 days. Freeze drying or brief air drying obviously killed or inhibited germination of many spores of *R. luteolus*.

Basidiospores of *P. tinctorius* have also been mixed with the pelletizing matrix (clay and adhesive) of encapsulated pine seeds and used to form *Pisolithus* ectomycorrhizae on container-grown and bare-root seedlings (Marx, unpublished data). In early tests, clay and adhesive with spores were applied too heavily (increased seed diameter by 1.5 mm) and inhibited seed germination by nearly 40 percent. When the quantity of clay and adhesive was decreased (seed diameters increased by only 1 mm), inhibition of seed germination was lowered to 20 percent. Seed encapsulation without spores did not significantly reduce seed germination. There was no spore rate: degree of ectomycorrhizae relationship with encapsulated pine seeds. One mg of spores/seed was effective in forming *Pisolithus* ectomycorrhizae as were 2 to 8 mg of spores/seed. All basidiospore collections of *P. tinctorius* obtained from mature basidiocarps with ruptured peridia contain yeast, bacteria, and an array of other fungi. These microorganisms may cause damage to seeds and seedlings. To decrease this possibility various fungicides were added to encapsulated seeds. Two grams of captan or benomyl (50% WP) or 5.45 grams of thiram (42% EC in latex) were added per 2,000 encapsulated pine seeds. In container tests with *P. taeda*, captan and thiram increased and benomyl decreased *Pisolithus* ectomycorrhizal development on 16-week-old seedlings. Thiram and benomyl also inhibited seed germination. In a companion bare-root seedling test, neither fungicide affected seed germination or *Pisolithus* ectomycorrhizal development on 9-month-old seedlings of *P. taeda*. Earlier, captan was shown to stimulate *Pisolithus* and *Thelephora* ectomycorrhizal development on pine in fumigated nursery soils (38,45). *Pisolithus* ectomycorrhizae also were formed from spore encapsulated seeds of *P. virginiana*, *P. elliottii* var. *elliottii*, *P. oocarpa*, *P. echinata*, and *P. caribaea*. Tests of spore encapsulated seeds in conventional nurseries, direct-seeding trials, and inclusion of other tree species are warranted.

There are both advantages and disadvantages to using spores of ectomycorrhizal fungi for inoculation. The major advantage is that spores require no extended growth phase under aseptic conditions like vegetative inoculum (see later section). Another advantage is that spore inoculum is very light. One gram of basidiospores of *Rhizopogon luteolus* or *Pisolithus tinctorius* contains over 1 billion potentially infective propagules (basidiospores). As mentioned earlier, large numbers of basidiospores can be collected from sporophores of Gastromycetes. Over 450 kg (1,000 pounds) of mature, dry basidiospores of *P. tinctorius* were collected from under pine on coal spoils near Birmingham, Alabama, by personnel of the International Forest Seed Company in approximately 75 man-days. This one collection represents sufficient basidiospores to produce over 225 million pine seedlings with *Pisolithus* ectomycorrhizae if we assume that 1 mg of spores/seed or seedling is used. This quantity of seedlings is about one-fifth of the number grown in nurseries in the United States each year. It would be nearly impossible to collect this quantity of spores from any other ectomycorrhizal fungus, especially from fungi belonging to the Agaricales or Aphyllophorales. Another advantage of spores, at least those of certain fungi, is their ability to survive storage from one season to the next. Storability is important since spores collected in the summer or early autumn must be stored until the following spring if they are to be used to inoculate spring-sown nursery beds. The precise storage conditions for large quantities of basidiospores have not been worked out. Frequently, if the spores of *P. tinctorius* have a moisture content exceeding 13 to 15 percent (determined by drying at 85°C for 48 hr), growth of mold fungi will occur in cold storage (5°C) after a few months. Moldy spore collections are not very effective as inoculum. Shortly after *P. tinctorius* spores are collected they should be placed 1 to 2 cm deep in trays with large surface areas and air-dried at 22 to 26°C and 40 to 50 percent relative humidity. Weight loss of as much as 10 to 30 percent can occur within 60 hrs if the spores are initially moist. During drying, the spore layer in the tray should be carefully mixed every 3 to 4 hrs. Individuals processing dry spores should wear a filter mask or respirator to avoid inhaling the spores.

One of the major disadvantages of spore inoculum obtained from many ectomycorrhizal fungi is the lack of standard laboratory tests to determine spore viability. Several workers (14, 22, Marx, unpublished data) have tried a variety of physical, chemical, and biological stimuli in attempts to germinate basidiospores of *P. tinctorius* without reproducible results. At this time, the only reliable means of determining spore viability is through ectomycorrhizal synthesis tests. Another disadvantage is that sufficient sporophores of many fungi required to inoculate nurseries may not be available every year. For example, 450 kg of spores were obtained from numerous sporophores of P. tinctorius produced from August through December 1980 on coal spoils around Birmingham, Alabama. From these same spoils, only 100 kg of spores were obtained during these months in 1979. The area experienced a severe drought in 1979 and in 1980 there was adequate rainfall. One would need ideal storage conditions to maintain a large inventory of spores in order to insure a constant supply of spore inoculum from year to year.

Formation of ectomycorrhizae by basidiospores usually takes 3 to 4 weeks longer than vegetative inoculum of the same fungus (40,77). This can be a disadvantage because, during this period, pathogenic fungi (26) or other ectomycorrhizal fungi often colonize the roots and reduce the effectiveness of the introduced spore inoculum. Also, seedlings experiencing a delay in ectomycorrhizal formation lose whatever growth stimulation ectomycorrhizae may give them during this period. It should be pointed out, however, that in parts of the world where the natural occurrence of ectomycorrhizal fungi is erratic or deficient, this delay may not have a significant effect on the final amount of mycorrhizae developed on tree seedlings from spore inoculations.

Perhaps the most significant problem in using spore inoculum is the lack of genetic definition. Although basidiospores of P. tinctorius collected from different sporophores in northeast Georgia did not vary significantly in their capacity to form ectomycorrhizae on pine (29), they may have varied in other traits. Genetic variation would be greater if basidiospores from sporophores collected from many geographic areas and from different tree hosts were combined into a single inoculum. Marx (32,34) demonstrated that vegetative isolates of P. tinctorius from different world locations and tree hosts varied significantly in their abilities to form ectomycorrhizae on pine and oak and to grow in pure culture at different temperatures. These same traits and undoubtedly others could vary in combined basidiospore collections. With vegetative cultures, on the other hand, certain traits can be controlled and propagated.

Vegetative Inoculum

Pure mycelial or vegetative inoculum of ectomycorrhizal fungi has been repeatedly recommended (1,33,51,69,83) as the most biologically sound method of inoculation. Unfortunately, ectomycorrhizal fungi as a group are difficult to grow in the laboratory. Many species have never been isolated and grown in pure culture. Some species grow slowly, others often die after a few months in culture. Most of these fungi require specific growth substances, such as thiamine and biotin, in addition to simple carbohydrates; most are sensitive to growth inhibitory substances (58).

The first and most important step in any inoculation program of tree seedlings is the selection of the fungi. The physiological differences are great among ectomycorrhizal fungi. These differences can be used as criteria for their selection. Host specificity is one physiological trait important to consider in the selection process. The consistent association of certain fungi with only a few specific tree hosts is well documented in the literature. Many other fungi are associated with a great number of different tree hosts (30,71,79). Any candidate fungus should exhibit the physiological capacity to form ectomycorrhizae on the desired hosts; the more hosts the better. Isolate variability within any candidate symbiont is another criterion to consider. Several isolates from different tree hosts and geographic regions should be used, at least initially, to determine the amount of variation that exists between isolates. This point has been stressed by Moser (56) and demonstrated with isolates of Rhizopogon luteolus (77), Pisolithus (32,34,52), and Paxillus involutus (18). For example, isolates of P. tinctorius from various pines were reported to form abundant ectomycorrhizae on P. taeda and Quercus rubra, but isolates from various oak species formed few ectomycorrhizae on pine or oak. Some oak isolates formed no ectomycorrhizae on either host (32,34) as did other pine isolates from Australia and Brazil (34). Another criterion is the ability of the selected fungus to grow in pure culture and withstand manipulation. A variety of culture media (55,71,79) and methods of isolation (58) can be used to obtain pure cultures of the selected fungus. Ideally, the fungus should be able to grow rapidly. A good rate of growth in petri dish culture is a mycelial colony 6 to 8 cm in diameter after 15 to 20 days. Procedures for isolation, type of culture media, and conditions necessary for isolate storage are described elsewhere in this Manual. Once the growth characteristics of a fungus have been confirmed, it is important to determine its capacity to withstand physical, chemical, and biological manipulations. Producing large quantities of vegetative inoculum of a fungus is of little value if the inoculum cannot survive the rigors of various manipulations, such as physical processing (blending, leaching, or drying) and soil incorporation. The inoculum must also be able to survive a minimum of 4 to 6 weeks between soil inoculation and the production of short roots by the seedlings. During this period, it must survive fluctuations of soil moisture, temperature, and microbial competition.

Another criterion is the adaptation of the selected fungus to the major type of site on which the seedlings are to be planted. Of equal importance is the ability of the fungus to survive and grow under cultural conditions used in nurseries. According to Trappe (83), the ecological adaptability of an ectomycorrhizal fungus hinges on the metabolic pathways it has evolved to contend with environmental variation. Extremes of soil and climatic factors, antagonism from other soil organisms including other ectomycorrhizal fungi, pesticide application, physical disruption of mycelium from nursery operation (undercutting and root pruning), and the abrupt physiological adjustment from a well fertilized and irrigated nursery soil to an uncultivated, low fertility planting site with all its stresses are only a few of the environmental variations to which the selected fungus must adapt. The effect of temperature on different species and ecotypes of ectomycorrhizal fungi is perhaps the most widely researched environmental factor. Upper and lower temperature limits of the candidate fungi should be determined. Moser (56) studied the ability of fungi to survive long periods (up to 4 months) of freezing at -12°C and to grow at 0 to 5°C. He found that high elevation ecotypes of Suillus variegatus survived freezing for 2 months, but valley ecotypes were killed after freezing for only 5 days. The ability of certain fungi to survive freezing was not correlated with their ability to grow at low temperature, however. Generally, he did find that mountain ecotypes and species had much lower temperature optima than lowland ones. Pisolithus tinctorius can grow at 40 to 42°C (16,53) and has a hyphal thermal death point of 45°C (19). It not only survives and grows well at high temperatures but grows at 7°C (36, 42) and can overwinter in frozen soil (37). Reaction of candidate fungi to soil moisture, organic matter, and pH are also important traits to consider. Cenococcum geophilum is not only drought tolerant but forms ectomycorrhizae in natural soils ranging in pH from 3.4 to 7.5 (83). Unfortunately, the drought tolerant characteristics of C. geophilum also make it difficult to establish on pine seedlings in irrigated nurseries where it can be supplanted by Thelephora terrestris (44). Pisolithus ectomycorrhizae on pine and oak have been observed in drought-prone coal spoils ranging in pH from 2.6 to as high as 8.4. Suillus bovinus (24) and Paxillus involutus (18) form abundant ectomycorrhizae on seedlings in nurseries with high organic matter, but these ectomycorrhizae are supplanted by others after seedlings are outplanted on sites having low organic matter.

The production of hyphal strands and sclerotia are also important traits in candidate fungi. Uptake of nutrients, especially phosphorus (2), and translocation of carbon compounds (61) take place through hyphal strands. In Australia, one of the initial criteria for selection of fungi for inoculation programs is their ability to produce hyphal strands. Although research data is lacking, abundant hyphal strand production by P. tinctorius apparently enhances nutrient absorption and increases its survival potential under adverse conditions. Yellow-gold hyphal strands of this fungus, easily visible to the naked eye, have been traced through coal spoils as far as 4 meters from seedlings to sporophores by Schramm (68) and others (31). The production of sclerotia by P. tinctorius (4,46) and C. geophilum

(81) in soil or container rooting media should enhance the abilities of these fungi to survive under harsh soil conditions and, therefore, are also favorable traits in candidate fungi.

All the criteria mentioned are meaningless unless the candidate fungus is aggressive and can form abundant ectomycorrhizae on seedlings as soon as short roots are produced. It should be able to maintain superiority on seedling roots over naturally occurring fungi in the nursery. In order for tree seedlings to obtain any measurable benefit from a specific ectomycorrhizal association, there is a threshold amount of the specific ectomycorrhizal association that must be present on seedling roots. Southern pine seedlings derive little growth benefits from Pisolithus ectomycorrhizae unless half or more of all their ectomycorrhizae are Pisolithus ectomycorrhizae at outplanting on routine reforestation sites (17, 41,66). In these reports, increased growth was determined by comparing the seedlings with Pisolithus ectomycorrhizae to control seedlings with comparable amounts of T. terrestris ectomycorrhizae.

All of these criteria for selecting a candidate fungus are important, however, one must be reminded that unknown numbers of ectomycorrhizal fungi have been introduced, usually in soil inoculum, into various parts of the world to establish exotic, man-made forests(33,51). Although many species probably died, numerous fungi are currently thriving in areas halfway around the world from their original habitat. Apparently, either individually or as a group, these fungi have a tremendous capacity to adapt to different environments. Once pure culture inoculation techniques have been perfected, the value of a specific fungus should be tested over a wide range of environmental conditions. Even though the effect of a fungus on seedlings may only be temporary until it is supplanted by other fungi, the brief advantages may make the difference between survival or death of newly planted seedlings.

Several researchers in various parts of the world have developed cultural procedures for producing vegetative inoculum of a variety of fungi for research purposes. Unfortunately, large-scale nursery applications of pure mycelial cultures, even involving only a few million tree seedlings, have been severely hampered by the lack of sufficient quantities of inoculum. It is relatively simple to produce a sufficient volume of inoculum, i.e., 30 to 40 liters, for research studies carried out in small containers, pots, microplots, or even small nursery plots, but it is difficult to produce a sufficient quantity of vegetative inoculum for large-scale nursery inoculation in a practical program. A considerable quantity of vegetative inoculum would be needed to inoculate even a few southern nurseries, which together produce nearly 1 billion pine seedlings annually on some 1,400 acres of nursery soil.

Moser (54,55,56) in Austria was the first to make a serious attempt at producing vegetative inoculum of ectomycorrhizal fungi. His pioneering research has furnished the necessary basic information used by others to modify and expand upon. Moser's purpose was to develop techniques for inoculating seedlings of Pinus cembra with low temperature strains of Suillus plorans in the nursery. These strains were absent from the warmer nursery soils in the valley and in alpine meadows, but they proved to be highly beneficial to pine seedlings for reforestation of the colder high elevation sites near the timberline.

For production of inoculum, mycelium of S. plorans was first grown in Moser's (54) nutrient solution in small flasks for several days. These mycelial cultures were then decanted into 10 liter tanks containing the same nutrient solution and aerated for 2 to 3 hours daily for 3 to 4 months. The mycelium and liquid were transferred to 5 liter flasks containing sterilized peat moss and fresh nutrient solution. During the next 2 to 4 months at laboratory temperatures, mycelium of S. plorans permeated the peat moss substrate. The inoculum was removed from the culture tanks, packaged in sterile plastic bags, transported to the nursery, and used within 3 days. Although attempts were made to maintain these cultures in aseptic condition during the 5 to 8 months incubation, contamination by Penicillium, Mucor, and bacteria often occurred. Moser (57) refers to this inoculum as "half-pure" cultures and states that on certain occasions it proved more effective in forming ectomycorrhizae than pure cultures. He speculated that these contaminants produced a rhizosphere effect beneficial to ectomycorrhizal development and seedling growth. He tested various organic materials as the final inoculum substrate and found that forest litter or sawdust were not as effective as peat moss. He also found that agar-mycelial inoculum or mycelial suspension were not effective inocula in the nursery. Other workers (7,23,51, Marx, unpublished data) have used agar-mycelial inoculum or mycelial suspension with varying degrees of success. Moser (57) used the above technique to produce vegetative inoculum of Suillus placides, S. grevillei, S. aeruginascens, Paxillus involutus, Phlegmacium glaucopus, Amanita muscaria, and Lactarius porninsis. More recently in Austria, Göbl (9) modified Moser's technique and used mycelium to inoculate 1 liter bottles of cooked and sterilized cereal grains such as wheat or white millet. Calcium sulfate (0.4 to 0.5 g/100 g of grain) was added to improve the growth of certain fungi. These cultures were shaken lightly each week, and after 4 weeks at 20 to 22°C the grains were usually thoroughly colonized by the fungi. These grain cultures were successfully stored for up to 9 months at 4 to 6°C. During storage, they were periodically checked for microbial purity. The final inoculum was produced by adding the grain cultures to sterile peat moss enriched with nitrogen and carbohydrates (ammonium tartrate, asparagine, soybean meal, blood meal, malt extract, glucose), as well as inorganic nutrients. The kinds and amounts of these enrichments varied according to the species of fungus grown. Usually 7 to 10 grams of glucose per liter of peat moss was a standard. Ten to 15 liters of sterile, enriched peat moss were placed in large, transparent plastic bags and inoculated with 1 to 3 bottles of grain cultures. The plastic bags were plugged with cotton to provide aeration and were shaken occasionally during storage at 20 to 22°C. After 3 to 6 weeks incubation the inoculum was ready. Contaminated cultures were apparently discarded. Göbl has used this method to produce inoculum of Suillus plorans, S. grevillei, Boletinus cavipes, Amanita muscaria, and Hebeloma crustuliniforme.

Takacs in Argentina modified Moser's technique to produce inoculum for new pine nurseries established in formerly treeless areas lacking ectomycorrhizal fungi. Takacs (72,73,74) obtained isolates from sporophores and then grew them in liquid culture for several days. The mycelium was used to inoculate sterilized, germinated grains of cereals (such as barley), the cereal chaff, a mixture of grain and chaff, or peat moss. Peat moss was more commonly used than other substrates. All substrates were enriched with a liquid medium. The inoculum, regardless of substrate, was used after 1 to 2 months incubation at room temperature. Inoculum of Amanita verna, Suillus granulatus, S. luteus, Hebeloma crustuliniforme, a Russula sp., Scleroderma verrucosum, and S. vulgare have been produced using this technique. Details for a large-scale nursery inoculation in Argentina were described by Mikola (50). Usually five 200-ml flasks of each of four different fungi contained in either peat moss or grain-chaff inoculum were each mixed with 4 to 10 kg of sterilized (heat or fumigated) soil or forest litter at the nursery. These mixtures were kept moist and incubated for 3 weeks in small piles. Using this method, twenty 200 ml "starter" cultures were used to produce 100 to 200 kg of inoculum, an amount sufficient to inoculate 500 m^2 of nursery soil. Since quantitative data are not available for this work it is difficult to evaluate the success of this inoculation method. However, based on our current knowledge, it is difficult to believe that these fungi grew saprophytically in the initially sterile (probably only partially sterile) soil or litter from the cereal grain food base. This mycelium had to grow in the presence of competitive microorganisms and in the absence of essential carbohydrates. If the starter inoculum survived the soil inoculation at the nursery, perhaps all that was accomplished was a dilution of the "starter" cultures.

In Canada, Park (59) also grew pure mycelial cultures of Suillus granulatus and Cenococcum geophilum in cereal grains. Five grams of wheat grains, 100 mg of $CaCO_3$ and 8 ml of water were autoclaved in test tubes. These tubes (master cultures) were then inoculated with a fungus and incubated at 25°C for 2 to 3 weeks.

To produce mass inoculum, wheat grains were soaked in water, boiled for 30 min, and drained. Four to 5 grams of $CaCO_3$/100 g of grain were added. The grain in flasks was autoclaved, cooled, and inoculated with master cultures. After several weeks, growth of the fungi in the grain was apparent. Unfortunately, Park failed to report whether this inoculum would form ectomycorrhizae on tree hosts.

In Australia, vegetative inoculum of Rhizopogon luteolus was produced using techniques similar to those of Moser. Inoculation with R. luteolus was used to correct the deficiency of ectomycorrhizal fungi in some Australian soils and to produce seedlings of Pinus radiata with a root system having a greater capacity to absorb phosphorus from forest soil. Pure cultures were produced in a 10:2:1 ratio of vermiculite, chaff, and corn meal moistened with a liquid medium (75). The fungus was placed in bottles containing about 80 grams of medium and incubated for 1 month at 21°C. Later, Theodorou and Bowen (77) grew Suillus granulatus, S. luteus, Cenococcum geophilum, and R. luteolus in the above substrate for 3 weeks and found that it formed ectomycorrhizae on seedlings of P. radiata in pot culture.

In the United States, tests to artificially introduce pure mycelial cultures of ectomycorrhizal fungi into soil were begun by Hatch (12,13). He found that seedlings of Pinus strobus grown in prairie soil for 3 months in a filtered chamber were nonmycorrhizal and severely stunted. Half of the seedlings were inoculated with agar-mycelial inoculum of S. luteus, Boletinus pictus, Lactarius deliciosus, L. indigo, and C. geophilum. After 5 months, root evaluations revealed that S. luteus and L. deliciosus formed ectomycorrhizae and stimulated seedling growth. Noninoculated seedlings remained nonmycorrhizal and stunted. Hacskaylo and Vozzo (11) initiated a series of inoculation experiments in Puerto Rico with pure mycelial cultures of various fungi to correct a deficiency of these fungi on the island. Following Moser's (57) general technique, they grew C. geophilum, Corticium bicolor, Rhizopogon roseolus, and Suillus cothuranatus on agar medium and then in liquid medium. The mycelium from liquid culture was added to polypropylene cups containing a 2:1 ratio of sterile peat moss and vermiculite moistened with a glucose-ammonium tartrate nutrient solution (pH 3.8). After 16 weeks of incubation at room temperature in a Maryland laboratory the inoculum was transported to Puerto Rico. In the nursery, one-half cup of inoculum per seedling was placed against the nonmycorrhizal roots of 4-month-old Pinus caribaea seedlings growing in plastic bags. After 10 months, all fungi except C. geophilum formed ectomycorrhizae to varying degrees.

Marx and colleagues (33) tested various methods for producing mycelial inoculum of Pisolithus tinctorius, Thelephora terrestris, and Cenococcum geophilum. P. tinctorius was chosen as a candidate fungus because: 1) it was considered valuable in reclamation of coal spoils and other adverse sites with pines due to its natural occurrence and apparent ecological adaptation to adverse sites (46), 2) it grew rapidly in pure culture, 3) it had a broad tree host range (30), and 4) the ectomycorrhizae it formed were easily recognized and quantified on roots because of their mustard-yellow color. T. terrestris was chosen because: 1) it occurred naturally in many tree nurseries and was adapted to the good tilth and fertility of nursery soils, 2) it grew rapidly in pure culture, and 3) it had a broad tree host range (45,84). Unfortunately, Thelephora ectomycorrhizae were not very distinctive and were easily confused with the ectomycorrhizae formed by other fungi. C. geophilum was chosen because: 1) it tolerated drought and high temperature (47,43,67) and should improve seedling growth on hot, droughty sites, 2) it produced a distinctive jet-black, easily identifiable ectomycorrhizae, and 3) it had a broad tree host range (80). Unfortunately, isolates of C. geophilum grew very slowly in pure culture; a mycelial colony on agar medium of 2 to 3 cm in diameter after 3 weeks at 25°C was considered normal.

Attempts to produce effective inoculum of these fungi in wheat grains, using Park's and Takacs' techniques, failed (33). Grain cultures were added to sterilized soil in a 1:15 ratio and planted to seed of Pinus taeda in a special growth room (27). After 5 months, none of the inoculated and control seedlings had ectomycorrhizae. Microscopic examination and plating procedures revealed that the grain cultures were colonized by saprophytic fungi and bacteria as early as 3 weeks after soil inoculation. A great deal of damping-off of pine seed was also observed. The high nutritive value of the wheat grains probably contributed to rapid colonization by saprophytes which, in turn, killed the ectomycorrhizal fungi. Macro- and microscopic examination of the grain inoculum before soil inoculations revealed that mycelium of the ectomycorrhizal fungi completely permeated the endosperm of the grain.

Vermiculite and peat moss moistened with a modification of Melin-Norkrans medium (MMN) was found to be an excellent substrate for the production of mycelial inoculum of these fungi. For decades, Melin-Norkrans medium has been the basic nutrient used in aseptic ectomycorrhizal synthesis tests. The modified formulation is simply an enrichment of this nutrient medium (25). The formulation of MMN is 0.05 g $CaCl_2$, 0.025 g NaCl, 0.5 g KH_2PO_4, 0.25 g $(NH_4)_2HPO_4$, 0.15 g $MgSO_4 \cdot 7H_2O$, 1.2 ml of 1% $FeCl_3$, 100 μg Thiamine HCl, 3 g malt extract, and 10 g glucose in distilled water to equal one liter. Fifteen grams of agar/liter are added for agar formulation. After autoclaving, the pH of both the liquid and agar formulations is 5.5 to 5.7. This medium has proven to be as good or better than most other media for the growth of many ectomycorrhizal fungi. A ratio of 28:1 vermiculite and peat moss substrate moistened with a volume of MMN liquid medium equal to approximately half the volume of the dry substrate has proved to be the best substrate. An example is 1400 ml of vermiculite and 50 ml of peat moss mixed thoroughly and then moistened with 750 ml of MMN liquid (37). Horticultural grade No. 4 vermiculite should have all fine particles removed by screening it through a fine mesh screen (window screen). Peat moss should be finely divided and passed through the same mesh screen; the screened peat moss is mixed with the vermiculite. After autoclaving the substrate, the pH should be within the range of 4.5 and 5.5 (39).

The size of the culture vessel is not critical. One-liter Erlenmyer flasks can be used as well as 50-liter carboys. Autoclavable plastic containers of various sizes have also been used successfully. The only characteristic important in the culture vessel for these stationary, solid substrate cultures is that its dimensions should allow the vermiculite-peat moss to stay moist. Vermiculite-peat moss in a tall cylinder-like container will be dry at the top and excessively moist at the bottom after a few weeks. These extremes in moisture conditions in the medium inhibit rapid growth of ectomycorrhizal fungi. Various means of inoculating the substrate can be used. Mycelial discs from agar plate cultures or mildly blended mycelium from liquid culture work well. The more starter inoculum placed in the substrate the more rapid will be mycelial colonization of the substrate. Four to six evenly spaced agar-mycelial discs/liter of substrate are recommended. Length of incubation after inoculation with mycelial discs depends on the growth rate of the fungi. Typically, fast-growing fungi such as _Thelephora terrestris_ and _Pisolithus tinctorius_ incubated at room temperature will thoroughly colonize the substrate in 2-liter containers in 2 to 3 months. Slower-growing fungi, such as _Cenococcum geophilum_, take up to 8 to 10 months to colonize the same amount of substrate. Blended mycelial starter inoculum mixed thoroughly in the substrate will reduce the time of incubation by half. After incubation, if the fungus has completely colonized the substrate and is free of microbial contamination, the inoculum is ready for use. The latter feature can be easily confirmed by plating out portions of the substrate onto various agar media. Various other substrates, such as perlite, and 0.6-cm-diameter particles of peanut hulls, corn cobs, or pine bark, are not as suitable as vermiculite-peat moss. These fungi fail to grow in peanut, corn, or bark substrates because growth inhibitors are released during autoclaving. They will grow in perlite substrates, but the mycelium tends to grow around, rather than into, the particles which is undesirable.

Vermiculite-peat moss-nutrient inoculum as described above, when taken directly from the container and mixed into fumigated nursery soil, becomes rapidly colonized by saprophytic microorganisms. Heavy colonization reduces the effectiveness of this inoculum to form ectomycorrhizae on pine. Leaching the inoculum before inoculation to remove most nonassimilated nutrients reduces this colonization and increases inoculum effectiveness (33). Leaching is simply done by wrapping 4 to 6 liters of inoculum in several layers of cheesecloth and irrigating this for 2 to 3 min under cool tapwater. Excess water is then removed by squeezing the inoculum in the cheesecloth by hand. Leaching removes over 65 percent of the original glucose and reduces the original inoculum volume by one-third. Leached inoculum has been used by many workers (33), but it has a very high bulk density (over 600 g/ℓ), a high moisture content (up to 91 percent), and a physical consistency of a sticky paste. These problems are corrected by drying the inoculum (38). Leached inoculum is placed 5 to 6 cm deep on wood framed wire screens (window screen) and dried at 20 to 26°C and 35 to 45 percent relative humidity. During drying the inoculum is mixed every 2 to 3 hrs to minimize excessive drying of the surface particles. Inoculum is dried at the IMRD in a room (22.6 m^3 volume) constructed of clear polyethylene plastic containing a small air conditioner, two dehumidifiers, and a small heater. Over 150 liters of dried inoculum can be processed in this room in 100 uninterrupted hours of drying. Final bulk densities range from 320 to 390 g/liter, and moisture contents from 20 to 65 percent. Final pH of inoculum ranges from 4.4 to 5.2 (45,46). The original volume of inoculum is reduced by nearly 60 percent after leaching and drying.

Dried inoculum of *P. tinctorius*, *T. terrestris*, and other fungi can be mixed more thoroughly in soil than the wetter, nondried formulations and can be effectively used at lower rates (33). Dried inoculum of *P. tinctorius* stored at 5°C for as long as 9 weeks and at room temperature for 5 weeks was still effective in forming ectomycorrhizae (33,45,46). However, we recommend that inoculum be dried and used as soon as possible.

A commercial formulation of mycelial inoculum of *P. tinctorius* has been recently developed by the USDA, Forest Service and Abbott Laboratories (45,46). This inoculum, which has been trademarked MycoRhiz®, is also grown in the vermiculite-peat moss nutrient medium. Liquid culture of *Pisolithus tinctorius* has recently been reviewed (70). The medium used for all liquid culture is Melin-Norkrans modified by Marx (25). Liquid culture is grown in shake flasks, 14-liter fermentors and large industrial fermentors (Plate 7) and is used as inoculum for the vermiculite-peat substrate. All liquid culture is grown with continuous agitation and high aeration rates (one volume change per minute) at temperatures of 28 to 32°C. The incubation time is dependent on inoculum level and ranges from 7 to 14 days.

Beds of vermiculite-peat moss-nutrient medium are sterilized by injection of high pressure steam in either deep-tank or rotary drum fermentors. This medium is inoculated with an actively growing liquid culture of *Pisolithus tinctorius*. The level of inoculum varies considerably but is usually in the range of 5-20 percent of the fermentor volume. The medium is mixed and incubated with or without agitation for 1 to 4 weeks, depending on the rapidity of growth and colonization of the substrate.

The colonized substrate is harvested by floatation in water and filtration; this also serves to wash the product form. The wet product is dried with an Aeromatic fluid bed dryer (type STS-60; Aeromatic Ltd. Muttenz-Basle, Switzerland) to a moisture level of 20-25 percent. MycoRhiz® normally has a bulk density of 240 to 310 g/liter and a pH of 5.1 to 5.6. It is packaged in 50 liter units and stored at 5°C. The shelf life under these conditions is 5 to 6 weeks. Using this fermentation technology, MycoRhiz® can be produced in sufficient volumes for large scale, practical inoculation of tree nurseries. Recently, a tractor-drawn machine was developed that will inoculate soil with MycoRhiz® and seed nursery beds in one operation (3).

During the research and development on MycoRhiz®, various assays were developed to characterize the inoculum in an attempt to predict the eventual effectiveness of the inoculum in forming *P. tinctorius* ectomycorrhizae in nursery soil or container substrate. One assay involved determination of propagules of *P. tinctorius* in inoculum. One cc of dried inoculum (final product form) was spread onto petri dishes containing 10 ml of MMN agar medium fortified with 25 mg/liter of benomyl and 10 mg/liter of erythromycin phosphate. The plates, usually 10 per inoculum batch, were incubated at 30°C for 5 days, growth centers of the yellow-gold mycelium of P. tinctorius were counted, and counts were expressed as propagules/g of inoculum. In early trials with MycoRhiz®, certain batches had a wide range of vermiculite particle sizes (0.5 mm to 7 mm). Survival of the fungus in the smaller particles was questioned. Thus, another propagule assay was developed to determine viability of the fungus in the smaller particles. This assay involved screening inoculum through a No. 6 sieve, and placing 30 to 40 of the particles that collected on a No. 8 sieve on the surface of fortified MMN agar. These were incubated and counted as in the other propagule assay. Since microbial contamination of inoculum caused problems in previous work at the IMRD, an assay for microbial contamination of the inoculum was developed. One gram of inoculum was blended with 100 ml of sterile water for 3 min; this blend was serially diluted by factors of 10 and 1 ml lots of the diluted material were placed onto the surface of Difco trypticase soy agar. After plates, usually five per treatment, were incubated for 5 days at 30°C the bacterial and fungal colonies were counted. The dilution yielding less than 10 colonies of bacteria or fungi per plate was considered the contamination level. Since leaching of nonassimilated nutrients from the inoculum was shown to improve inoculation effectiveness in soil, a low concentration of glucose in the leached and dried inoculum was considered beneficial to inoculum effectiveness. Residual glucose (mg glucose/g of oven-dried inoculum) was determined, therefore, with an autoanalyzer by a modification of Hoffman's (15) technique.

Although MycoRhiz® effectiveness in nursery tests could not be predicted from the results of any single assay, a pattern did emerge. The most effective formulations were those with: 1) abundant hyphae of P. tinctorius inside the vermiculite particles (inoculum niche), 2) pH between 4.5 and 6.0 by having a 5 to 10 percent volume of peat moss as a component of substrate, 3) low levels of bacterial and fungal contaminants which apparently colonized inoculum during leaching and drying, and 4) low amounts of residual glucose (<16 mg/g inoculum).

CONCLUSIONS

There are a variety of methods available to ensure the development of ectomycorrhizae on forest tree seedlings for the establishment of man-made forests. Certain methods have more advantages than others. Pure mycelial inoculum has the greatest biological advantage. There is sufficient information to conclude that pure cultures of certain fungi, such as Suillus granulatus, Rhizopogon luteolus, Thelephora terrestris, and Pisolithus tinctorius, can be used to improve survival and growth of tree seedlings on a variety of sites (33,46). These results represent only the beginning of a universal practical program, however. When one considers the millions of hectares of potential exotic forests which could be established in Third World nations on former treeless lands, as well as the millions of hectares of former forested lands awaiting artificial regeneration throughout the world, the importance of the selection, propagation, manipulation, and management of superior strains or species of ectomycorrhizal fungi as a forest management tool is paramount. Research so far has only revealed the tip of the iceberg in regards to potential use of specific ectomycorrhizal fungi in world forestry. There still remains a tremendous reservoir of basic and practical information which must be revealed if these fungi are to be managed and, therefore, fully utilized in forestry. Hopefully, the techniques to produce ectomycorrhizal fungus inoculum described in this chapter will be used by challenged and foresighted plant scientists to further define and expand the value of ectomycorrhizae in forest regeneration.

LITERATURE CITED

1. Bowen, G. D. 1965. Mycorrhiza inoculation in forestry practices. Aust. For. 29:231-237.

2. Bowen, G. D. 1973. Mineral nutrition of ectomycorrhizae. Pages 151-205, in; Ectomycorrhizae: their ecology and physiology, G. C. Marks and T. T. Kozlowski, eds., Academic Press, New York. 444 pp.
3. Cordell, C. E., Marx, D. H., Lott, J. R. and Kenney, D. S. 1981. The practical application of Pisolithus tinctorius ectomycorrhizal inoculum in forest tree nurseries. Pages 38-42, in; Forest Regeneration - Proc. Amer. Soc. Agr. Engineers, Symp. on Engineering Systems for Forest Regeneration, Raleigh, N. C. 376 pp.
4. Dennis, J. J. 1980. Sclerotia of the Gasteromycete Pisolithus tinctorius. Can. J. Microbiol. 26:1505-1507.
5. Donald, D. G. M. 1975. Mycorrhizal inoculation of pines. J. S. Afr. For. Assoc. 92:27-29.
6. Donald, D. G. M. 1979. Nursery and establishment techniques as factors in productivity of man-made forests in Southern Africa. S. Afr. For. J. 109:19-25.
7. Ekwebelam, S. A. 1973. Studies of pine mycorrhizae at Ibadan. Res. Paper, For. Serv., No. 18. Fed. Dept. For. Res., Nigeria. 10 pp.
8. Gerdemann, J. W. and Trappe, J. M. 1974. The Endogonaceae in the Pacific Northwest. Mycologia Mem. 5. 76 pp.
9. Göbl, F. 1975. Erfahrungen bei der Anzucht von Mykorrhiza-Impfmaterial. Cbl. Gesamte Forstwesen. 92:227-237.
10. Hacskaylo, E. 1971. Metabolic exchanges in ectomycorrhizae. Pages 175-196, in; Mycorrhizae, E. Hacskaylo, ed., USDA Forest Serv. Misc. Publ. No. 1189. 255 pp.
11. Hacskaylo, E. and Vozzo, J. A. 1967. Inoculation of Pinus caribaea with pure cultures of mycorrhizal fungi in Puerto Rico. in; Proc. 14th Int. Union Forest. Res. Organ. Munich. Vol. 5:139-148.
12. Hatch, A. B. 1936. The role of mycorrhizae in afforestation. J. For. 34:22-29.
13. Hatch, A. B. 1937. The physical basis of mycotrophy in Pinus. Black Rock For. Bull. No. 6:1-168.
14. Hile, N. and Hennen, J. F. 1969. In vitro culture of Pisolithus tinctorius mycelium. Mycologia 61:195-198.
15. Hoffman, W. S. 1937. A rapid photoelectric method for the determination of glucose in blood and urine. J. Biol. Chem. 120:51-55.
16. Hung, L. L. and Chien, C. Y. 1978. Physiological studies on two ectomycorrhizal fungi, Pisolithus tinctorius and Suillus bovinus. Trans. Mycol. Soc. Japan 19:121-127.
17. Kais, A. G., Snow, G. A. and Marx, D. H. 1981. The effects of benomyl and Pisolithus tinctorius ectomycorrhizae on survival and growth of longleaf pine seedlings. South. J. Appl. For. 5:189-195.
18. Laiho, O. 1970. Paxillus involutus as a mycorrhizal symbiont of forest trees. Acta For. Fenn. 106:1-73.
19. Lamb, R. J. and Richards, B. N. 1971. Effect of mycorrhizal fungi on the growth and nutrient status of slash and radiata pine seedlings. Aust. For. 35:1-7.
20. Lamb, R. J. and Richards, B. N. 1974. Inoculation of pines with mycorrhizal fungi in natural soils. I. Effect of density and time of application of inoculum and phosphorus amendment on mycorrhizal infection. Soil Biol. Biochem. 6:167-171.
21. Lamb, R. J. and Richards, B. N. 1974. Inoculation of pines with mycorrhizal fungi in natural soils. II. Effects of density and time of application of inoculum and phosphorus amendment on seedling yield. Soil Biol. Biochem. 6:173-177.
22. Lamb, R. J. and Richards, B. N. 1974. Survival potential of sexual and asexual spores of ectomycorrhizal fungi. Trans. Br. Mycol. Soc. 62: 181-191.

23. Levisohn, I. 1956. Growth stimulation of forest tree seedlings by the activity of free-living mycorrhizal mycelia. Forestry 29:53-59.
24. Levisohn, I. 1965. Mycorrhizal investigations. Pages 228-235, in; Experiments on nutrition problems in forest nurseries, B. Benzian, Ed., For. Comm. Bull. No. 37, Vol. 1. HMSO, London.
25. Marx, D. H. 1969. The influence of ectotrophic mycorrhizal fungi on the resistance of pine roots to pathogenic infections. I. Antagonism of mycorrhizal fungi to root pathogenic fungi and soil bacteria. Phytopathology 59:153-163.
26. Marx, D. H. 1972. Ectomycorrhizae as biological deterrents to pathogenic root infections. Ann. Rev. Phytopathol. 10:429-454.
27. Marx, D. H. 1973. Growth of ectomycorrhizal and nonmycorrhizal shortleaf pine seedlings in soil with Phytophthora cinnamomi. Phytopathology 63:18-23.
28. Marx, D. H. 1975. Mycorrhizae of exotic trees in the Peruvian Andes and synthesis of ectomycorrhizae on Mexican pines. Forest Sci. 21:353-358.
29. Marx, D. H. 1976. Synthesis of ectomycorrhizae on loblolly pine seedlings with basidiospores of Pisolithus tinctorius. Forest Sci. 22:13-20.
30. Marx, D. H. 1977. Tree host range and world distribution of the ectomycorrhizal fungus Pisolithus tinctorius. Can. J. Microbiol. 23:217-223.
31. Marx, D. H. 1977. The role of mycorrhizae in forest production. Pages 151-161, TAPPI Conf. Pap., Annu. Meet., Feb. 14-16, Atlanta, Ga.
32. Marx, D. H. 1979. Synthesis of ectomycorrhizae by different fungi on northern red oak seedlings. USDA Forest Serv. Res. Note SE-282, 8 pp.
33. Marx, D. H. 1980. Ectomycorrhizal fungus inoculations: a tool for improving forestation practices. Pages 13-71, in; Tropical Mycorrhiza Research, P. Mikola, ed., Oxford Univ. Press, London. 270 pp.
34. Marx, D. H. 1981. Variability in ectomycorrhizal development and growth among isolates of Pisolithus tinctorius as affected by source, age, and reisolation. Can. J. Forest. Res. 11:168-174.
35. Marx, D. H. and Barnett, J. P. 1974. Mycorrhizae and containerized forest tree seedlings. Pages 85-92, in; Proc. N. Amer. Containerized For. Tree Seedling Symp., R. W. Tinus, W. E. Stein and W. E. Balmer, eds., Great Plains Agric. Counc. Publ. 68. 458 pp.
36. Marx, D. H. and Bryan, W. C. 1971. Influence of ectomycorrhizae on survival and growth of aseptic seedlings of loblolly pine at high temperature. Forest Sci. 17:37-41.
37. Marx, D. H. and Bryan, W. C. 1975. Growth and ectomycorrhizal development of loblolly pine seedlings in fumigated soil infested with the fungal symbiont Pisolithus tinctorius. Forest Sci. 21:245-254.
38. Marx, D. H. and Rowan, S. J. 1981. Fungicides influence growth and development of specific ectomycorrhizae on loblolly pine seedlings. Forest Sci. 27:167-176.
39. Marx, D. H. and Zak, B. 1965. Effect of pH on mycorrhizal formation of slash pine in aseptic culture. Forest Sci. 11:66-75.
40. Marx, D. H., Bryan, W. C. and Cordell, C. E. 1976. Growth and ectomycorrhizal development of pine seedlings in nursery soils infested with the fungal symbiont, Pisolithus tinctorius. Forest Sci. 22:91-100.
41. Marx, D. H., Bryan, W. C. and Cordell, C. E. 1977. Survival and growth of pine seedings with Pisolithus ectomycorrhizae after two years on reforestation sites in North Carolina and Florida. Forest Sci. 23:363-373.
42. Marx, D. H., Bryan, W. C. and Davey, C. B. 1970. Influence of temperature on aseptic synthesis of ectomycorrhizae by Thelephora terrestris and Pisolithus tinctorius on loblolly pine. Forest Sci. 16:424-431.
43. Marx, D. H., Mexal, J. G. and Morris, W. G. 1979. Inoculation of nursery seedbeds with Pisolithus tinctorius spores mixed with hydromulch increases ectomycorrhizae and growth of loblolly pines. South. J. Appl. For. 3:175-178.

44. Marx, D. H., Morris, W. G. and Mexal, J. G. 1978. Growth and ectomycorrhizal development of loblolly pine seedlings in fumigated and nonfumigated soil infested with different fungal symbionts. Forest Sci. 24:193-203.
45. Marx, D. H., Cordell, C. E., Kenney, D. S., Mexal, J. G., Artman, J. D., Riffle, J. W. and Molina, R. J. 1982. Commercial vegetative inoculum of Pisolithus tinctorius and inoculation techniques for development of ectomycorrhizae on bare-root seedlings. Forest Sci. 28: (In Press)-
46. Marx, D. H., Ruehle, J. L., Kenney, D. S., Cordell, C. E., Riffle, J. W., Molina, R. J., Pawuk, W. H., Navratil, S., Tinus, R. W. and Goodwin, O. C. 1982. Commercial vegetative inoculum of Pisolithus tinctorius and inoculation techniques for development of ectomycorrhizae on container-grown tree seedlings. Forest Sci. 28(In press).
47. Mexal, J. G. and Reid, C. P. P. 1973. The growth of selected mycorrhizal fungi in response to induced water stress. Can. J. Bot. 51:1579-1588.
48. Meyer, F. H. 1964. The role of the fungus Cenococcum graniforme (Sow.) Ferd. et Winge in the formation of mor. Pages 23-31, in; Soil Microbiology, E. A. Jongerius, ed., Elsevier, Amsterdam.
49. Meyer, F. H. 1973. Distribution of ectomycorrhizae in native and man-made forests. Pages 79-105, in; Ectomycorrhizae: their ecology and physiology, G. C. Marks and T. T. Kozlowski, eds., Academic Press, New York. 444 pp.
50. Mikola, P. 1969. Afforestation of treeless areas. Unasylva (Suppl.) 23:35-48.
51. Mikola, P. 1973. Application of mycorrhizal symbiosis in forestry practice. Pages 383-411, in; Ectomycorrhizae: their ecology and physiology, G. C. Marks and T. T. Kozlowski, eds., Academic Press, New York. 444 pp.
52. Molina, R. J. 1979. Ectomycorrhizal inoculation of containerized Douglas-fir and lodgepole pine seedlings with six isolates of Pisolithus tinctorius. Forest Sci. 25:585-590.
53. Momoh, Z. O. and Gbadegesin, R. A. 1980. Field performance of Pisolithus tinctorius as a mycorrhizal fungus of pines in Nigeria. Pages 72-79, in; Tropical Mycorrhiza Research, P. Mikola, ed., Oxford Univ. Press, London. 270 pp.
54. Moser, M. 1958. Die künstliche Mykorrhizaimpfung an Forstpflanzen. I. Erfahrungen bei der Reinkultur von Mykorrhizapilzen. Forstw. Cbl. 77:32-40.
55. Moser, M. 1958. Die künstliche Mykorrhizaimpfung an Forstpflanzen. II. Die Torfstreukultur von Mykorrhizapilzen. Forstw. Cbl. 77:273-278.
56. Moser, M. 1958. Die Einfluss tiefer Temperaturen auf das Washstum und die Lebenstätigkeit höherer Pilze mit spezieller Berücksichtigung von Mykorrhizapilzen. Sydowia 12:386-399.
57. Moser, M. 1963. Die Bedeutung der Mykorrhiza bei Aufforstungen unter besonderer Berücksichtigung von Hochlagen. Pages 407-424, in; Mykorrhiza W. Rawald and H. Lyr, eds., Fischer, Jena. 482 pp.
58. Palmer, J. G. 1971. Techniques and procedures for culturing ectomycorrhizal fungi. Pages 32-36, in; Mycorrhiza, E. Hacskaylo, eds., USDA Forest Serv. Misc. Publ. 1189. 255 pp.
59. Park, J. Y. 1971. Preparation of mycorrhizal grain spawn and its practical feasibility in artificial inoculation. Pages 239-240, in; Mycorrhiza, E. Hacskaylo, ed., USDA Forest Serv. Misc. Publ. 1189. 255 pp.
60. Pryor, L. D. 1956. Chlorosis and lack of vigour in seedlings of Renantherous species of Eucalyptus caused by lack of mycorrhizae. Proc. Linn. Soc. N.S.W. 81:91-96.
61. Reid, C. P. P. and Woods, F. W. 1969. Translocation of ^{14}C-labeled compounds in mycorrhizae and its implications in interplant nutrient cycling. Ecology 50:179-187.
62. Ruehle, J. L. 1980. Growth of containerized loblolly pine with specific ectomycorrhizae after 2 years on an amended borrow pit. Reclam. Rev. 3:95-101.

63. Ruehle, J. L. 1980. Inoculation of containerized loblolly pine seedlings with basidiospores of Pisolithus tinctorius. USDA Forest Serv. Res. Note SE-291. 4 pp.
64. Ruehle, J. L. 1982. Field performance of container-grown loblolly pine seedlings with specific ectomycorrhizae on a reforestation site in South Carolina. South. J. Appl. For. 6: (In Press).
65. Ruehle, J. L. and Marx, D. H. 1977. Developing ectomycorrhizae on containerized pine seedlings. USDA Forest Serv. Res. Note SE-242. 8 pp.
66. Ruehle, J. L., Marx, D. H., Barnett, J. P. and Pawuk, W. H. 1981. Survival and growth of container-grown and bare-root shortleaf pine seedlings with Pisolithus and Thelephora ectomycorrhizae. South. J. Appl. For. 5:20-24.
67. Saleh-Rastin, N. 1976. Salt tolerance of the mycorrhizal fungus Cenococcum graniforme (Sow.). Ferd. Eur. J. For. Pathol. 6:184-187.
68. Schramm, J. R. 1966. Plant colonization studies on black wastes from anthracite mining in Pennsylvania. Trans. Am. Phil. Soc. 56:1-194.
69. Shemakhanova, N. M. 1962. Mycotrophy of woody plants. U.S. Dep. Comm. Transl. TT66-51073 (1967), Washington, D.C. 329 pp.
70. Smith, R. A. 1982. Nutritional study of Pisolithus tinctorius. Mycologia 74: (In Press).
71. Stevens, R. B., ed., 1974. Mycology guidebook. Univ. Washington Press, Seattle. 703 pp.
72. Takacs, E. A. 1961. Inoculación de especies de pinos con hongos formadores de micorrizas. Silvicultura 15:5-17.
73. Takacs, E. A. 1964. Inoculación artificial de pinos de regiones subtropicales con hongos formadores de micorrizas. Idia. Suplemento Forestal 12:41-44.
74. Takacs, E. A. 1967. Producción de cultivos puros de hongos micorrizógenos en el Centro Nacional de Investigaciónes Agropecuarias, Castelar. Idia. Suplemento Forestal 4:83-87.
75. Theodorou, C. 1967. Inoculation with pure cultures of mycorrhizal fungi of radiata pine growing in partially sterilized soil. Aust. For. 31:303-309.
76. Theodorou, C. 1971. Introduction of mycorrhizal fungi into soil by spore inoculation of seed. Aust. For. 35:23-26.
77. Theodorou, C. and Bowen, G. D. 1970. Mycorrhizal responses of radiata pine in experiments with different fungi. Aust. For. 34:183-191.
78. Theodorou, C. and Bowen, G. D. 1973. Inoculation of seeds and soil with Basidiospores of mycorrhizal fungi. Soil Biol. Biochem. 5:765-771.
79. Trappe, J. M. 1962. Fungus associates of ectotrophic mycorrhizae. Bot. Rev. 28:538-606.
80. Trappe, J. M. 1964. Mycorrhizal hosts and distribution of Cenococcum graniforme. Lloydia 27:100-106.
81. Trappe, J. M. 1969. Studies on Cenococcum graniforme. I. An efficient method for isolation of sclerotia. Can. J. Bot. 47:1389-1390.
82. Trappe, J. M. 1971. Mycorrhiza-forming Ascomycetes. Pages 19-37, in; Mycorrhiza, E. Hacskaylo, ed., USDA Forest Serv. Misc. Publ. No. 1189. 255 pp.
83. Trappe, J. M. 1977. Selection of fungi for ectomycorrhizal inoculation in nurseries. Ann. Rev. Phytopathol. 15:203-222.
84. Weir, J. R. 1921. Thelephora terrestris, T. fimbriata, and T. caryophylla on forest tree seedlings. Phytopathology 11:141-144.

ECTOMYCORRHIZAL INOCULATION PROCEDURES FOR GREENHOUSE AND NURSERY STUDIES

Jerry W. Riffle and Dale M. Maronek

INTRODUCTION

Ectomycorrhizal fungi are ubiquitous in forest soils, but are absent or deficient in prairie and other grassland soils which have never supported trees, in new tree nurseries established in areas devoid of adequate wind-dispersed inoculum, in fumigated soils, and in nurseries where heavy fertilization is used to obtain rapid seedling growth (8, 16, 42, 50). Seedling survival on planting sites lacking ectomycorrhizal fungi depends greatly upon adequate ectomycorrhizal development in planting stock. Dramatic improvements in seedling performance have been observed following introduction of ectomycorrhizal fungi to planting stock in Australia, the Russian Steppes, parts of Africa, the Caribbean Islands, and the United States (5).

Ectomycorrhizal fungi show strong physiological variability both within and between species; some species appear more ecologically adapted to certain sites than do other species. To gain maximum benefit from mycorrhizal inoculations, seedlings should be inoculated with fungal symbionts which are best suited to specific tree species, and which are well adapted to the environmental conditions of the planting site.

A discussion of all facets of ectomycorrhizal inoculations is beyond the scope of this paper. The inoculation procedures presented here should be helpful to investigators involved in improving quantity and quality of tree seedlings for use in forestry and horticultural programs. Marx (21), Mikola (34), and Trappe (48) give detailed reviews of ectomycorrhizal inoculations.

INOCULATION TECHNOLOGY

Successful production of ectomycorrhizal seedlings is contingent upon type and age of inoculum used, timing of inoculation, inoculum density, inoculum placement in the growing medium, and a number of host and fungus interactions. Mycorrhizal inocula consisting of soil duff from natural stands and plantations, mycorrhizal seedlings and roots, sporocarps and spores, and pure cultures of ectomycorrhizal fungi have been used to inoculate seedlings. Details on the production of these inocula, and advantages and disadvantages in their use are considered elsewhere in this manual.

There are three times when seedlings can be inoculated with mycorrhizal inoculum: i) before seeds are sown, ii) when seeds are sown, and iii) after seedlings emerge. The most efficient times to inoculate seedlings are before or when seeds are sown. An efficient time to inoculate cuttings is at the time of propagation (19). These inoculation times are based on economic considerations and on the ecology of ectomycorrhizal fungi. From the economic standpoint, inoculation before or when seeds are sown, or when cuttings are propagated, requires the least amount of inoculum per volume of growing medium. Additionally, the newly developing rootlets of seedlings or cuttings are receptive to mycorrhizal infections because they initially are nonlignified. Also, mycorrhizal fungi have increased the rooting of cuttings and increased root development during propagation (11). Finally, it is more cost efficient to develop a mechanized inoculation system for use when seeds are sown or when cuttings are propagated than at other stages of plant production. Inoculation of seedlings at the time of field planting is time consuming, requires more inoculum, and the introduced fungus must be compatible with native microorganisms and climatic conditions of the planting site.

A major goal of ectomycorrhizal inoculation programs is to develop efficient techniques to minimize the amount of inoculum needed to obtain a required percentage of ectomycorrhizae on roots of seedlings. Two common inoculation methods are surface broadcasting of inoculum followed by incorporation into the root zone, and banding, or layering of inoculum in the root zone. Broadcasting and incorporation of inoculum require large quantities of inoculum for uniform coverage of the surface of seedbeds. In contrast, the layer technique concentrates small quantities of inoculum near developing root systems of seedlings.

ECTOMYCORRHIZAL INOCULATION PROCEDURES

Broadcast Inoculations

Broadcast inoculation involves spreading a known quantity of mycorrhizal inoculum over a given area of soil surface and then mixing the inoculum into the soil to a depth of 10 to 20 cm before seeding. Several inocula, including pine duff, sporocarps and spores, and pure culture vegetative mycelium have been applied in this manner to obtain development of ectomycorrhizae on seedlings in nursery beds. These inocula may also be broadcast and mixed into container growing medium.

Duff inoculum consisting of mycorrhizal tree roots and fungus propagules was effective in promoting development of ectomycorrhizae on pine seedlings in Great Plains nurseries during the 1930's and 1940's (33, 39). The inoculum was broadcast at rates of 2 to 4 kg per m^2 on nursery beds and plowed or disked into the upper 10 cm of soil prior to seeding. Application of this inoculum was simple, and the inoculation was most effective when combined with soil acidification and fertilization. Similarly, pine duff (10 and 20% by volume) mixed in a rotary mixer with moistened 1:1 vermiculite-peat growing medium before seeding was successful for development of ectomycorrhizae on container-grown *Pinus ponderosa* and *P. sylvestris* seedlings (40).

Pure culture inoculations have been used when a desirable fungus species is lacking or when endemic nursery fungi are not well adapted to planting sites. In 1958 Moser (38) obtained best mycorrhizal development on *Pinus cembra* with *Suillus plorans* by using 8 to 10 l of inoculum per m^2 of soil surface mixed 10 cm deep in the soil before transplanting seedlings. Since then investigators have successfully used similar inoculation techniques with vermiculite-based vegetative inoculum of ectomycorrhizal fungi for development of mycorrhizae on tree seedlings in nurseries (20, 27, 31, 48, 49, Marx et al, unpublished data).

Laboratory and commerical vegetative inocula of *Pisolithus tinctorius* have been broadcast evenly over the surface of fumigated nursery soil and mixed into the upper 20 cm of soil for development of ectomycorrhizae on *Quercus rubra* and conifer species in 33 nurseries located in 25 states during a four year period (Marx et al., unpublished data). Different rates (0.27, 0.54, 1.08, 1.62, and 2.16 l per m^2) of the vermiculite-based inoculum were used. Inoculum broadcast at a rate of 1.08 l per m^2 (100 ml per ft^2) of soil surface was as effective in forming *P. tinctorius* ectomycorrhizae as higher rates.

Laboratory and commercial vegetative inocula of *P. tinctorius* were mixed at rates of 3, 6, and 12 percent by volume in various growing media before sowing with *Quercus macrocarpa* or numerous conifer species in 21 container tests conducted at seven research laboratories (32). Inoculum at the six or 12 percent levels was effective in forming ectomycorrhizae on the tree species. Container tree seedlings produced in various growing media have been effectively inoculated with other ectomycorrhizal fungi by use of similar procedures (37, 40, 43).

Banding of Inoculum Below Seeds

Inoculum placement below the seed in a layer or band is an effective and efficient inoculation method. Most types of mycorrhizal inoculum can be applied using the banding technique which facilitates concentration of the inoculum near developing roots. A banding technique for inoculation of nursery seedbeds with commercially produced vegetative inoculum of *P. tinctorius* has been developed by scientists of USDA Forest Service (2). The technique involves simultaneous application of inoculum and seeds with a modified nursery tree seed planter (Plate 8A). The modified seeder was successfully tested in four southern nurseries in 1980; row injected inoculum was effective in forming *P. tinctorius* ectomycorrhizae on *Pinus taeda* seedlings. A major advantage of the banding inoculation technique is that it requires only one-third as much inoculum as the broadcast method. Other advantages include substantial savings in time and labor. The need for an additional machine is a disadvantage of the technique; however, most nursery seeders can be modified at moderate cost to apply mycorrhizal inoculum. Use of this machine applicator represents a practical and economic method of mycorrhizal inoculation in bareroot nursery seedling production.

For inoculation of container seedlings, the technique of banding mycorrhizal inoculum is more effective than the technique of incorporating inoculum throughout the container growing media (18). Inoculation of container-grown seedlings has been improved by a banding unit which inoculates a large number of seedlings in one operation (Plate 8B and 8C). For a 65 ml volume container, banding 10 ml of vegetative inoculum in the center of the container has been adequate for production of mycorrhizal seedlings (D. Maronek and D. Crowley, unpublished data).

Slurry Dips

Slurries of various forms of mycorrhizal inoculum can be prepared by simply mixing mycorrhizal inoculum with water and, if needed, a carrier such as clay or soil. Barerooted or container-grown seedlings are inoculated by dipping them into the slurry before planting (20).

A technique has been developed to assess the mycorrhizal development potential of laboratory and commercially produced inoculum of *P. tinctorius* applied to *Pinus taeda* seedlings by the inoculum slurry-root dip technique (32). Ten-week-old seedlings, initially grown in a mycorrhizal fungus-free growth room (24), are immersed in a 1:1 inoculum-water slurry for 5 seconds to cover the root system with inoculum. The slurry is drained from roots for a few seconds, and then seedlings are transplanted in containers with a 1:1 vermiculite-peat growing medium. The seedlings are grown in a greenhouse under a 16-hour photoperiod, and after four weeks, root systems are assessed for ectomycorrhizal development.

Basidiospore Inoculations

Successful production of mycorrhizal seedlings by basidiospore inoculation has been reported for many tree species (10, 20, 30, 41, 45, 47). Several inoculation techniques have been used, such as suspending spores in water and leaching into nursery soil (20), mixing dry spores directly into soil or container medium (10, 20, 41), coating tree seeds with spores prior to sowing (46, 47), dusting spores on roots of nonmycorrhizal seedlings (20), or mixing spores in a carrier and applying them after sowing seeds (30). Vermiculite, kaolin, sand, hydromulch, or water may be used as physical carriers of basidiospores for soil infestation. Spores applied to soil or to container growing medium should be thoroughly incorporated into the medium. Incorporation can be done by irrigation or by mechanical means. However, incorporation by irrigation seldom results in even distribution of spores throughout the root zone of seedlings.

A specially designed chamber and compressed air have been used to deposit spores of *P. tinctorius* uniformly on the surface of a container growing medium (41). Five-day-old *P. taeda* seedlings were inoculated with spores 0, 2, 4, 6, or 8 weeks after transplanting; the growing medium was misted with water for 2 minutes to wash spores into the medium. Seedlings inoculated with spores 0 or 2 weeks after transplanting had significantly more *P. tinctorius* ectomycorrhizae than those inoculated later. Naturally occurring ectomycorrhizal fungi from wind-dispersed spores colonized all seedlings and were probably responsible for a lower rate of *Pisolithus* colonization on seedlings inoculated more than 2 weeks after transplanting. Ruehle (41) suggests that spore inoculation should be done at time of transplanting to develop ectomycorrhizae on the greatest number of seedlings.

Five methods of introducing basidiospores of *P. tinctorius* into fumigated soil were tested in an Oklahoma nursery to determine their effectiveness in forming ectomycorrhizae on *P. taeda* seedlings (30). The methods were: i) basidiospores mixed into hydromulch and applied after sowing, ii) basidiospores dusted onto soil after sowing, iii) basidiospores injected into soil before sowing, iv) basidiospores dusted onto seedlings 6 weeks after sowing, and v) basidiospores drenched onto seedlings 6 weeks after sowing. The first two methods were most effective. About 75 percent of the seedlings treated with spores mixed in hydromulch and applied after sowing formed *Pisolithus* ectomycorrhizae. When spores were dusted onto the soil after sowing, one-third of the seedlings formed *Pisolithus* ectomycorrhizae.

Basidiospores as inoculum remains advantageous, but there is need for more information on spore transport and storage, on spore viability, on spore inoculum concentrations, and on application methods (48).

Pelletizing Seed

Introduction of ectomycorrhizal fungi into soil or container growing media on seeds coated with spores has been investigated (44, 46, 47). A technique is being developed to incorporate basidiospores of *P. tinctorius* in the external matrix of encapsulated pine seed (26). Abundant *P. tinctorius* ectomycorrhizae were formed on pine seedlings grown in containers and in a bareroot nursery when seeds encapsulated with basidiospores at 1, 2, or 3 mg per seed were used. In some tests the degree of *P. tinctorius* ectomycorrhizal development on pine seedlings from encapsulated seeds was identical to that formed with vegetative ectomycorrhizal inoculum. In a bareroot nursery test with *Pinus taeda*, the fungicides captan, arasan, and benlate had no effect on development of *P. tinctorius* ectomycorrhizae on seedlings when mixed with spore encapsulated seeds prior to sowing (26). Such results are significant because these fungicides are used in bareroot nurseries to control damping-off pathogens.

Ectomycorrhizal Seedlings and Roots

Transplanting mycorrhizal seedlings in nursery beds has been a successful inoculation method where application of soil duff mixtures has not been effective (35). This inoculation method was first used in Indonesia and was reliable for production of Merkus pine (*Pinus merkusii*). Mycorrhizal seedlings are planted at 1- to 2-m intervals, and from there mycelia spread to adjacent seedlings. At time of lifting, some seedlings are left in beds to serve as inoculum for the next crop.

Incorporation of mycorrhizal roots into nursery beds has also been successfully used for inoculation of seedlings (9). This inoculation procedure is nonselective for fungus symbionts. Roots with abundant mycorrhizae should be selected as inoculum, and should be incorporated while fresh and before sowing seeds.

Other Inoculation Procedure

Some ectomycorrhizal fungi grow slowly in pure culture and appear unsuitable for artificial inoculation of seedlings in nurseries (31). Recent findings indicate that *Cenococcum geophilum*, a slow growing fungus, is adaptable to inoculation of container *Pseudotsuga menziesii* and *Picea sitchensis* (4, 37, 43). In one study *P. menziesii* was inoculated with three isolates of *C. geophilum* at time of seeding and at 1, 2, and 3 months after emergence (4). For post-emergence inoculations, folding book containers were carefully opened to expose surfaces of the root ball, and 25 ml of vegetative inoculum per seedling was spread evenly over each side of the root ball. Inoculation of 1-, 2-, and 3-mo-old seedlings resulted in more ectomycorrhizae than inoculation at time of seeding. The investigators suggest that delayed inoculation maximizes the interaction between fresh, viable inoculum and short roots. When short roots were available, fresh inocula of all the isolates were equally effective in mycorrhizal formation.

FACTORS INFLUENCING ECTOMYCORRHIZAL INOCULATIONS

Simple addition of mycorrhizal inoculum to growing medium is in itself no guarantee that ectomycorrhizae will develop on a host plant. Interactions between the fungus and host in growing medium are complex and influenced by a number of interrelated biochemical, physiological, and environmental processes (20). Hence the mycorrhizal formation process may be affected by a number of factors, including length of time fungal isolates have been in culture, age of vegetative inoculum, inoculation method, and the host species (14, 16, 17, 22). Significant interaction between age of vegetative inoculum and inoculation method has been reported (18).

A number of cultural practices, environmental conditions (moisture, mineral nutrients, pollutants), and biological factors (interaction with other soil organisms) affect short root development and ectomycorrhizal formation. Since mycorrhizal fungi exhibit ecological selectivity (14, 16, 23, 48), production cultural practices may need to be adjusted to provide an environment conducive for fungal host interactions. For example, most mycorrhizal fungi are adapted to low fertility levels. Current production practices often involve heavy fertilization which may inhibit mycorrhizal development (3, 29). The rate of fertilizer release may be critical in controlling mycorrhizal formation. Ectomycorrhizal plants have been produced at an adequate or accelerated growth rate by use of slow release fertilizer (12, 13, 14, 15, 17). Studies have shown that slow release fertilizer (4.5 kg per m^3) applied at recommended rates was not inhibitory to formation of ectomycorrhizae by *P. tinctorius*. Poor mycorrhizal development may not always indicate nutritionally poor soils; it may also indicate an inbalance of nutrients. Marx et al (unpublished data) analyzed 20 nursery cultural practices and soil conditions in 19 bareroot nurseries to see how they related to ectomycorrhizal formation on 1-yr-old *Pinus taeda* seedlings inoculated with laboratory produced inoculum of *P. tinctorius*. No single practice or condition accounted for significant differences in ectomycorrhizal development.

The effect of photoperiod and light intensity on mycorrhizal formation is unclear. Although light is essential for carbohydrate production and plant carbohydrate levels are known to influence mycorrhizal formation, there is conflicting evidence regarding the light intensity needed for mycorrhizal formation. Mycorrhizal formation increases at both low and high light intensities (1, 7). These variations may be related to host fungus specificity or to a photoperiodic response. Additional work is needed to determine the association between fungal-plant carbohydrate relationships and subsequent mycorrhizal formation.

Mycorrhizal formation is influenced by temperature. Most mycorrhizal fungi have an optimum temperature for establishment of the symbiotic relationship and for survival of the mycorrhizal condition. There is considerable variation in the temperature-range tolerance of individual fungi. Marx et al (28) found that *Thelephora terrestris* had a considerably lower tolerance to a wide temperature range than *P. tinctorius*. *Thelephora terrestris* formed ectomycorrhizae on 45 percent of the feeder roots of *Pinus taeda* at 14, 19 and 24° C, but formed only 30 percent at 29° C and none at all at 34° C. In contrast, *P. tinctorius* formed mycorrhizae in increasing numbers up to a maximum of 80 percent at 34° C. This fungus has a high survival rate even at 40° C (25).

Mycorrhizal development in inoculated nursery soils and container growing medium may be suppressed or prevented by microorganisms (fungi and plant-parasitic nematodes) present in these substrates. Sterilization of the soil or container medium with fumigants or with low temperature aerated steaming may be used to eliminate feeder root pathogens that normally damage or destroy sites for mycorrhizal development. Autoclaving and high temperature aerated steaming should be used with caution because of the possibility of toxin formation (51). Vegetative mycelia of *P. tinctorius*, *T. terrestris*, and *C. geophilum*, and basidiospores of *P. tinctorius* are more effective in forming ectomycorrhizae on roots of *P. taeda* in fumigated soil than in nonfumigated soil (27, 31). Similarly, *Rhizopogon luteolus* introduced on spore-inoculated *Pinus radiata* seed promote greater seedling height growth and mycorrhizal infection in methyl bromide fumigated soil than in nonfumigated soil (46). Stimulation of seedling growth in fumigated soil apparently is a nutrient response; inoculation of fumigated soil promotes mycorrhizal development which in turn increases seedling efficiency in absorption of nutrients released by fumigation (46).

Production of mycorrhizal seedlings with specific fungal species can be affected by competing ectomycorrhizal fungi. Greenhouse- and field-grown plants often become mycorrhizal by natural dispersal of spores. Such dispersal usually occurs in areas where native ectomycorrhizal host plants are found. For instance, in the South most nursery seedlings are naturally infected by spores of *T. terrestris*. A special growth room has been constructed to remove spores by filtration because *T. terrestris* rapidly colonizes many conifer root systems, and diminishes the effectiveness of other slower growing ectomycorrhizal fungi used in inoculation experiments (24).

Contamination of experimental seedlings can also occur as a result of poor sanitation and insect control practices. Many ectomycorrhizal fungi will produce sporocarps after development of ectomycorrhizae on seedlings in the greenhouse (Plate 8D). Unharvested sporocarps can mature, and their spores can be dispersed by air, insects, and splashing water.

SUGGESTIONS

Research on ectomycorrhizal fungi is intensifying, but there is little information on many fungus species. Roots of ectomycorrhizal trees support a large number of fungal symbionts, but only a few symbionts have been cultured and used as vegetative inoculum. Additional information is needed to grow symbionts more effectively in culture. Use of vegetative mycelium offers the opportunity to select proven, highly beneficial fungi which are well adapted to particular planting sites. To fully realize the potential of pure cultue inoculations, further information is needed on the biology and ecology of ectomycorrhizal fungi, and on the benefits specific fungus species and ecotypes lend to their hosts.

Most ectomycorrhizal research for the immediate future will necessarily involve fungi that are easily cultured. Use of these fungi will facilitate the development of inoculation technology. However, it will be essential to identify the mycorrhizal fungi in specific crop systems and to determine how specific fungi benefit a particular crop. In so doing, researchers must pay strict attention to experimental methods. Ultimately, field testing of mycorrhizal seedlings will be necessary to determine if a fungus can produce a desired effect. Several researchers (17, 23, 36, 48) have already emphasized the need to examine fungal species and ecotypes of symbionts for their suitability for nursery and container inoculation, and for their performance in field plantings.

Standardization of procedures whenever possible and initial avoidance of large multivariate studies are desirable because of possible inherent variability among fungal-host associations. Initial work should deal with one or two variables. Because adjustment of fertility regimes may be essential to produce mycorrhizal seedlings, efficient nutrient control systems must be developed. Nutrient concentration should be standardized to express the ions actually absorbed by plants. This standardization could be accomplished by expressing nutritional data in milliequivalents per liter. This method would take into account the types of ions in the growing medium and the types absorbed by plants and fungi. Nutrient concentrations could be readily determined regardless of the situation (6).

Many ectomycorrhizal inoculation procedures have been developed and are available for inoculation of plants. The information presented in this paper is intended to serve only as a guide to ectomycorrhizal inoculation technology. Because the practical use of ectomycorrhizal fungi can be of major significance in forest regeneration and horticultural programs, existing inoculation methodology will likely be improved and additional techniques will likely be developed.

LITERATURE CITED

1. Bjorkman, E. 1970. Mycorrhiza and tree nutrition in poor forest soils. Stud. For. Suec. 83:1-24.
2. Cordell, C. E., Conn, J. P., Marx, D. H. and Kenney, D. S. 1981. Practical machine application of ectomycorrhizal fungus inoculum in forest tree nurseries. Page 74, in: Program and Abstracts, 5th No. Amer. Conf. Mycorrhizae, Laval Univ., Quebec, Canada. 83 pp.
3. Crowley, D. E., Maronek, D. M. and Hendrix, J. W. 1981. Fertility mycorrhizal interactions on container-grown *Pinus echinata*. Hort. Sci. 16:430.
4. Graham, J. H. and Linderman, R. G. 1981. Inoculation of containerized Douglas-fir with the ectomycorrhizal fungus *Cenococcum geophilum*. Forest Sci. 27:27-31.
5. Hacskaylo, E. 1972. Mycorrhizae: the ultimate in reciprocal parasitism. Bioscience 22:577-583.
6. Hanan, J. J., Holley, W. D. and Goldsberry, K. L. 1978. Greenhouse management. Springer Verlag, New York. 530 pp.
7. Harley, J. L. and Waid. J. S. 1955. The effect of light on the roots of beech and its surface population. Plant and Soil 7:96-112.
8. Hatch, A. B. 1937. The physical basis of mycotrophy in *Pinus*. Black Rock Forest Bull. 6:1-168.
9. Hernandez Gil, R. and Garcia, F. F. 1976. Ensayo de inoculacion con micorrizas en *Pinus pseudostrobus* Lindl. Revista Forestal Venezolana 17:61-66.
10. Hodson, T. J. 1979. Basidiospore inoculation of soil: the effect of application timing on *Pinus elliottii* seedling development. South Afr. For. J. 108:10-15.
11. Linderman, R. G. and Call, G. A. 1977. Enhanced rooting of woody plant cuttings by mycorrhizal fungi. J. Amer. Soc. Hort. Sci. 102:529-532.
12. Maronek, D. M. 1977. Mycorrhizae and plant growth. Proc. Int. Plant Prop. Soc. 27:382-388.
13. Maronek, D. M. and Hendrix, J. W. 1979. Slow release fertilizer for optimizing mycorrhizal production in pine seedlings of *Pisolithus tinctorius*. in: Program and Abstracts, 4th No. Amer. Conf. Mycorrhizae, Colorado State Univ., Fort Collins, Colo.
14. Maronek, D. M. and Hendrix, J. W. 1979. Growth acceleration of pin oak seedlings with a mycorrhizal fungus. Hort. Sci. 4:627-628.
15. Maronek, D. M. and Hendrix, J. W. 1980. Synthesis of *Pisolithus tinctorius* ectomycorrhizae on seedlings of four woody species. J. Amer. Soc. Hort. Sci. 105:823-825.
16. Maronek, D. M., Hendrix, J. W. and Kierman, J. 1981. Mycorrhizal fungi and their importance in horticultural crop production. Hort. Rev. 3:172-213.
17. Maronek, D. M., Hendrix, J. W. and Stevens, C. D. .1981. Fertility-mycorrhizal interactions in production of containerized pin oak seedlings. Scientia Hortic. 15:283-289.

18. Maronek, D. M., Crowley, D. E. and Hendrix, J. W. 1981. Inoculum banding, a new technique for production of ectomycorrhizal containerized seedlings. Hort. Sci. 16:429. (Abstract 224).

19. Maronek, D. M., Hendrix, J. W. and Kierman, J. 1981. Adjusting nursery practices for production of mycorrhizal seedlings during propagation. Proc. Int. Plant Prop. Soc. 31: (in press).

20. Marx, D. H. 1976. Synthesis of ectomycorrhizae on loblolly pine seedlings with basidiospores of *Pisolithus tinctorius*. Forest Sci. 22:13-20.

21. Marx, D. H. 1981. Ectomycorrhizal fungus inoculations: a tool for improving forestation practices. Pages 13-71. in: Tropical Mycorrhizae Research. P. Mikola, ed. Clarendon Press, Oxford, London. 270 pp.

22. Marx, D. H. 1981. Variability in ectomycorrhizal development and growth among isolates of *Pisolithus tinctorius* as affected by source, age, and reisolation. Can. J. For. Res. 11:168-174.

23. Marx, D. H. and Artman, J. D. 1979. The significance of *Pisolithus tinctorius* ectomycorrhizae to survival and growth of pine seedlings on coal spoils in Kentucky and Virginia. Reclamation Rev. 2:23-31.

24. Marx, D. H. and Bryan, W. C. 1969. Studies on ectomycorrhizae of pine in an electronically air-filtered, air conditioned, plant growth room. Can. J. Bot. 27:1903-1909.

25. Marx, D. H. and Bryan, W. C. 1971. Influence of ectomycorrhizae on survival and growth of aseptic seedlings of loblolly pine at high temperatures. Forest Sci. 17:37-41.

26. Marx, D. H., Bell, R. W. and Jarl, K. 1981. Use of seed encapsulated with basidiospores of *Pisolithus tinctorius* to form ectomycorrhizae on pine. Page 38. in: Program and Abstracts, 5th No. Amer. Conf. Mycorrhizae, Laval Univ., Quebec, Canada. 83 pp.

27. Marx, D. H., Bryan, W. C. and Cordell, C. E. 1976. Growth and ectomycorrhizal development of pine seedlings in nursery soils infested with the fungal symbiont *Pisolithus tinctorius*. Forest Sci. 22:91-100.

28. Marx, D. H., Bryan, W. C. and Davey, C. B. 1970. Influence of temperature on aseptic synthesis of ectomycorrhizae by *Thelephora terrestris* and *Pisolithus tinctorius* on loblolly pine. Forest Sci. 16:424-431.

29. Marx, D. H., Hatch, A. B. and Mendicino, J. F. 1977. High soil fertility decreases sucrose content and susceptibility of loblolly pine roots to ectomycorrhizal infection by *Pisolithus tinctorius*. Can. J. Bot. 55: 1569-1574.

30. Marx, D. H., Mexal, J. G. and Morris. W. G. 1979. Inoculation of nursery seedbeds with *Pisolithus tinctorius* spores mixed with hydromulch increases ectomycorrhizae and growth of loblolly pines. So. J. Appl. For. 3:175-178.

31. Marx, D. H., Morris, W. G. and Mexal, J. G. 1978. Growth and ectomycorrhizal development of loblolly pine seedlings in fumigated and nonfumigated nursery soil infested with different fungal symbionts. Forest Sci. 24: 193-203.

32. Marx, D. H., Ruehle, J. L., Kenney, D. S., Cordell, C. E., Riffle, J. W., Molina, R. J., Pawuk, W. H., Navratil, S., Tinus, R. W. and Goodwin, O. C. 1982. Commercial inoculum of *Pisolithus tinctorius* for ectomycorrhizal development on containerized tree seedlings. Forest Sci. 28: (in press).

33. McComb, A. L. 1938. The relation between mycorrhizae and the development and nutrient absorption of pine seedlings in a prairie nursery. J. For. 36: 1148-1154.

34. Mikola, P. 1970. Mycorrhizal inoculation in afforestation. Int. Rev. For. Res. 3:123-196.

35. Mikola, P. 1973. Application of mycorrhizal symbiosis in forestry practice. Pages 383-411. in: Ectomycorrhizae Their Ecology and Physiology. Marks, G. C. and T. T. Kozlowski. eds. Academic Press, N.Y. 444 pp.

36. Molina, R. 1979. Ectomycorrhizal inoculation of containerized Douglas-fir and lodgepole pine seedlings with six isolates of Pisolithus tinctorius. Forest Sci. 25:585-590.
37. Molina, R. 1980. Ectomycorrhizal inoculation of containerized western conifer seedlings. USDA Forest Service Research Note PNW-357. 10 pp.
38. Moser, M. 1958. Die künstliche Mykorrhizaimpfung an Forstpflanzen. I. Erfahrungen bei der Reinkulture von Mykorrhizapilzen. Forstw. Chl. 77:32-40.
39. Riffle, J. W. 1977. Ectomycorrhizal inoculation of nursery seedbeds and container growing media. Proc. Intermountain Nurserymen's Assoc. 1977:73-85.
40. Riffle, J. W. and Tinus, R. W. 1982. Ectomycorrhizal characteristics, growth and survival of artificially inoculated ponderosa and Scots pine in a greenhouse and plantation. Forest Sci. 28: (in press).
41. Ruehle, J. L. 1980. Inoculation of containerized loblolly pine seedlings with basidiospores of Pisolithus tinctorius. USDA Forest Service Research Note SE-291. 4 pp.
42. Ruehle, J. L. and Marx, D. H. 1979. Fiber, food, fuel and fungal symbionts. Science 206:419-422.
43. Shaw, C. G. III and Molina, R. 1980. Formation of ectomycorrhizae following inoculation of containerized Sitka spruce seedlings. USDA Forest Service Research Note PNW-351. 8 pp.
44. Shemakhanova, W. M. 1962. Mycotrophy of woody plants. Washington, D.C. U.S. Dept. Commer.Transl. TT66-51073 (1967). 329 pp.
45. Sinclair, W. A., Cowles, D. P. and Hee, S. M. 1975. Fusarium root rot of Douglas-fir seedlings: Suppression by soil fumigation, fertility management, and inoculation with spores of the fungal symbiont Laccaria laccata. Forest Sci. 21:390-399.
46. Theodorou, C. 1971. Introduction of mycorrhizal fungi into soil by spore inoculation of seed. Aust. For. 35:23-26.
47. Theodorou, C. and Bowen, G. D. 1973. Inoculation of seeds and soil with basidiospores of mycorrhizal fungi. Soil Biol. and Biochem. 5:765-771.
48. Trappe, J. M. 1977. Selection of fungi for ectomycorrhizal inoculation in nurseries. Annu. Rev. Phytopathol. 15:203-222.
49. Vozzo, J. A., and Hacskaylo, E. 1971. Inoculation of Pinus caribaea with ectomycorrhizal fungi in Puerto Rico. Forest Sci. 17:239-245.
50. Wilde, S. A. 1965. Mycorrhizal fungi: their distribution and effect on tree growth. Soil Sci. 78:23-31.
51. Zak, B. 1971. Detoxication of autoclaved soil by a mycorrhizal fungus. USDA Forest Service Research Note PNW-159. 4 pp.

QUANTITATIVE MEASUREMENT OF ECTOMYCORRHIZAE ON PLANT ROOTS

L. F. Grand and A. E. Harvey

INTRODUCTION

Early research with ectomycorrhizae, as is true of the history of most areas of life sciences, emphasized the descriptive biology of organisms and their interactions. However, as research elucidated the role and function of ectomycorrhizae it became clear that quantification would be necessary to establish the contributions of the symbionts to plant and/or ecosystem development. Simple means of estimating numbers of ectomycorrhizae soon evolved into standardized (and repeatable) procedures then into complex but reliable measurements.

The Need

Reliable quantification of ectomycorrhizae is required for most studies in this field. As research delved into the effects of ectomycorrhizae on plant growth, nutrient uptake and resistance to root pathogens, researchers needed to determine statistical differences in quantity of ectomycorrhizae between their own experimental treatments and for biological and interpretive comparison with results from other studies.

Forest tree nursery workers have recognized the importance of a good ectomycorrhizal root system on conifer seedlings to insure successful nursery culture (48) and in some cases for survival of outplanted stock (38). This requires a quantitative evaluation of ectomycorrhizal development present on the root systems.

More recently, interest in the ecology of forest ecosystems (5,6,10-14,17,19, 35,36,42,49,50) required quantification of ectomycorrhizae to understand contributions to biomass production, nutrient recycling and the effects of fire, harvesting practices and other site disturbances on soil-ectomycorrhizal relationships.

The Problem

Sampling.--The choice of a sampling technique is usually dictated by the size and location of the plant(s) under study. Entire root systems can be examined on seedlings; however, this is impossible for larger plants. The number and size of samples (root or soil) required for valid statistical and biological comparisons can vary considerably. Most often this must be determined for each study.

Preparing Root Samples.--Perhaps one of the most critical phases of quantification of ectomycorrhizae is preparation of the roots for examination. Experience has shown that the cleaner the material the faster and more accurate the quantification. Difficulty arises in separation of organic debris from the roots without major damage to the ectomycorrhizae. Improper separation may mask or damage ectomycorrhizae and lead to erroneous data.

Similarly, roots growing in a soil high in clay content are often difficult to clean without serious damage to the ectomycorrhizae. Adhering soil may also make it difficult to distinquish important morphological features, especially color and surface features. This is particularly important when the quantification of different types of ectomycorrhizae is desired.

Timing of Sampling.--Populations of ectomycorrhizae can vary considerably during seasons of the year. Several studies have shown a distinct change in numbers of ectomycorrhizae from spring to fall (6,10-12,19,35,36,50). Consideration must be given to the objective(s) of the study to determine what time of the year samples should be taken and if more than one sample period should be included in the study.

Rooting Patterns.--The nature of the rooting patterns of the plant species under investigation must be taken into consideration when sampling. While most studies indicate that the majority of roots on which ectomycorrhizae are found occur in the upper 15-30 cm of the soil profile, this situation may vary with the plant species and soil type (5,13,19). Plants growing in light soils frequently have roots which are able to penetrate much deeper in the soil profile than the same species growing in a heavy soil.

On certain sites, roots can occur as a thick mat so tangled and interwoven as to preclude the separation required to insure accurate counts with most estimation procedures.

Contaminating Roots.--With a little experience, the recognization of roots from plants of different species is usually not a problem (18,50). Occasionally roots from understory vegetation, often unfamiliar to the researcher, can cause confusion.

Differing Morphological Forms.--The variation in branching habit of ectomycorrhizae is considerable. Branching can vary from simple monopodial to coralloid, with a number of intermediate forms (Plate 7, A-D). Any single type may or may not be a single fungus-host combination. This can further confuse the issue, particularly if one is interested in a particular combination. Even if only the number of ectomycorrhizal short roots and not the individual identity is desired (a commonly used procedure), Fig. 2B, C, D, E would receive the same value though Fig. 2E obviously has more volume, surface area, ectomycorrhizal tips and presumably, more weight than Fig. 2B, C, D.

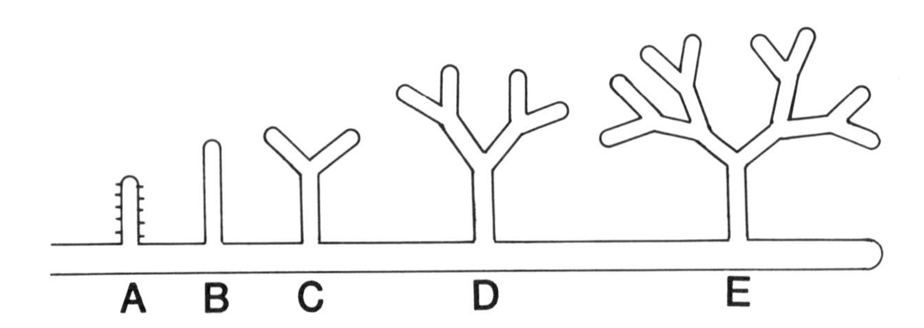

Fig. 2. Schematic diagram of a nonmycorrhizal short root (A) and ectomycorrhizae (B-E).

Further confusing this matter are the tuberculate forms (Plate 7, D) reported in the literature (3,7,47,52). Presumably these result from the transformation of a single short root to a single tubercule that may contain up to 150 ectomycorrhizal tips.

Choosing a Quantification Method.--The nature of the research, in many instances, dictates the quantification procedure. Studies that require large numbers of samples can seldom "afford" the time (and expense) necessary to use the most detailed system of estimation. In contrast, studies involving a limited number of samples frequently do use a detailed system of estimation.

In the final analysis, the choice of a quantification procedure may be determined by the objective(s) of the study, the time factor, the sampling procedure, the preparation procedure, plant species, rooting habit, predominant ectomycorrhizal types, and, although we might not like to admit it, the cost of the study. It is no easy choice!

The Approach

Sample Types.--Studies of ectomycorrhizae involving quantification can be conveniently separated into two sample categories; those based on the entire root system (or major parts thereof) and those based on soil cores. Sampling procecures using root systems almost exclusively involve pure culture synthesis of ectomycorrhizae and/or seedling studies. Sampling procedures using soil cores are used chiefly for studies involving large plants in field situations.

Root Systems.--Entire plants (seedlings) are carefully removed from containers (pure culture synthesis) or from soil (pots or nursery beds) and placed in a suitable container until processed, or into a container of water if examination is done within a short period of time.

Seedlings grown in nursery beds are often undercut at a certain depth (10-25 cm) prior to removal by hand (28,30-33). This procedure results in the loss of some roots from the root system; however if all plants in the study are sampled in the same manner, the comparison between or among treatments remains valid.

Soil Cores.--The type of device used to take soil cores is usually a standard metal soil sampler (35,36,50) or a modification of one (5,10-14). However, in cases where soils contain large rocks or roots, specific designs of samplers may be required (17). In studies requiring larger samples, cubes of soil have been used (19).

The size of the core and depth to which the sample is taken can vary substantially depending, to a large degree, on the nature of the soil and the rooting habit of the plant(s) under investigation. The diameters of soil cores reported in the literature ranged from 1.2 cm to 10 cm and the depth of sampling ranged from 7.5 cm to 90 cm. Several studies (5,10,12-14,19,20,50), which included root and ectomycorrhizae distribution in the soil profile, sampled at depths of 10, 30, 50, 70, and 90 cm. The majority of these studies sampled to a depth of 30 cm. Fogel (5) has presented a summary of the distribution of ectomycorrhizae in various soil profiles.

Number of Samples.--The number of samples necessary to provide sufficient material for accurate estimates of ectomycorrhizae is dependent on the objective(s) of the research, size of the study plots, distribution of plants, size of the soil core, and other considerations. These factors, together with the statistical design of the study, determine the "optimum" number of samples to be taken.

Preparation of Samples.--Studies involving ectomycorrhizae on seedlings generally use water to remove soil or potting mixtures from the roots (21-23,26-28,39,42,45). Although most studies indicated roots were gently washed in water, one study (29) indicated vigorous agitation for up to one hour was needed. The duration and vigor of washing should be determined for each plant species being studied; unnecessary breakage of ectomycorrhizae must be avoided. In studies where vermiculite is used in the potting mixture, the expanded mica particles are often difficult to remove from the ectomycorrhizae and they may require removal by hand under a dissecting microscope. Presoaking soil cores in water or deflocculating agents such as polyoxyethylene sorbitan monolaurate, sodium hexametaphosphate or sodium pyrophosphate (6) in water may be necessary to facilitate removal of soil and/or debris from ectomycorrhizal roots. Generally, soil cores are presoaked at least 12 hours prior to further preparation. Soils high in clay content may require even longer periods of presoaking (up to 36 hours) to facilitate separation (6,19,35,36). Preparation by ultrasonics has also been successfully used (40).

Pretreated or nontreated soil cores are placed onto a soil sieve, or nested sieves, after removal of litter, large roots, stones, or organic debris and cleaned in running tap water or distilled water (6,10-14). Harvey, *et al.* (10-14) shook each sample for five minutes in a 2-mm sieve and gently crumbled various aggregates prior to washing. Nylon mesh screen has been used in place of soil sieves in several studies (35,36,50) and appears to work satisfactorily. Most studies collected ectomycorrhizae and tips of ectomycorrhizae on either 2-mm sieves or 1- or 2-mm nylon mesh screens (6,10-14,19,20,35,36,50). Large roots with secondary roots having ectomycorrhizae were removed by hand prior to sieving. When nested sieves were used, soil cores were placed directly into sieves of a larger mesh size. Large roots and root segments with ectomycorrhizae were removed by hand and ectomycorrhizae and tips of ectomycorrhizae were collected on a second sieve (10-14).

Menge, et al. (35,36) combined several steps in the cleaning procedure. Soil cores were placed into a wide mouth jar fitted with 1-mm mesh screen, soaked in water overnight, and rinsed vigorously under running tap water directly through the screen until roots and soil separated. The contents of the jars, fragments of roots and ectomycorrhizae, were swirled and poured onto a 0.147-mm sieve.

Following the cleaning procedure(s), material captured on the small mesh screens or sieves is placed in appropriate containers, usually Petri plates, in water and the ectomycorrhizae or fragments of ectomycorrhizae examined with a dissecting scope at 4-40X.

Estimation Methods

Visual Estimates.--Most early workers used visual estimations to report quantities of ectomycorrhizae. These were subjectively determined using a rating system. Some of the most subjective systems were those which rated ectomycorrhizae as "absent", "rare", "frequent", or "very common" or -(absent), +(present), ++(present in small numbers), +++(present in large numbers), and ++++(mass incidence) (3,4,37). Other studies (24,25,49) improved on this type of quantification visually estimating the % of short roots ectomycorrhizal and assigning values to a rating system such as + (< 1%), ++ (< 10%), +++ (< 30%), and ++++ (> 30%), or excellent (75-100% short roots ectomycorrhizal), good (50-74% short roots ectomycorrhizal), moderate (24-49% short roots ectomycorrhizal) or poor (1-24% short roots ectomycorrhizal).

The obvious limitations are the highly subjective nature of the system and the difficulty in comparing results among studies. The majority of studies using this type of visual estimation dealt with seedlings.

The visual quantification scheme proposed and used by Richards and Wilson (43) and Marx and co-workers and others (1,9,21,22,25,27,28,30-33,35,42,44-46) for seedling evaluation has received considerable use. Entire root systems are examined visually and the estimation of ectomycorrhizae, expressed as a % of the total number of short roots/seedling, determined. Roots are then removed from a number of randomly selected seedlings and the accuracy of the initial estimation determined with a dissecting microscope by counting all ectomycorrhizae and short roots. In addition, ectomycorrhizae are selected from these samples, sectioned, and examined to verify Hartig net and mantle formation. Repeated experience with this scheme results in accurate quantification. This scheme is especially useful when large numbers of seedlings are involved in a study.

Recently, Marx (22) developed an ectomycorrhizal index for quantitative assessment of ectomycorrhizae formed by *Pisolithus tinctorius* (Pers.) Coker & Couch (Pt) using the formula a x b/c where a = % of seedlings with any amount of Pt ectomycorrhizae, b = average % of feeder roots with Pt ectomycorrhizae, and c = average % of total ectomycorrhizae formed by Pt and other fungi. Based on 100, this index provides a single value which represents the aggressiveness of the isolate of the effectiveness of the inoculum.

Direct Counts of Entire Root System.--Studies involving relatively small numbers of seedlings frequently count ectomycorrhizae and short roots on all roots (2,16,34,51). The length of each root is also measured. This scheme has been widely used in pure culture syntheses studies. Results may be presented as the total number of ectomycorrhizae/seedling, number of ectomycorrhizae/unit length of root or % of short roots ectomycorrhizal. The user of percent based data should be forewarned that, in our experience, % ectomycorrhizal short roots may not correlate well with growth (height, weight, etc.). This is apparently because factors such as size of seedling, number of roots, soil fertility, and others can have a much greater impact. As a result, the % figure may be statistically overwhelmed. In most cases this does not happen with the ectomycorrhizal tip method of quantification.

Direct Counts of Selected Roots.--In some studies involving seedlings, the number of seedlings and the amount of roots to be examined preclude the counting of ectomycorrhizae on all roots. In such cases roots are randomly selected from seedlings in some manner and the number of ectomycorrhizae counted. The number of roots (subsample size) selected from each seedling and the number of seedlings from which roots are selected varies with the study (8,24,41). Variations have included: random selection of a predetermined number of roots from each seedling; random selection of roots until a predetermined total length from each seedling is attained; selection of the major lateral root; and ectomycorrhizae counted on a predetermined length of the selected root.

Results from these quantification procedures are usually expressed as the number of ectomycorrhizae/unit length of root or the % of short roots ectomycorrhizal. Twarowski (49) developed an index of abundance for "genera" of ectomycorrhizae based on the sum of the percentage contributions of a given "genus" on a monthly or yearly basis and an index of frequency of occurrence of "genera" of ectomycorrhizae based on the number of months a given "genus" occurred.

Counts of Ectomycorrhizal Tips.--Research requiring quantification procedures that more accurately reflect the function of ectomycorrhizae was addressed by Hatch (15) and Richards and Wilson (43), but not adapted by them. Recent studies have quantified ectomycorrhizae by counting tips (10-14,20) or counting tips and calculating the average surface area/tip (19,35,36). Most studies using these quantification procedures dealt with ectomycorrhizae of large plants and used a soil core sampling technique. Results may be expressed as the number of living or active ectomycorrhizal tips/unit volume of soil (12), surface area of ectomycorrhizae/unit volume of soil, number of living ectomycorrhizal tips/soil fraction, or number of living ectomycorrhizal tips/unit weight of soil. Again a caution! When dealing with many large samples (cores, cubes, etc.) representing forest stand conditions, the volume (and number) of dead or inactive ectomycorrhizal roots is too large for it to be practical to count them to determine live or dead status. In such cases counting active (12) types has proven practical and should represent the current host situation as it relates to ectomycorrhizae. On the other hand, if nutrient capital, cycling, turnover rates, etc. are to be determined, it is important to consider all tips, including dead or inactive (5).

Recent studies dealing with biomass and nutrient cycling have used ectomycorrhizal tips and weight to determine the average biomass of an ectomycorrhizal tip (5,6,50). Results may be expressed as weight of ectomycorrhizal tips/unit area or weight of ectomycorrhizal tips/unit volume of soil.

Studies of the ectomycorrhizae of large plants which use soil cores as the sampling technique have employed several, often elaborate, means of handling to provide accurate quantification. Frequently the quantity of material (ectomycorrhizae and tips) from processing soil core samples is so great that a means of subsampling is necessary. Subsampling is also necessary if portions of the initial sample are required to determine weight or surface area of the ectomycorrhizae or tips, nutrient analysis or similar data (6,20,50). Marks, et al. (20) developed a 10-inch square perspex tray which was divided into inch squares and 0.1 inch subdivisions. Material from sieved soil cores was uniformly spread over the tray in 200 ml of water and ectomycorrhizae and tips counted in 7 randomly selected squares with a dissecting microscope at 7X. Fogel and Hunt (6) modified this procedure using a 25-cm tray and counted ectomycorrhizal fragments from 7 randomly selected 6.5-cm^2 squares.

Some Final Words

None of the quantification methods described here can be considered "all purpose" or suitable to achieve "all ends". One of the more interesting aspects of research involving ectomycorrhizal systems is the sometimes complexity in dealing with decisions regarding quantification systems that satisfy particular objectives. Despite the fact that ectomycorrhizae are easily seen and counted (usually!), the morphology of the ectomycorrhizal structure, host root system and soil frequently make appropriate quantification difficult. This is particularly true when dealing with morphological types and/or many individual species of ectomycorrhizal fungi. It seems appropriate at this point to raise a caution... don't be fooled! Despite the myriad of quantification systems available, we are still working for adequate methods to deal with some of our problems. So... use these herein described only if they are appropriate to "your ends", if not, good luck in developing new ones that will!

LITERATURE CITED

1. Berry, C. R. and Marx, D. H. 1977. Growth of loblolly pine seedlings in strip-mined kaolin spoil as influenced by sewage sludge. J. Environ. Qual. 6:379-381.
2. Bjorkman, E. 1970. Mycorrhiza and tree nutrition in poor forest soils. Studia Forestalea Suecica. No. 83. 24 p.
3. Dominik, T. and Ferchav, H. 1958. Study on mycotrophy of forest trees. Prace Instytutu Badawczego Lesnictwa, No. 179:85-99.
4. Dominik, T. and Madej. 1963. Investigation into mycotrophy in the pine association, Starzenina Forest District. Prace Instytutu Badawczego Lesnictwa, No. 259:61-69.
5. Fogel, R. 1980. Mycorrhizae and nutrient cycling in natural forest ecosystems. New Phytol. 86:199-212.
6. Fogel, R. and Hunt, G. 1979. Fungal and arboreal biomass in a western Oregon Douglas-fir ecosystem: distribution patterns and turnover. Can. J. For. Res. 9:245-256.
7. Grand, L. F. 1971. Tuberculate and *Cenococcum* mycorrhizae of *Photinia* (Rosaceae). Mycologia 63:1210-1212.
8. Grand, L. F. and Ward, W. W. 1969. An antibiotic detected in conifer foliage and its relation to *Cenococcum graniforme* mycorrhizae. For. Sci. 15:286-288.
9. Harley, J. L. 1969. The biology of mycorrhiza. Leonard Hill, London. 334 p.
10. Harvey, A. E., Jurgensen, M. F., and Larsen, M. J. 1978. Seasonal distribution of ectomycorrhizae in a mature Douglas-fir/larch forest soil in western Montana. For. Sci. 24:203-208.
11. Harvey, A. E., Jurgensen, M. F., and Larsen, M. J. 1980. Clearcut harvesting and ectomycorrhizae:survival of activity on residual root and influence on a bordering forest stand in western Montana. Can. J. For. Res. 10: 300-303.
12. Harvey, A. E., Larsen, M. J., and Jurgensen, M. F. 1976. Distribution of ectomycorrhizae in a mature Douglas-fir/larch forest soil in western Montana. For. Sci. 22:393-398.
13. Harvey, A. E., Larsen, M. J., and Jurgensen, M. F. 1979. Comparative distribution of ectomycorrhizae in soils of three western Montana forest habitat types. For. Sci. 25:350-358.
14. Harvey, A. E., Larsen, M. J., and Jurgensen, M. F. 1980. Partial cut harvesting and ectomycorrhizae: early effects in Douglas-fir-larch forests of western Montana. Can. J. For. Res. 10:436-440.

15. Hatch, A. B. 1937. The physical basis of mycotrophy in plants. Black Rock Forest Bull. 6. 168 p.
16. Hatch, A. B. and Doak, K. D. 1933. Mycorrhizal and other features of the root systems of *Pinus*. J. Arnold Arbor. 14:85-99.
17. Jurgensen, M. F., Larsen, M. J., and Harvey, A. E. 1977. A soil sampler for steep, rocky sites. U. S. Dept. Agri., For. Ser. Res. Pap. INT-217. Intermt. Forest and Range Exp. Sta., Ogden, Utah, 5 p.
18. Leaphart, C. D. 1958. Root characteristics of western white pine and associated tree species in a stand affected with pole blight of white pine. USDA For. Serv. Res. Pap. INT-52. Intermt. For. and Range Exp. Sta., Ogden, Utah, 10 p.
19. Lorio, P. L., Jr., Howe, V. K., and Martin, C. N. 1972. Loblolly pine rooting varies with microrelief on wet sites. Ecology 53:1134-1140.
20. Marks, G. C., Ditchburne, N., and Foster, R. C. 1968. Quantitative estimates of mycorrhiza populations in radiata pine forests. Aust. For. 32:26-38.
21. Marx, D. H. 1975. Synthesis of ectomycorrhizae by different fungi on northern red oak seedlings. U. S. Dept. Agr., For. Res. Note SE-282. 8 p.
22. Marx, D. H. 1981. Variability in ectomycorrhizal development and growth among isolates of *Pisolithus tinctorius* as affected by source, age, and reisolation. Can. J. For. Res. 11:168-174.
23. Marx, D. H. and Artman, J. D. 1978. Growth and ectomycorrhizal development of loblolly pine seedlings in nursery soil infested with *Pisolithus tinctorius* and *Thelephora terrestris* in Virginia. U. S. Dept. Agr., For. Ser. Res. Note SE-256. 4 p.
24. Marx, D. H. and Bryan, W. C. 1969. Studies on ectomycorrhizae of pine in an electronically air-filtered, air-conditioned, plant-growth room. Can. J. Bot. 47:1903-1909.
25. Marx, D. H. and Bryan, W. C. 1970. Pure culture synthesis of ectomycorrhizae by *Thelephora terrestris* and *Pisolithus tinctorius* on different conifer hosts. Can. J. Bot. 48:639-643.
26. Marx, D. H. and Bryan, W. C. 1971. Influence of ectomycorrhizae on survival and growth of aseptic seedlings of loblolly pine at high temperature. For. Sci. 17:37-41.
27. Marx, D. H. and Bryan, W. C. 1975. Growth and ectomycorrhizal development of loblolly pine seedlings in fumigated soil infested with the fungal symbiont *Pisolithus tinctorius*. For. Sci. 22:245-254.
28. Marx, D. H., Bryan, W. C., and Cordell, C. E. 1976. Growth and ectomycorrhizal development of pine seedlings in nursery soils infested with the fungal symbiont *Pisolithus tinctorius*. For. Sci. 22:91-100.
29. Marx, D. H., Bryan, W. C., and Davey, C. B. 1970. Influence of temperature on aseptic synthesis of ectomycorrhizae by *Thelephora terrestris* and *Pisolithus tinctorius* on loblolly pine. For. Sci. 16:424-431.
30. Marx, D. H., Hatch, A. B., and Mendicino, J. F. 1977. High soil fertility decreases sucrose content and susceptibility of loblolly pine roots to ectomycorrhizal infection by *Pisolithus tinctorius*. Can. J. Bot. 55: 1569-1574.
31. Marx, D. H., Mexal, J. G., and Morris, W. G. 1979. Inoculation of nursery seedbeds with *Pisolithus tinctorius* spores mixed with hydromulch increases ectomycorrhizae and growth of loblolly pines. S. J. Appl. For. 3:175-178.
32. Marx, D. H., Morris, W. G., and Mexal, J. G. 1978. Growth and ectomycorrhizal development of loblolly pine seedlings in fumigated and nonfumigated nursery soil infested with different fungal symbionts. For. Sci. 24:193-203.
33. Marx, D. H. and Rowan, S. J. 1981. Fungicides influence growth and development of specific ectomycorrhizae on loblolly pine seedlings. For. Sci. 27:167-176.
34. Marx, D. H. and Zak, B. 1965. Effect of pH on mycorrhizal formation of slash pine in aseptic culture. For. Sci. 11:66-75.
35. Menge, J. A. and Grand, L. F. 1978. Effect of fertilization on production of epigeous basidiocarps by mycorrhizal fungi in loblolly pine plantations.

Can. J. Bot. 56:2357-2362.

36. Menge, J. A., Grand, L. F., and Haines, L. W. 1977. The effect of fertilization on growth and mycorrhizae numbers in 11-year-old loblolly pine plantations. For. Sci. 23:37-44.

37. Mikola, P. 1953. An experiment on the invasion of mycorrhizal fungi into prairie soil. Karstenia 2:33-34.

38. Mikola, P. 1970. Mycorrhizal inoculation in afforestation. Int. Rev. For. Res. 3:123-196.

39. Molina, R. 1979. Ectomycorrhizal inoculation of containerized Douglas-fir and lodge pole pine seedlings with six isolates of *Pisolithus tinctorius*. For. Sci. 25:585-590.

40. Neal, J. L., Jr., Trappe, J. M., Lu. K. C., and Bollen, W. B. 1968. Some ectotrophic mycorrhizae of *Alnus rubra*. *In* Biology of alder. *Ed*. by Trappe, J. M., Franklin, J. F., Tarrant, R. F., and Hansen, G. M. Northwest Sci. Assoc. 40th Ann. Meet., Symp. Proc. 1967. pp. 179-184.

41. Park, J. Y. 1970. Effects of field inoculation with mycorrhizae of *Cenococcum graniforme* on basswood growth. Can. For. Ser., Dept. Fish and For. BiMonthly Res. Notes 26:27-28.

42. Pawuk, W. H., Ruehle, J. L., and Marx, D. H. 1980. Fungicide drenches affect ectomycorrhizal development of container-grown *Pinus palustris* seedlings. Can. J. For. Res. 10:61-64.

43. Richards, B. N. and Wilson, G. L. 1963. Nutrient supply and mycorrhiza development in Caribbean pine. For. Sci. 9:405-412.

44. Ruehle, J. L. 1980. Inoculation of containerized loblolly pine seedlings with basidiospores of *Pisolithus tinctorius*. U. S. Dept. Agr., For. Ser. Res. Note SE-291. 4 p.

45. Ruehle, J. L. 1980. Ectomycorrhizal colonization of container-grown northern red oak as affected by fertility. U. S. Dept. Agr., For. Ser. Res. Note SE-297. 5 p.

46. Ruehle, J. L. and Marx, D. H. 1977. Developing ectomycorrhizae on containerized pine seedlings. U. S. Dept. Agr., For. Ser. Res. Note SE-242. 8 p.

47. Trappe, J. M. 1965. Tuberculate mycorrhizae of Douglas-fir. For. Sci. 11: 27-32.

48. Trappe, J. M. and Strand, R. F. 1969. Mycorrhizal deficiency in a Douglas-fir region nursery. For. Sci. 15:381-389.

49. Twarowski, Z. 1963. Investigations on the annual development dynamics of mycorrhizae in 40-year-old spruce stand. Prace Instytutu Badawczego Lesnictna. No. 260:72-102.

50. Vogt, K. A., Edmonds, R. L., and Grier, C. C. 1981. Seasonal changes in biomass and vertical distribution of mycorrhizal and fibrous-textured conifer fine roots in 23- and 180-year-old subalpine *Abies amabilis* stands. Can. J. For. Res. 11:223-229.

51. Whitney, R. D. 1965. Mycorrhiza-infection trials with *Polyporus tomentosus* and *P. tomentosus* var. *circinatus* on white spruce and red pine. For. Sci. 11:265-270.

52. Zak, B. 1971. Characterization and classification of mycorrhizae of Douglas fir. II. *Pseudotsuga menziesii* + *Rhizopogon vinicolor*. Can. J. Bot. 49:1079-1084.

EVALUATION OF PLANT RESPONSE TO INOCULATION

A. HOST VARIABLES

W. A. Sinclair and D. H. Marx

In this section, examples from literature are cited primarily for illustrative value rather than for documentation of fact or first use of a technique.

PARAMETERS THAT ARE MEASURED OR DERIVED

Survival and Population Density

Mycorrhizal fungi influence the survival of tree seedlings, especially when these grow under stressful conditions (55, 63). Population counts or estimates to evaluate survival are most useful at six stages of seedling development: immediately after germination, after one growth season, at the beginning of the second growth season, immediately after transplanting, at the end of one growth season after transplanting, and at the beginning of the second growth season after transplanting. Population changes after one year in a seedbed or in containers, or more than one year after transplanting usually continue previous trends.

The initial seedling census is a measure of germination and a basis for detecting population changes. If germination is asynchronous, two or more early stand counts may be needed to measure both ingrowth (newly emerged seedlings) and early mortality. The timing of mortality after the germination period during the first season is seldom important in relation to mycorrhizal variables. Seedling population at the end of one growth period or season, preferably expressed as percentage of initial population, reflects net changes during the period and serves as the baseline for estimates of winter survival. The least ambiguous measure of survival after winter is ability of seedlings to resume growth. The three population counts for survival after transplanting provide a baseline for comparisons, a measure of effectiveness of root regeneration (of bare-root seedlings) or root development outside the container mix, and a measure of winter hardiness. Survival data, if acquired by an adequate sampling scheme, also indicate variation in population density in seedbeds or stocking in young plantations (35).

Mass and Dimensions

Weights and dimensions of roots, stems and leaves are intercorrelated (53). Each parameter varies in usefulness as an indicator of mycorrhizal influence according to the stage of plant development.

<u>Seedlings in the first year of development</u>.--Mycorrhizal fungi may influence the weight of seedlings within as few as 8 wk after sowing seed, but 12 wk is the practical minimum for small-seeded species. Those with large seeds (e.g., oaks, nut trees) may require at least 20 wk (14) because their early growth depends greatly upon food mobilized from the seeds. The minimum time for detection of mycorrhizal influence on growth of such seedlings has not been well defined.

In young seedlings, dry weight is the most sensitive indicator. Weigh tops and roots separately because some fungi influence root and top development differentially (22, 64). Fungal tissue can contribute significantly to total root mass in mycorrhizal seedlings. This contribution is usually ignored or acknowledged without measurement (1), but can be quantified by assay for glucosamine (67). When weights of tops and roots reflect treatments similarly, report total weights and top:root weight ratios, with measures of variability.

Height (stem length) differences among mycorrhizal treatments are usually the first effects of inoculation visible above the soil line. Stem length = distance from root collar or soil line to base or apex of the apical bud or to the apical node. Neither root collar nor soil line is consistently marked by a morphological feature. Therefore the stem base for measurements is arbitrarily identified on each seedling. Variations of stem length measurements include length of hypocotyl, length above the cotyledonary node, and length at time 2 minus length at time 1 (33, 39).

Stem diameter at root collar or soil line is a useful indicator of growth differences at the end of one growth season because it can be measured nondestructively, is strongly correlated with dry weight (53), and can therefore serve as a surrogate for weight measurements in seedlings that are allowed to continue development. The dimensional parameter most strongly correlated with top weight is a volume index: (stem diameter at soil line)^2 x stem length (35).

Root dimensions are not often recorded because of cumbersome techniques and the usual need for destructive sampling. Length of the primary root may reflect premycorrhizal influences during early development, but after lateral roots form, the length of the longest root eventually becomes unrelated to other parameters and to mycorrhizal treatments. Total root length of young seedlings, however, is sensitive to mycorrhizal influence and can be measured directly or estimated electronically by a line-intersect technique (50).

Root volume can be measured by displacement. Absorbing area of roots can be measured indirectly on the basis of absorption of a dye such as methylene blue (55). Total root surface area can be estimated photometrically (44) or gravimetrically on the basis of adsorption of a viscous solution such as concentrated calcium nitrate (9). These techniques yield relative values that can be converted to estimates of actual surface area by means of standard curves that relate light transmittance or weight of adsorbed solution to area derived from measurements of root dimensions.

Seedlings one or more years old.--Measure fresh or dry weight according to convenience. Dry basis is preferred. Do not report both. The larger the plants, the more satisfactory are fresh weight data because water loss before weighing becomes unimportant in relation to total weight. Reserve the term "biomass" for reference to dry weight. Add a word such as "index" when fresh weight or a combination of weight and dimensions is used as a surrogate for dry mass.

Dry weights of current season's leaves and shoots are useful if annual growth increments are of interest, but are not more useful than dimensional data unless the weights are also used as bases for comparing mineral concentrations, rates of physiological processes, etc. (4).

Beginning in the second season of growth, and especially thereafter, the length of current season's shoots is more important than total plant height for evaluating long-term plant responses to inoculation. For this measurement the longest apical shoot is usually adequate, but if variability within treatments is great, the aggregate length of multiple shoots per plant may be a more useful parameter.

Stem diameter in the second and subsequent years of growth should be measured at soil line until trees are tall enough for useful measurements at breast height (1.4 m). Nondestructive measurements of height, shoot lengths and diameter at soil line can serve as bases for calculating volume indices or growth rates (35, 57).

Leaf length or area can be measured nondestructively (8, 16, 29). Portable meters that can measure leaf areas are commercially available (13).

Derived Values

Derived values are useful for integrating information about experimental plants or plant populations. Observe three criteria when selecting a derived parameter: (a) it should convey information not readily discernible in its predecessors; (b) it should preserve the original patterns of variability in the data or reflect the pattern of variability in a real but unmeasured parameter; (c) it should not contort standard definitions or widely understood concepts. For example, the conversion of seedling stem diameter to cross sectional area does not meet the first criterion but could be justified if a volume index is desired.

Some derived parameters, conditions of their usefulness, and examples of application follow.

Top:root ratio, preferably on dry weight basis, can reflect differential influence of mycorrhizal fungi on weight accretion by tops and roots (40, 43).

Plot volume index, the product of mean seedling volume index and number of plants, integrates seedling size and site occupancy (35). For any volume index, report values as whole integers without units of measure.

Plot weight index, the product of mean top weight and number of seedlings in a plot, reflects biomass production per unit area (35). Use a weight index to report plot productivity if mean weight is obtained from a subsample of the seedling population. If all seedlings are weighed, however, report aggregate weight rather than an index. As this index is a projection of real sample values, retain the unit of measure.

Use indices only as measures of plot occupancy or productivity, or for other purposes that can not be accomplished using real dimensional or weight data.

Mycorrhizal influence value.--The influences of mycorrhizal fungi on dimensions and mass of plants have been quantified in many ways. For comparisons among diverse plant-fungus combinations, values for mycorrhizal plants have been converted to proportions of the mean values of nonmycorrhizal or noninoculated controls (42). For such relative values, we propose the general term "mycorrhizal influence value" (MIV), which is derived for any given parameter by assigning a value of 100 to the mean for noninoculated plants and calculating treatment means as integers relative to 100. MIV's could be reported not only for dimensional and mass parameters, but also for chemical and physiological variables. As MIV is simply "percent of control", it would seem familiar to readers. For authors and editors it would be an economical term, requiring neither units nor repetitious symbols or phrases.

Form and Color

Ectomycorrhizal fungi can influence the form of both roots and tops of tree seedlings. Effects on root form and branching are indicated by rate of lateral root formation (17), total root tips per seedling, short roots per unit length of lateral root (38), or short roots per unit root weight (1). These effects of inoculation may be independent of mycorrhizal formation (26).

In tree seedlings, the number and size of leaves, degree of stem branching, duration of annual growth and number of growth periods per year have been related to ectomycorrhizal inoculation (1, 12, 14, 55). Conversely, the greenness of foliage, or chlorophyll content measured spectrophotometrically, may be subnormal, or other symptoms of mineral deficiency may appear in plants that do not enter effective associations with mycorrhizal fungi (32, 55, 63, 65). Influences of ectomycorrhizal fungi on color and form (with exception of top:root ratio) are seldom recorded because the associated influences on mass and dimensions are easier to document in a standard format.

Physiological Parameters

Rates of growth or carbon assimilation.--Growth rates of plants are seldom estimated or analyzed in mycorrhizal research although techniques for this are available (66). Instead, differing rates of growth attributed to treatments are inferred from dimensional or weight differences among seedlings. If growth rates are of interest, however, and destructive sampling is undesirable, a volume index can be used as a surrogate for mass (57). Investigators should distinguish between absolute and relative rates of growth (46, 61). If seedlings are grown in closed containers, the net rate of carbon assimilation is a basis for evaluating growth rates associated with mycorrhizal treatments (4).

Rates or products of other physiological processes.--Water relations of seedlings have been little studied in relation to inoculation with ectomycorrhizal fungi. Transpiration rate, measured gravimetrically; leaf conductance, measured by diffusive resistance porometer; and xylem pressure potential, measured with pressure bomb, have been considered (4, 13, 54). Mycorrhizal respiration has been studied in relation to mineral uptake (7,15) and translocation. Rates of root respiration (on weight basis) were unaffected by inoculation with ectomycorrhizal fungi (4, 55), although the respiration of ectomycorrhizae is apparently a major drain on current photosynthate. Carbohydrate types and distribution in roots as influenced by ectomycorrhizal fungi have been studied by histochemical techniques (27, 45), chromatographic and scintillation techniques (2, 25), and by enzymatic conversion (36). Changes in amino acid composition associated with ectomycorrhizal infection have been characterized using automated analytical procedures (21). Phenolics which accumulate in response to fungal symbionts may be localized by histochemical stains (28, 45). Chromatographic and spectrometric techniques have been used to show that fungal symbionts also influence the amounts and composition of volatile organic compounds in roots (20).

Mineral Uptake and Content

The amounts and sometimes the concentrations of minerals, especially phosphorus, in plants change after inoculation with ectomycorrhizal fungi. Ion uptake is usually studied by radiotracer techniques (7, 41, 47, 58), and element content by standard titrimetric, spectrometric or photometric techniques (4, 6, 18, 40). Polyphosphate granules may be detected using cytochemical stains (3).

Tolerance of Environmental Stress

Weight and height gains by seedlings have been used to evaluate tolerance of inoculated and noninoculated seedlings to high (40°C) versus moderate (25°C) temperatures (33). Effects of water stress have been evaluated in terms of survival, xylem pressure potential, rates of transpiration and carbon assimilation, leaf conductance, root elongation and increase in root volume during treatment periods (13, 54, 55). The greater tolerance of inoculated than noninoculated seedlings to acidity, alkalinity and ion excesses have been demonstrated in terms of survival, size, plot volume index, mineral nutrient content, foliar color, and concentrations of cations in tissues (10, 11, 32). Tolerance or detoxication of injurious chemicals in soil have been evaluated in terms of seedling size and color, short root development, and color of roots (56, 68). In one study, an arbitrary numerical scale of root development was useful; scale values were illustrated by diagrams of the root systems (56).

Prevention or Mitigation of Disease

Ectomycorrhizal fungi can protect mycorrhizal or premycorrhizal seedlings against pathogens. Protective influences have been detected in survival, foliar color, dimensions and weights, number of short roots, frequency of isolation of pathogenic fungi from roots, and degree of root invasion by pathogens as determined microscopically (19, 30, 31, 48, 60).

ANTICIPATING AND COPING WITH VARIABILITY

Intraspecific Variability

Most inoculations with mycorrhizal fungi have been done with seedlings, seldom with vegetatively propagated plants, and this will continue. Variability in seedling parameters, including intensity of mycorrhizal formation, has often caused ambiguous results. We recommend that investigators grow a practice crop and evaluate the variability in each parameter before designing the main experiment.

Some tree species vary so much in seedling characteristics, even within individual seedlots, that special precautions or selection for uniformity may be needed. If an experiment involves transplanting, for example, select seedlings for uniform size and appearance (52), or mark uniform subgroups that can later be segregated for analyses of data.

Effects of asynchronous germination can be minimized by allowing it to begin on a moist surface, and transferring germinants at the same stage of development into the experiment. An alternative is to plant a surplus of seed and later rogue the fastest and slowest germinants. Germination behavior of tree seeds becomes increasingly erratic with duration of seed storage if storage containers are repeatedly opened for removal of sublots.

Seed size can influence seedling size in young plants. To minimize this, sort seeds into size, weight or specific gravity classes. Even seed orientation may influence the speed of germination (59). Thus in closely controlled experiments, place each seed similarly in the seed bed or on the germination surface.

Interspecific Differences

If several tree species are to be included in an experiment, design it to provide not only interspecific comparisons, but also to yield special information about individual species. The strategy is to obtain as much information as possible by schedules and procedures common to all species, and to modify or expand the procedures for some species in recognition of their particular characteristics. As examples, consider four hypothetical situations and proposed actions.

1. Root weights of several species are to be determined. During early growth of one species, most of the root mass is characteristically in a taproot. The weight of roots of this species might be partitioned into "taproot" and "other roots", thus allowing detection of possible differential mycorrhizal influences on these root types.
2. One species germinates asynchronously. Classify its seedlings (e.g. by color coded marks on leaves) according to the interval between planting and germination. Subsequently, stratify the observations according to the differing lengths of growth period caused by asynchronous germination, or schedule observations so as to observe each germination class after the same length of growth period.
3. Two species grow at very different rates. Therefore schedule two harvests. When the fast-growing species is harvested, get interim data for the slow one either by making nondestructive measurements or by destructively sampling extra replications included for the purpose.
4. One species varies more than others in form or growth rate. Design the experiment with extra replications for that species only.

CULTURAL FACTORS THAT INFLUENCE RESPONSE TO MYCORRHIZAL FUNGI

Type of Cultural Unit

Seedling growth and responses to ectomycorrhizal fungi vary with the type of experimental unit. Small containers such as tubes and nursery containers of capacity 200 cc or less, whether closed or with open tops, quickly influence root form of seedlings and often become associated with stress as indicated by slow growth, few or no ectomycorrhizae, or non-response to ectomycorrhizal fungi. Although useful for many types of research including tests of mycorrhizal formation (43), they are usually not suitable for evaluation of growth responses to mycorrhizal fungi because the restricted root environment dominates seedling behavior. Also, if seedlings are grown singly in unconnected containers and inoculation is not uniformly successful, there is no opportunity for subsequent natural inoculation by growth of mycelium or of roots. Individually grown seedlings, even within containers of large capacity, may thus have disadvantages associated with variability and experimental inefficiency.

Flats or pots that each contain multiple plants are relatively efficient for tests of plant response to inoculation because environmental and biological treatments can be applied conveniently to sufficient plants for detection of modest growth differences. If seedling groups are grown too long in pots, however, growth differences among experimental treatments may be limited by the pot environment (62). Out of doors, microplots allow greatest control of experimental conditions (34).

Nature and Microbiological Status of Rooting Medium

In general, mycorrhizal inoculations are more uniformly successful in soil-free rooting media than in soil or mixtures that contain a large proportion of soil. Plants can be induced to grow rapidly in soil-free mixtures regardless of mycorrhizal inoculation. But in such mixtures, ectomycorrhizal fungi often do not significantly enhance the growth of seedlings in comparison to noninoculated controls (40). Presumably the reason is that nonmycorrhizal plants can exploit the nutrient resources of soil-free mixtures nearly as well as can mycorrhizal plants.

Growth stimulation by mycorrhizal fungi in soils is easiest to demonstrate if microbial populations have been altered by heat, radiation or biocidal chemicals (37). Inoculation with mycorrhizal fungi is more quickly and uniformly successful in such soils than in untreated soil because of less competition and antagonism from other microorganisms. Also, plant growth suppression by soil organisms is minimal in pasteurized or sterilized soil.

Seedlings Versus Transplants

This choice makes little difference in quickness of growth response. The selection of uniform seedlings for transplanting can, however, help control variability in weight and dimensions.

HINTS ABOUT EXPERIMENTAL DESIGN AND ANALYSIS OF DATA

Designs

For experiments with multiple treatments, common useful designs are factorials and randomized blocks (5, 23, 37, 43, 51, 63).

Types of Controls

Consider three types of controls on inoculation: 1) no inoculum, 2) killed inoculum, 3) inoculum material lacking the mycorrhizal fungus. If just one of these is included in the experiment, the choice should be based upon the result of a preliminary experiment showing no difference among these controls.

Randomization

During experiments in greenhouses or controlled environment chambers, frequent rerandomization of the locations of pots or other cultural units is advisable because light, air movement, and other environmental factors are neither uniform nor randomly variable. Rerandomization spreads the influence of the uncontrolled variables across the experimental plant population.

Duration

Eight weeks is the briefest interval we have observed between inoculation and significant growth response of tree seedlings to ectomycorrhizal fungi. Experiments of duration less than one year commonly run 12-20 weeks, the latter roughly equivalent to one growth season in the field in northern climates. Experiments in pots or other containers should be terminated before plant development becomes limited by the pot environment.

Whereas pots and containers tend to suppress seedling variation (60, 62) and also to minimize differences among treatments, seedlings in open soil of a nursery or field plot may become increasingly variable because of differences in individual growth rates (57). Initially significant differences among treatments then tend to become insignificant. Observations in nursery or field experiments should therefore begin as soon as preliminary sampling indicates significant differences among treatments.

Variable Mycorrhizal Frequency Within Treatments

If inoculation has caused inconsistent mycorrhizal formation, analyze data separately for mycorrhizal plants, nonmycorrhizal plants, and all plants. If most seedlings are mycorrhizal but the mycorrhizal frequency (ectomycorrhizae per unit weight or per hundred short roots, etc.) varies, consider analysis of covariance.

Replication

Literature from many sources indicates that for mass and dimensional data, treatment means that differ from control means by less than 20% of the control value are seldom significant at $P = 0.05$ (5, 24, 51, 63). To discriminate relative differences of this magnitude as significant, here are guidelines for replication: for seedlings grown singly in containers, flasks, etc., 20 seedlings per treatment (49); for experiments in pots, at least 6 pots per treatment and at least 5 seedlings per pot (63); for experiments in flats or microplots, at least 3 plots per treatment and at least 10 random seedlings observed in each plot (34); in nurseries, at least four plots per treatment and at least 10 seedlings observed per plot (37); in outplantings, five replicate plots per treatment and 20 seedlings observed per plot (5, 12, 35, 51). Plan extra replicates or plot borders to allow preliminary sampling that will not disturb the design of the experiment.

LITERATURE CITED

1. Alexander, I. J. 1981. The *Picea sitchensis* + *Lactarius rufus* mycorrhizal association and its effects on seedling growth and development. Br. Mycol. Soc. Trans. 76:417-423.
2. Ahrens, J. R. and Reid, C. P. P. 1973. Distribution of ^{14}C-labeled metabolites in mycorrhizal and nonmycorrhizal lodgepole pine seedlings. Can. J. Bot. 51:1029-1035.
3. Ashford, A. E., Lee, M. L. and Chilvers, G. A. 1975. Polyphosphate in eucalypt mycorrhizas: a cytochemical demonstration. New Phytol. 74:447-453.
4. Benecke, U. and Gobl, F. 1974. The influence of different mycorrhizae on growth, nutrition and gas-exchange of *Pinus mugo* seedlings. Plant Soil 40:21-32.
5. Berry, C. R. and Marx, D. H. 1978. Effects of *Pisolithus tinctorius* ectomycorrhizae on growth of loblolly and Virginia pines in the Tennessee copper basin. U.S. Dep. Agric. For. Serv. Res. Note SE-264. 6 pp.
6. Bowen, G. D. 1973. Mineral nutrition of ectomycorrhizae. Pages 151-205, *in*: Ectomycorrhizae: their ecology and physiology, Marks, G. C. and Kozlowski, T. T., eds. Academic Press, N.Y. 444 pp.
7. Bowen, G. D., Skinner, M. F. and Bevege, D. I. 1974. Zinc uptake by mycorrhizal and uninfected roots of *Pinus radiata* and *Araucaria cunninghamii*. Soil Biol. Biochem. 6:141-144.
8. Carbon, B. A., Bartle, G. A. and Murray, A. M. 1979. A method for visual estimation of leaf area. For. Sci. 25:53-58.
9. Carley, H. E. and Watson, R. D. 1966. A new gravimetric method for estimating root-surface areas. Soil Sci. 102:289-291.
10. Clement, A., Garbaye, J. and Le Tacon, F. 1977. Importance des ectomycorhizes dans la résistance au calcaire du pin noir (*Pinus nigra* Arn. ssp. *nigricans* Host). Oecol. Plant. 12:111-131.

11. Dale, J. and McComb, A. L. 1955. Chlorosis, mycorrhizae, and the growth of pines on a high lime soil. For. Sci. 1:148-157.
12. Dixon, R. K., Garrett, H. E., Cox, G. S., Johnson, P. S. and Sander, I. L. 1981. Container- and nursery-grown black oak seedlings inoculated with Pisolithus tinctorius: growth and ectomycorrhizal development following out-planting on an Ozark clear-cut. Can. J. For. Res. 11:492-496.
13. Dixon, R. K., Wright, G. M., Behrns, G. T., Teskey, R. O. and Hinckley, T. M. 1980. Water deficits and root growth of ectomycorrhizal white oak seedlings. Can. J. For. Res. 10:545-548.
14. Dixon, R. K., Wright, G. M., Garrett, H. E., Cox, G. S., Johnson, P. S. and Sander, I. L. 1981. Container- and nursery-grown black oak seedlings inoculated with Pisolithus tinctorius: growth and ectomycorrhizal development during seedling production period. Can. J. For. Res. 11:487-491.
15. Edmonds, A. S., Wilson, J. M. and Harley, J. L. 1976. Factors affecting potassium uptake and loss by beech mycorrhiza. New Phytol. 76:307-315.
16. Gist, C. S. and Swank, W. T. 1974. An optical planimeter for leaf area determination. Am. Midl. Nat. 92:213-217.
17. Graham, J. H. and Linderman, R. G. 1980. Ethylene production by ectomycorrhizal fungi, Fusarium oxysporum f. sp. pini, and by aseptically synthesized ectomycorrhizae and Fusarium-infected Douglas-fir roots. Can. J. Microbiol. 26:1340-1347.
18. Harley, J. L. 1969. The biology of mycorrhiza. Leonard Hill, London. 334 pp.
19. Hyppel, A. 1968. Effect of Fomes annosus on seedlings of Picea abies in the presence of Boletus bovinus. Stud. For. Suec., Stockh. No. 66. 16 pp.
20. Krupa, S., Andersson, J. and Marx, D. H. 1973. Studies on ectomycorrhizae of pine. IV. Volatile organic compounds in mycorrhizal and nonmycorrhizal root systems of Pinus echinata Mill. Eur. J. For. Pathol. 3:194-200.
21. Krupa, S., Fontana, A. and Palenzona, M. 1973. Studies on the nitrogen metabolism in ectomycorrhizae. I. Status of free and bound amino acids in mycorrhizal and nonmycorrhizal root systems of Pinus nigra and Corylus avellana. Physiol. Plant 28:1-6.
22. Laiho, O. 1970. Paxillus involutus as a mycorrhizal symbiont of forest trees. Acta. For. Fenn. No. 106. 72 pp.
23. Lamb, R. J. and Richards, B. N. 1974. Inoculation of pines with mycorrhizal fungi in natural soils -- I. Effects of density and time of application of inoculum and phosphorus amendment on mycorrhizal infection. Soil. Biol. Biochem. 6:167-171.
24. Lamb, R. J. and Richards, B. N. 1974. Inoculation of pines with mycorrhizal fungi in natural soils -- II. Effects of density and time of application of inoculum and phosphorus amendment on seedling yield. Soil Biol. Biochem. 6:173-177.
25. Lewis, D. H. and Harley, J. L. 1965. Carbohydrate physiology of mycorrhizal roots of beech. I. Identity of endogenous sugars and utilization of exogenous sugars. New Phytol. 64:224-237.
26. Linderman, R. G. and Call, C. A. 1977. Enhanced rooting of woody plant cuttings by mycorrhizal fungi. J. Am. Soc. Hort. Sci. 102:629-632.
27. Ling-Lee, M., Ashford, A. E. and Chilvers, G. A. 1977. A histochemical study of polysaccharide distribution in eucalypt mycorrhizas. New Phytol. 78:329-335.
28. Ling-Lee, M., Chilvers, G. A. and Ashford, A. E. 1977. A histochemical study of phenolic materials in mycorrhizal and uninfected roots of Eucalyptus fastigata Deane and Maiden. New Phytol. 78:313-328.
29. Marshall, J. K. 1968. Methods for leaf area measurement of large and small leaf samples. Photosynthetica 2:41-47.
30. Marx, D. H. 1973. Growth of ectomycorrhizal and nonmycorrhizal shortleaf pine seedlings in soil with Phytophthora cinnamomi. Phytopathology 63: 18-23.
31. Marx, D. H. 1973. Mycorrhizae and feeder root diseases. Pages 351-382, in: Ectomycorrhizae: their ecology and physiology, Marks, G. C. and Kozlowski, T. T., eds. Academic Press, N.Y. 444 pp.

32. Marx, D. H. and Artman, J. D. 1979. Pisolithus tinctorius ectomycorrhizae improve survival and growth of pine seedlings on acid coal spoils in Kentucky and Virginia. Reclam. Rev. 2:23-31.
33. Marx, D. H. and Bryan, W. C. 1971. Influence of ectomycorrhizae on survival and growth of aseptic seedlings of loblolly pine at high temperature. For. Sci. 17:37-41.
34. Marx, D. H. and Bryan, W. C. 1975. Growth and ectomycorrhizal development of loblolly pine seedlings in fumigated soil infested with the fungal symbiont Pisolithus tinctorius. For. Sci. 21:245-254.
35. Marx, D. H., Bryan, W. C. and Cordell, C. E. 1977. Survival and growth of pine seedlings with Pisolithus ectomycorrhizae after two years on reforestation sites in North Carolina and Florida. For. Sci. 23:363-373.
36. Marx, D. H., Hatch, A. B. and Mendocino, J. F. 1977. High soil fertility decreases sucrose content and susceptibility of loblolly pine roots to ectomycorrhizal infection by Pisolithus tinctorius. Can. J. Bot. 55: 1569-1574.
37. Marx, D. H., Morris, W. G. and Mexal, J. G. 1978. Growth and ectomycorrhizal development of loblolly pine seedlings in fumigated and nonfumigated nursery soil infested with different fungal symbionts. For. Sci. 24:193-203.
38. Marx, D. H. and Zak, B. 1965. Effects of pH on mycorrhizal formation of slash pine in aseptic culture. For. Sci. 11:66-75.
39. McComb, A. L. 1943. Mycorrhizae and phosphorus nutrition of pine seedlings in a prairie soil nursery. Iowa State Coll. Agric. Exp. Stn. Res. Bull. 314:581-612.
40. Mejstřík, V. K. 1975. The effect of mycorrhizal infection of Pinus sylvestris and Picea abies by two Boletus species on the accumulation of phosphorus. New Phytol. 74:455-459.
41. Mejstřík, V. K. and Krause, H. H. 1973. Uptake of ^{32}P by Pinus radiata roots inoculated with Suillus luteus and Cenococcum graniforme from different sources of available phosphate. New Phytol. 72:137-140.
42. Menge, J. A., Johnson, E. L. V. and Platt, R. G. 1978. Mycorrhizal dependency of citrus cultivars under three nutrient regimes. New Phytol. 81:553-559.
43. Molina, R. 1980. Ectomycorrhizal inoculation of containerized western conifer seedlings. U.S. Dep. Agric. For. Serv. Res. Note PNW-357. 10 pp.
44. Morrison, I. K. and Armson, K. A. 1968. The rhizometer -- a device for measuring roots of tree seedlings. For. Chron. 44:21-23.
45. Piché, Y., Fortin, J. A. and Lafontaine, J. G. 1981. Cytoplasmic phenols and polysaccharides in ectomycorrhizal and non-mycorrhizal short roots of pine. New Phytol. 88:695-703.
46. Radford, P. J. 1967. Growth analysis formulae -- their use and abuse. Crop Sci. 7:171-175.
47. Reid, C. P. P. and Bowen, G. D. 1979. Effect of water stress on phosphorus uptake by mycorrhizas of Pinus radiata. New Phytol. 83:103-107.
48. Richard, C. and Fortin, J. A. 1975. Role protecteur du Suillus granulatus contre le Mycelium radicis atrovirens sur des semis de Pinus resinosa. Can. J. For. Res. 5:452-456.
49. Richard, C., Fortin, J. A. and Fortin, A. 1971. Protective effect of an ectomycorrhizal fungus against the root pathogen Mycelium radicis atrovirens. Can. J. For. Res. 1:246-251.
50. Rowse, H. R. and Phillips, D. A. 1974. An instrument for estimating the total length of root in a sample. J. Appl. Ecol. 11:309-314.
51. Ruehle, J. L. 1980. Growth of containerized loblolly pine with specific ectomycorrhizae after two years on an amended borrow pit. Reclam. Rev. 3:95-101.
52. Ruehle, J. L., Marx, D. H., Barnett, J. P. and Pawuk, W. H. 1981. Survival and growth of container-grown and bare-root shortleaf pine seedlings with Pisolithus and Thelephora ectomycorrhizae. So. J. Appl. For. 5:20-24.

53. Rutter, A. J. 1955. The relation between dry weight increase and linear measures of growth in young conifers. Forestry 28:125-135.
54. Sands, R. and Theodorou, C. 1978. Water uptake by mycorrhizal roots of radiata pine seedlings. Aust. J. Plant Physiol. 5:301-309.
55. Shemakhanova, N. M. 1962. Mycotrophy of woody plants. Transl. from Russian by S. Nemchonok, 1967. Israel Progr. Sci. Transl., Jerusalem. 329 pp.
56. Sherwood, J. L. and Klarman, W. L. 1980. IAA involvement in fungal protection of Virginia pine seedlings exposed to methane. For. Sci. 26:172-176.
57. Sinclair, W. A. 1973. Development of weight variation in Douglas-fir seedlings. For. Sci. 19:105-108.
58. Smith, F. A. 1972. A comparison of the uptake of nitrate, chloride and phosphate by excised beech mycorrhizas. New Phytol. 71:875-882.
59. Sorensen, F. C. and Campbell, R. K. 1981. Germination rate of Douglas-fir [*Pseudotsuga menziesii* (Mirb.) Franco] seeds affected by their orientation. Ann. Bot. 47:467-471.
60. Stack, R. W. and Sinclair, W. A. 1975. Protection of Douglas-fir seedlings against Fusarium root rot by a mycorrhizal fungus in the absence of mycorrhiza formation. Phytopathology 65:468-472.
61. Sweet, G. B. and Wareing, P. F. 1966. The relative growth rates of large and small seedlings in forest tree species. Pages 110-117 in: Physiology in forestry, Supplement to Forestry, Society of Foresters, Great Britain. 126 pp.
62. Terman, G. L., Bengston, G. W. and Allen, S. E. 1970. Greenhouse pot experiments with pine seedlings: methods for reducing variability. For. Sci. 16:432-441.
63. Theodorou, C. and Bowen, G. D. 1970. Mycorrhizal responses of radiata pine in experiments with different fungi. Aust. For. 34:183-191.
64. Trappe, J. M. 1977. Selection of fungi for ectomycorrhizal inoculation in nurseries. Annu. Rev. Phytopathol. 15:203-222.
65. Trappe, J. M. and Strand, R. F. 1969. Mycorrhizal deficiency in a Douglas-fir region nursery. For. Sci. 15:381-389.
66. Van den Driessche, R. 1968. Growth analysis of four nursery-grown conifer species. Can. J. Bot. 46:1389-1395.
67. Wu, L. and Stahmann, M. A. 1975. Chromatographic estimation of fungal mass in plant materials. Phytopathology 65:1032-1034.
68. Zak, B. 1971. Detoxication of autoclaved soil by a mycorrhizal fungus. U.S. Dep. Agric. For. Serv. Res. Note PNW-159. 4 pp.

EVALUATION OF PLANT RESPONSE TO INOCULATION

B. Environmental Variables

C. P. P. Reid and E. Hacskaylo

INTRODUCTION

In this section we examine the major soil and atmospheric abiotic factors considered of importance in the growth of tree seedlings and mycorrhizal development. A primary problem in treating this subject is the many variables which are involved in the interactions of plants with their environment. It is important to recognize which variables or abiotic factors are of primary importance and not include more than are necessary. Emphasis is placed on the methods of measuring selected abiotic factors. For convenience, abiotic factors are grouped into those related to the root environment, and those related to the shoot environment.

Seedling vigor and inoculation success may often be directly related to the processes of photosynthesis and respiration which in turn respond to energy and gas exchange. Radiation, air temperature, substrate temperature, wind speed, and humidity are factors that affect the exchange of energy, while CO_2 concentration, O_2 concentration, and humidity affect the exchange of gases. Additionally, soil factors of nutrient status, pH, texture, and water availability affect mycorrhizal development and plant growth. Selection of the most appropriate technique to measure specific abiotic factors is often determined by one's objectives and the particular circumstances. We have endeavored to select those techniques which are generally accepted and can be used in a variety of growing environments. We cite key references for particular methods rather than including extensive detail.

THE ROOT ENVIRONMENT

The growing medium or rooting substrate is very important, whether it is a natural soil, or various artificial mixes such as vermiculite, perlite, or peat. Soil conditions affect growth of the plant and influence the growth of mycorrhizal fungi and their successful infection of the root. The primary soil factors discussed below are: the level and balance of inorganic nutrients, pH, organic matter, texture, moisture and temperature.

Inorganic Nutrients

Major and Minor Elements.--Higher plants require fifteen mineral elements in varying amounts for normal growth and development. Of these, N and P are often emphasized because of their great demand by the plant and their apparent direct involvement in the formation of mycorrhizae. In addition to the needs for essential mineral nutrients in soil, high concentrations of toxic elements such as Pb, Zn, and Cu, the effect of soil pH, and the excess of salt in some soils can present problems for the growth of woody plants. In alkaline soils, Fe and P deficiencies are common; in acid soils, K and Mg may be frequently deficient and the probability of Al and Mn toxicity is increased.

Specific information on the levels of nutrients required in solution for optimum growth of a number of tree species can be found in Tinus and McDonald (28). Their chapter on mineral nutrition provides useful information on the formulation and preparation of nutrient solutions and on nutrient deficiencies in forest trees. However, recommendations on nutrient levels for optimum tree seedling growth must be interpreted with some caution, since such recommendations rarely consider the establishment of the mycorrhizal association. Although very high levels of N and P, and large imbalances in these elements should be avoided to assure good mycorrhiza formation, there is as of yet no general recommendation that can be made as to the nutrient levels required for optimum mycorrhiza formation. It is generally reported that low soil fertility enhances mycorrhiza formation. Although in some circumstances high levels of soil N and P inhibit the formation of mycorrhizae, the interpretation that high levels of these elements always inhibit mycorrhiza formation should be done with caution. The balance between N and P, as well as the relative level of other possibly limiting nutrients, must be considered. This also applies to other environmental conditions that impact the growth of the tree. Investigators should examine the available literature relative to specific tree species of interest to determine fertilization regimes that appear appropriate for good mycorrhiza formation.

There are many approaches for monitoring the levels of soil nutrients, however, we strongly recommend that researchers examine the "Handbook on Reference Methods for Soil Testing," published by the Council on Soil Testing and Plant Analysis in 1980 (8). This reference places the more commonly used procedures in a standard format and documents ramifications of these procedures. The Handbook also has a specific section on greenhouse soil analysis. Detailed step-by-step procedures are presented with appropriate references. We indicate below what we believe to be the more appropriate procedures to use for both alkaline and acidic soils. Since extraction of soils for nutrient analysis depends on their pH, we suggest different extraction procedures for soils above pH 6.0 and for those below pH 6.0. Once a soil sample is extracted by the proper technique, the same analytical techniques can be used for determining the actual level of various elements.

For "alkaline" soils (above pH 6.0) use the NH_4HCO_3-DTPA extraction procedures for available nutrients and trace elements (26, 27). This procedure simultaneously extracts P, K, Zn, Fe, Cu and Mn with 1M NH_4HCO_3 and 0.005M DTPA at pH 7.6. The method can also be used to evaluate the availability and toxicity of trace elements such as Pb, Ni, Cd, Mo, B, As, and Se. Although it is probably feasible to extract Mg by this method, it has not been thoroughly examined. Mg can be extracted by pure water or by ammonium acetate. For satisfactory extraction of the exchangable ammonium, nitrate, and nitrite in a wide variety of soils (both acidic and alkaline) soil samples can be shaken with 2N KCl for 1 hr using 10 ml of reagent per gram of soil (4). In acidic soils (below pH 6.0) we suggest that the Mehlich No. 1 (Double Acid) extraction method be used for P, K, Ca, Mg, and Zn. Details of this method using 0.05N HCl and 0.025N H_2SO_4 can be found in the Handbook on Reference Methods for Soil Testing (8). Specific reference to this method in forest soils can be found in Lea, Ballard and Wells (15).

Once soils have been extracted by appropriate means, P may be determined by the ascorbic acid method of Watanabe and Olsen (29). The cationic elements, K, Ca, Mg, Zn, Fe, Cu, Mn, can be individually determined by atomic absorption spectroscopy or simultaneously by the new method of inductively-coupled plasma optical emission spectrometry. Total N can be determined by the Kjeldahl wet-oxidation procedure (5). The determination of nitrate-N can be done colorimetrically by using chromotropic acid (30). Ammonium-N soil extracts can be readily determined by use of the ammonia specific-ion electrode method (6).

Conductivity as Indication of Salt Accumulation.--Electrical conductivity as measured by a conductivity bridge and cell can indicate salt accumulation in soils. Conductivity is measured by two basic approaches (8). First, specific conductance is measured in an extract which is filtered and collected under vacuum from a saturated soil paste. In the second method, soils (especially from more humid regions) are prepared as a 1:2 soil:water extract. Specific conductance is measured on the supernatant from the soil-water mixture. Other methods which involve measuring solute potential are discussed in the below section on soil moisture.

Soil pH

The pH of solutions of most soils ranges from 4 to 8. Although some trees are adapted for growth at extremes of the pH range, most trees grow well in a pH of 5 to 7. The importance of the pH of soil solutions, or added nutrient media, is through its influence on the oxidation-reduction equilibrium, the solubility of several constituents, and the ionic form of a number of mineral elements. For example, the availability of such ions as Fe and P are greatly influenced by pH.

The most common method for measuring soil pH employs a glass electrode pH meter in a soil-water mix (8). Five grams of air-dried soil are mixed with 5 ml of water for 5 sec and allowed to stand for 10 min. The soil-water suspension is then swirled as the electrodes from a calibrated pH meter are lowered into the solution. If it is necessary to measure the pH in a variety of soils and maintain an approximate constant ionic-strength, 0.01M $CaCl_2$ can be added to the soil instead of water (8). The advantage of measurement in the presence of $CaCl_2$ is the tendency to eliminate interferences from variable salt contents such as fertilizer residues and from suspension effects. In this method, 5 ml of 0.01M $CaCl_2$ are added to 5 g of soil and allowed to stand for 30 min with intermittent stirring. The pH of the soil-$CaCl_2$ suspension is then measured with a calibrated pH meter.

Soil Organic Matter and Texture

The amount of soil organic matter and the texture of soil mixes affect both the physical and chemical status of the rooting media. Moisture, aeration, pH control, and cation exchange are strongly influenced by these two factors. Total soil organic matter can be estimated routinely by measuring organic carbon content using a modification of the Walkley-Black potassium dichromate wet-oxidation procedure (2). The procedure is rapid and is routinely used for the determination of organic matter in mineral soils. To determine soil texture, the Bouyoucos hydrometer method for particle-size analysis is recommended (9).

Soil Moisture

Soil moisture can be viewed as part of the total continuum of water that exists between the soil, the plant, and the atmosphere. It is important in terms of its availability for uptake by the plant and the need to maintain a satisfactory water status of plant tissue. However, soil moisture is also indirectly important through its effect on nutrient availability, aeration of the soil, root development, and microorganism activity. It is important to distinguish between water content (weight or volume basis) and chemical potential of water. As defined thermodynamically, total water potential (Ψ) is a measure of the capacity of water at a particular point in soil to do work, as compared with free pure water. The difference in potential between two points, i.e. the potential gradient, defines the magnitude of the driving force and the direction for water movement. Water moves from areas of high potential to areas of low potential, although the actual flux of water is modified by the resistances to flow encountered by water as it moves through the soil, the plant, and the atmosphere. Where possible, water availability is considered in the standard terminology of *chemical potential of water*, where the total potential of a plant cell (Ψ_{cell}) is the sum of the solute (Ψ_s), pressure (Ψ_p), and matric (Ψ_m) potentials. *Soil water potential* (Ψ_{soil}) consists of Ψ_s, Ψ_m, Ψ_p, and an additional potential term due to gravitation (Ψ_g). The terms Ψ_g and Ψ_p (positive pressure) are often disregarded because of their minor contribution to Ψ_{soil} under most situations. In soils with water contents less than saturation, Ψ_{soil} approximately equals Ψ_m (adsorption and capillarity) except under saline conditions where Ψ_s may be of considerable importance. The usual pressure unit of Ψ is the Pascal (0.1 Mega Pascal (MPa) = 1 bar = 100 J kg^{-1}). Below we discuss both direct measurements of water in the soil and indirect measures of water availability to the plant by using the plant as an indicator.

Direct Measurement (Water Content).--The traditional gravimetric method of measuring the water content of soil by mass consists of removing a sample from the substrate and determining its "moist" and "dry" weights (the latter after drying the sample to constant weight in an oven at 105°C). The gravimetric water content (or mass water content) is the ratio of the weight loss in drying to the dry weight of the sample (mass and weight being proportional). Occasionally, however, mass ratio of the water to the wet soil is used. The water content on a dry basis (w_{md}) and on a wet basis (w_{mw}) can be interconverted by the following expressions:

$$w_{md} = \frac{w_{mw}}{1 - w_{mw}} \qquad w_{mw} = \frac{w_{md}}{1 + w_{md}}$$

To obtain the volumetric water content (Θ) from the gravimetric determination, a separate measurement must be made of the bulk density (ρ_b) and the following expression used:

$$\Theta = \frac{\rho_b}{\rho_w} w_{md},$$

where ρ_w is the density of liquid water. Bulk density measurement is usually difficult and subject to errors. Although the gravimetric method entails inherent errors because of sampling, transporting, and weighing, these can be reduced by increasing the size and number of samples.

The relationship between water content and soil water potential (mainly Ψ_m) is often determined by means of a tension plate apparatus at high soil-water potentials (above -0.1 MPa) or by pressure plate or pressure membrane apparatus at lower soil water potentials (20). A soil moisture retention curve can then be described by examining the amount of water left in soil samples after being subjected to different suction or matric potentials. Since the relationship between matric potential and water content exhibits a hysteresis upon the wetting up (sorption) or drying down (desorption) of a soil, the curve determined by desorption is ordinarily used in connection with processes that involve drainage, evaporation or plant extraction of soil moisture (11). Recommended procedures for determining soil moisture characteristic curves can be found in Richards (20).

Direct Measurement (Soil Water Potential).--There are basically two methods by which to measure water potential directly in soils, tensiometry and thermocouple psychrometry, each of which have their own particular shortcomings. *Tensiometers* basically consist of a tube sealed at one end by a porous ceramic cup which is in contact with the soil (21). The other end of the tube is sealed with a removable cap after the tube has been completely filled with water. If the soil is drier, that is, has a lower water potential than the water in the porous ceramic cup, water will be withdrawn from the tensiometer through the porous cup and out into the soil. As water is withdrawn, a partial vacuum is created in the tensiometer and registered by the vacuum gauge. Since the porous cup of the tensiometer is permeable to both water and solutes, only matric potential is determined. Tensiometers work quite well in wet soils (0 to -0.1 MPa), but become unusable when Ψ_m declines to near -0.1 MPa, and the water column fails as air enters the porous cup. Tensiometers can be useful where conditions are maintained near field capacity on a continuous basis. Although most tensiometers are left in place after insertion into the soil, there are portable tensiometers available which can be inserted into a cored hole and left for only 2-3 min to obtain a reading (see for example, equipment available from Soil Moisture Equipment Corporation).

The basic principle behind *thermocouple psychrometry* is, as in other psychrometric methods, the determination of a wet bulb depression in a closed system which is in vapor equilibrium with a sample of interest. The wet bulb depression is calibrated empirically against the wet bulb depression obtained over salt solutions of known solute potential. The wet bulb depression, which is dependent on the rate of evaporation from a thermocouple junction, is controlled by the vapor pressure of the sample. Although there are several types of thermocouple psychrometers, the one most readily available for use in the measurement of soil water potential uses the principle of Peltier cooling as the means of wetting the thermocouple junction with pure water (Wescor, Inc., EMCO). Thermocouple psychrometer sensors used for measurement of soil water consist of a thermocouple, usually enclosed in a ceramic cup or fine wire mesh that readily allows vapor equilibration between the sensor within the psychrometer cup and the soil. Peltier cooling of the sensing junction results in the condensation of water on the junction as its temperature is lowered below the dew point. As soon as the cooling is stopped, water begins to evaporate from the junction at a rate determined by vapor pressure of its environment. The lowering of temperature by evaporation from the sensing junction results in a wet bulb depression, usually measured in microvolts, and is compared to a calibration curve for conversion to total water potential. Soil psychrometers can be placed directly in the soil and left for periods of weeks to months. However, each sensor should be periodically calibrated against known solute potential solutions. Care must be taken to avoid the accumulation of salts on the porous ceramic bulb or fine mesh screen or on the thermocouple itself. Psychrometers can be used in the measurement of soil water over a wide range of potentials, and are especially useful in drier soils (Ψ_{soil} as low as -7.0 MPa). However, their usefulness in wetter soils, near field capacity, is limited because of inaccuracy at potentials greater than about -0.1 MPa. Because the microvolt output of the thermocouple junction is very small, it is extremely sensitive to slight temperature gradients, capacitance, or other electrical interference. This sensitivity often limits its reliable use to conditions where temperatures do not change rapidly.

Thermocouple psychrometer chambers are also available for measuring small samples removed from the soil. Samples are placed in small planchets which are then sealed into a vapor-tight chamber. The samples are then allowed to remain in the chamber for a sufficient amount of time to insure vapor equilibration between the sample and the thermocouple sensor. In order to obtain reliable measurements, care must be taken in handling soil samples to prevent loss of water by evaporation or major changes in bulk density that would lead to erroneous measurements once placed in the chamber. Since the instrumentation required for the use of thermocouple psychrometry (a potentiometer sensitive in the microvolt range) is relatively expensive, serious consideration must be given to the level of precision and accuracy required in soil moisture determinations. Specific details on the use of thermocouple psychrometry can be found in the symposium volume edited by Brown and Van Haveren (7), and in the publication by Wiebe et al. (31).

Direct Measurement (Solute Potential).--The osmotic component (solute potential) has little effect on the movement of liquid water within soils because there is no membrane barrier and usually no change of phase in the flow path. Therefore, soil matric potential has usually been considered the significant measurement to make when considering water movement within soils. However, when water moves in a soil-plant-atmosphere continuum, there is a membrane in the flow path at the root-soil interface, and under these circumstances soil solute potential becomes a relevant component in the water potential gradient. As discussed above, psychrometric techniques are technically demanding and are inappropriate for wet soils. Frequently solute potential has been considered small enough (in non-saline soils) to ignore. However, solute potential can be significant, particularly in dry soils or where salts accumulate from frequent fertilization. Sands and Reid (24) evaluated the techniques of dilution, pressure membrane extraction, displacement, in situ electrical conductivity for measuring solute potential to test their validity. They concluded that the displacement technique was the most suitable for calibrating soil osmotic potential against soil water content, because it was simple, inexpensive in materials and time, and probably the least subject to error. The displacement technique is one of the oldest, yet probably most reliable techniques for extracting soil solution in an unaltered form. Sampled soil is carefully packed into glass tubes. Distilled water is added to the top surface of the soil in the tube and then several aliquot volumes of soil water displaced through the bottom surface of the soil are collected and the solute potential measured with an osmometer. A more detailed description including optimum displacing solutions, column sizes, and column packing techniques can be found in Adams (1).

Indirect Measurement - The Plant as an Indicator.--Determination of the water status of the tree seedling can often be an easier way to assess soil moisture availability than direct soil measurements. However, since the development of water stress in a plant shoot results from the loss of water by transpiration exceeding the uptake by the root system, water stress can develop quite independently of soil moisture conditions if the evaporative demand of the atmosphere is high. Since the usual objective of measuring soil moisture is to serve as an indication of plant response, direct measurement of plant water status is often more desirable. Plant growth is often directly related to the chemical water potential of the foliage, since such processes as photosynthesis, respiration, and cell enlargement are both directly and indirectly affected by water potential of the leaf. With a high atmospheric evaporative demand, even well-watered plants will exhibit stress as a diurnal change in leaf water potential. Therefore, any measure of plant water status should take into account this temporal variation. The measurement of leaf water potential can be measured directly by thermocouple psychrometry or estimated by measuring xylem water potential with the pressure chamber method. The measurement of stomatal conductance will also be discussed because of its effect on photosynthesis and transpiration.

The *pressure chamber method* is one of the most rapid and fool-proof procedures for estimating leaf water potential in seedlings. Details on the use of the pressure chamber can be found in Kaufmann (12) and Ritchie and Hinkley (22). A small twig or leaf is cut from the seedling and quickly sealed in a steel chamber so the petiole or stem of the twig protrudes from the lid of the chamber. Pressure is then applied to the leaf in the sealed chamber until sap in the xylem appears at the cut surface of the petiole or stem. The pressure required to return the sap to the cut surface is related to the water potential of leaf cells by $\Psi_{leaf} = P + \Psi_s^{xylem}$, where P represents the negative component of the water potential of the xylem sap measured as a positive pressure in the pressure chamber. Since the solute potential of the xylem sap (Ψ_s^{xylem}) is generally higher than -0.3 MPa, estimates of leaf water potential may be obtained without accounting for Ψ_s^{xylem}. Measurements of xylem water potential by this method are generally very reproducible but should be considered relative since the relationship between actual leaf water potential and xylem water potential varies with species. McDonald and Running (17), suggest ways pressure chamber readings can be used to monitor seedling water stress for use in irrigation of western forest tree nurseries.

Thermocouple psychrometers (as discussed earlier) can be used for direct measurement of leaf water potential. Small leaf or needle samples can be removed from a seedling and placed in a thermocouple psychrometer sample chamber as described earlier for soil. As with soil, care must be taken to prevent the loss of water by evaporation from the samples once they are removed from the seedling. Since sample planchets for use in thermocouple psychrometer chambers are small, leaf samples usually have to be cut to fit into the planchets. Therefore, appropriate corrections must be made for the leakage of cell solutes from the cut surfaces and their effect on measurement of leaf water potential. Compared to pressure chamber measurements, thermocouple psychrometer measurements are very time consuming. Small leaf samples may require 30 to 60 min for vapor equilibration in the sample chamber. Very often, thermocouple psychrometry under controlled conditions is used for calibration of leaf water potential against xylem water potential as measured by the pressure chamber.

Stomatal opening, as measured by *stomatal conductance* (the reciprocal of resistance) is responsive primarily to the environmental factors of air temperature, humidity, radiation, and soil water supply as mediated through leaf water potential. The measurement of stomatal conductance can be useful in assessing the impact of various cultural practices on photosynthesis and transpiration since conductance is an indication of gas exchange between the leaf and the atmosphere.

There are a number of commercially available diffusion porometers which can be used to estimate stomatal conductance in leaves. These fall into three basic categories: 1) non-ventilated, 2) ventilated, and 3) ventilated null-balance (steady-state). Basically, all three of these instruments work on the same principle. The first two categories of instruments consist of a cup or cuvette containing a humidity sensor which is sensitive over a narrow range of relative humidity. In operation, the cuvette air is first dried by silica gel or other dehydrator. The cuvette is then placed over the leaf and as transpiration proceeds the time required for a prescribed range of humidity change is determined. Transpiration rate is assumed to be constant during the period of measurement and inversely proportional to the time required for a given change in humidity. Calibration is generally carried out by determining the amount of time required for a change in humidity over a prescribed range when the porometer is located over a variety of perforated plates whose diffusive resistance (or conductance) can be calculated from diffusion theory. Calibration by this method is extremely difficult for leaves of irregular geometry such as conifer needles. In non-ventilated porometers, water moves from the leaf to the humidity sensor by diffusion and often involves very long transit times if stomata are nearly closed or the area of transpiring surface is small. In ventilated porometers, the transfer of water vapor from the leaf to the sensor is considerably speeded up by a fan which provides a constant air-speed across the leaf surface. Since both types of instruments are subject to large errors because of the adsorption-desorption of water vapor on the surface of the cuvette, past history and preconditioning of the cuvette are very important in obtaining reliable readings. The advantage of both types of these instruments is their low cost and portability. With careful use they can give good relative indications of stomatal conductance. The null-balance porometer reduces the above errors considerably, but at a substantial expense.

In the null-balance technique, rather than monitoring the time required for a given humidity change on a sensor, the amount of water vapor lost by the leaf is determined by monitoring the flow of dry air required to maintain a given constant relative humidity in the cuvette. Since the cuvette can initially be "nulled" at air ambient humidities, the problem of adsorption-desorption of water vapor is minimal. An added advantage in the use of null-balance porometers is that leaf geometry is of little importance and the porometer can be easily calibrated for use with a variety of tree foliage types. The primary disadvantage of the instrument is its expense and relatively large size. All three types of porometers are commercially available (see for example those available from Li-Cor, Inc.). Specific details on the precautions in calibration and use of non-ventilated diffusion porometers can be found in Morrow and Slatyer (18). Details on the use of a ventilated porometer with tree species can be found in Kaufmann and Eckard (13). The use of the null-balance diffusion porometer is discussed by Beardsell, Jarvis and Davidson (3).

Soil Temperature

Monitoring of soil temperature is important for several reasons. Excessive surface temperatures can reduce seedling establishment by direct injury to succulent tissues in young germinating seedlings. Bulk soil temperature has a regulatory effect on root metabolism and can directly affect the uptake of water and nutrients by the root system. Soil temperature also influences growth of mycorrhizal fungi and the infection process. Production and transport of root-synthesized growth hormones is also impacted by soil temperature. Soil temperature measurement can be accomplished by a variety of instruments, ranging from mercury thermometers to thermistors and thermocouples (23) (see the section below on air temperature). The remote recording two- and three-point thermograph (c.f. Weathermeasure Corp.) is specifically designed for measuring and continuously recording soil temperature. Temperature sensors are mercury-filled bulbs encased in stainless steel probes with capillaries connected to a Bourdon tube. The sensors can be placed up to 32 ft from the spring or battery-driven clock mechanism used for continuous chart recording. This type of instrument is particularly durable, portable and reliable with minimal maintenance.

THE SHOOT ENVIRONMENT

The physical environment effect on plant growth is most easily recognized in the response of the shoot. The plant shoot functions within a fairly restrictive range of most environmental factors, and is subject to greater extremes of those factors controlling energy exchange than the root system. Physiological processes such as respiration, photosynthesis and cell enlargement are temperature dependent, and to some extent, light dependent. Assessment of the suitability of a particular set of environmental conditions is often judged by shoot response, whether it be shoot biomass, height growth, foliage color, needle length, or other observable characteristics. This observed response is basically related to the transfer of energy between the environment and the organism. The following consideration of radiation, temperature, humidity, atmospheric CO_2, are all directly or indirectly related to this energy exchange.

Radiation

Solar radiation provides two very basic needs of plants: 1) the establishment of a satisfactory thermal environment, and 2) the light energy required for photosynthesis. Additionally, the periodicity of the radiation environment (photoperiod) contols certain developmental patterns in woody plants, such as dormancy and cold hardiness. There are essentially three factors of solar radiation that are important to the plant: 1) the intensity (amount of energy per unit time), 2) duration, and 3) quality (wavelength). To relate radiant energy to photosynthesis, instruments which measure the energy in the photosynthetically active spectrum (380 nm to 710 nm) are preferable. If thermal effects of solar radiation are of primary concern, it may be necessary to use an instrument which measures a considerably greater spectrum, i.e. including both ultraviolet and infrared radiation. A detailed discussion on radiation and instrumentation can be found in Rosenberg (23) and Reifsynder and Lull (19).

Photosynthetic Active Radiation Measurement.--Photosynthetically active radiation (PAR) is best measured by quantum-flux meters, whose sensitivity is proportional to the number of quanta per unit wave-length. Therefore, PAR values are accurate regardless of spectral distribution. Quantum sensors usually measure PAR units as microeinsteins m^{-2} sec^{-1} or Watts m^{-2}. Full sun plus sky PAR is approximately 2,000 microeinsteins m^{-2} sec^{-1}. Such instruments are available from Li-Cor, Inc. and Cintra, Inc.

Traditionally, plant scientists have used measures of illuminance as an indication of the radiation environment. Such units as foot candle, lumen, and lux represent illuminance, a measure of visible light. Since such illuminance measurement devices indicate radiation in the visible spectrum, they have been very useful for estimating the amount of energy available for photosynthesis. The amount of illuminance received on a horizontal surface at sea level with sun at its zenith is about 10,000 foot candles (107,600 lux). One lumen ft^{-2} is equivalent to one foot candle, whereas one lumen m^{-2} equals one lux. The use of illuminance devices for measuring the visible spectrum is less satisfactory than using PAR-type meters. However, instruments for measuring illuminance are fairly inexpensive and are available from a number of manufacturers (c.f. United Detectors Technology, Photo Research Corp., Li-Cor, Inc.).

Total Incoming Shortwave Radiation.--It may be of interest to measure the total incoming shortwave radiation rather than just the visible spectrum when considering plant energy balance studies or the amount of energy available for transpiration and evaporation. A number of instruments are available for measuring direct beam solar radiation plus diffuse sky radiation, usually including the wavelengths from about 300 to 3,000 nm. These instruments are termed pyranometers. The Robitzch bi-metallic-type pyranometer (Belfort Instrument Co., Weathermeasure Corp.) is typically used for greenhouse and field measurements. The principle of operation is setting up a differential temperature between two identical bi-metallic strips, one of which is blackened while the other is highly reflective. The glass domed-covered sensors are placed horizontally and provide continuous recording on a chart-covered drum rotated by clock movement. Accuracy of these instruments is usually not better than ± 5 to 10% for daily totals, and instantaneous values are only reliable when radiation is changing slowly since the sensing strips have a large time-lag. However, they are rugged, durable, and require little attention except to change the chart. They are also useful for providing a daily record of daily fluctuations in the radiation environment.

Photoperiod Establishment.--The length of the photoperiod (or length of dark period) is important in controlling dormancy of most woody plants. If plants are to be kept actively growing beyond the normal growing season, i.e., in greenhouse or growth chamber conditions, then photoperiod must be controlled by some system of artificial lighting. For recommendations on lighting systems, photoperiod lengths, and specific requirements for a number of tree species see Tinus and McDonald (28).

Temperature

Temperature effects plant growth, both directly and indirectly. In some cases, the most significant effects of temperature are on physiological processes, while in others the effects of temperature are through strictly physical changes in the environment, for example, loss of water by evaporation and transpiration. Since plants are poikilothermic organisms, that is, their temperature tends to approach that of their surrounding environment, it is important to consider that plant tissue temperature is often not the same as that of its surrounding air. For this reason it is convenient to discuss air temperature separately from that which is called "test-body" temperature, or temperature of particular objects, such as plant leaves.

Air.--The measurement of air temperature can range from the use of very simple alcohol or mercury thermometers, up to very sophisticated electronic systems. Since the purpose of measuring air temperature is to indicate the mean kinetic energy of the air molecules, it is important to insure that the sensing element does not indicate elevated temperatures because of direct radiation absorption by the sensor itself. Shielding of sensing elements tends to minimize radiant heating while still allowing free movement of the air past the sensor. Detailed information on the use and installation of different kinds of temperature measuring instruments can be found in the Fire-Weather Observer's Handbook, a U.S. Forest Service publication (10).

Test Body.--The test body temperature of importance usually will be leaf temperature. The temperature of a leaf governs its vulnerability to frost damage or heat scorch and also affects the rates of transpiration and photosynthesis. The leaf is very active in partitioning energy by such processes as radiation, convection, and evapotranspiration. Measurements of leaf temperature, when coupled with air temperature measurements, give an indication of the vapor pressure gradient which controls the water loss from the leaf. Two basic requirements which must be fulfilled by an instrument used for measuring leaf temperature are: 1) lack of disturbance of the heat balance of the leaf, i.e., transfer by convection or radiation, and 2) the measurement of the real temperature of the tissue. Two basic types of instruments have found wide usage, thermistors, semi-conductors which produce a large signal with change of temperature, and thermocouples, a transducer which produces a low output voltage on change of temperature (23). Thermocouples are generally very stable and accurate but require a reference junction and relatively sophisticated electronics. Thermistors tend to be less sensitive but have the advantage of requiring less expensive equipment for use. Thermistors and thermocouples can be made small enough to be inserted into fleshly leaves. For thin leaves, it is often possible to have surface contact of the sensing element, usually on the lower surface of the leaf. As with any temperature sensing device, it is necessary to avoid self-heating by direct radiation. For a description and use of a thermistor leaf thermometer see Linacre and Harris (16). For large masses of tree foliage, temperature can be remotely sensed by the use of infrared scanners (23). Infrared scanners detect electromagnetic energy and wavelengths longer than the visible spectrum. They can operate over quite wide ranges of temperature and can detect a difference as little as 0.2°C. These instruments are relatively expensive and hence, may find limited use.

Humidity

The amount of water vapor or humidity in the air often gives an indication of the potential for plants to undergo desiccation. Air humidity may be expressed absolutely, either as partial water vapor pressure in pressure units such as millibars, or as a mass of water vapor per unit volume of moist air, e.g. kg cm^{-3}. Relative humidity is the ratio of the actual partial vapor pressure of the air to the partial vapor pressure of air saturated with water vapor at the same temperature, and is dimensionless.

Relative Humidity.--Psychrometric methods are the simplest and most reasonably accurate methods for determining relative humidity. The temperature difference between the dry and wet bulb of a pair of thermometers is measured. The temperature of the wet thermometer is lowered below that of the dry thermometer by evaporative cooling and equilibrates at the point at which the intake of heat equals the loss of heat by evaporation. The values of the wet and dry bulb temperature measurements can then be entered into psychrometric tables to determine relative humidity. Wet and dry bulb psychrometers are found in a number of configurations. Portable battery-powered fan psychrometers, such as the Bendix Psychron, are designed for accurate determination of relative humidity and dew point in a variety of locations. Sling psychrometers consist of a pair of wet and dry bulb thermometers ventilated by whirling in a circular pattern. Relative humidity can also be measured by the standard hygrothermograph in which humidity change is sensed by a change in the length of bundles of human hairs or by change in dimension of a biomembrane and transmitted to a clock-wound chart through a mechanical linkage system. The reliability of humidity measurements from hygrothermographs depend directly on calibration and maintenance procedures. Generally, hygrothermographs record too low at high humidities and too high at low humidities. Detailed information on the use of the various types of humidity sensors can be found in the Fire-Weather Observer's Handbook (10).

Absolute Humidity, Vapor Pressure, Vapor Pressure Deficit.--Since relative humidity (RH) is related to the partial vapor pressure (p) and the vapor pressure of air saturated with water vapor at the same temperature (p_o), RH = 100 p/p_o, absolute humidity can be calculated from measures of relative humidity and values of saturation vapor pressures determined from table values for specific temperatures. The rate of evaporation from a water source is proportional to the vapor pressure deficit (VPD) of the air, where VPD is defined as the difference between saturation vapor pressure (p_o) and the actual partial vapor pressure of the air (p). In transpiration, leaves are usually assumed to be at saturation and VPD is the difference between the saturation vapor pressure of the leaf and the partial vapor pressure of the air above that leaf. Since leaf temperatures can often be several degrees higher or lower than air temperature, both air and leaf temperature must be known to make accurate calculations of VPD for transpiration. Other than using vapor pressure, absolute humidity deficits can be calculated in the same manner where water vapor is expressed as mass per unit volume rather than vapor pressure. The calculated absolute humidity deficit ($\mu g\ cm^{-3}$) can then be used in conjunction with leaf stomatal conductance units ($cm\ sec^{-1}$) to indicate a transpiration flux density ($\mu g\ cm^{-2}\ sec^{-1}$) from leaves.

Atmospheric CO_2

Water and mineral nutrients are seldom limiting in culturing woody plants under greenhouse and other experimental conditions. However, it is possible for CO_2 concentration in the atmosphere surrounding the seedlings to become limiting to growth. There are a number of observations which indicate that an increase in CO_2 concentration can be accompanied by an increase in tree growth (14). In the culture of containerized seedlings, nursery managers often add CO_2 to the greenhouse environment by use of CO_2 generators. Although precise measurements of CO_2 concentration can be conducted by use of infrared gas analyzers, inexpensive CO_2 testors are commercially available (c.f. Unico, Inc.) with sufficient accuracy for general monitoring of CO_2 levels in greenhouses. Slavik and Catsky (25) describe several simple devices for measuring the CO_2 concentration in air by use of colorimetric means.

LITERATURE CITED

1. Adams, F. 1974. Soil solution. Pages 441-481 *in* E. W. Carson, ed. The Plant Root and its Environment. Univ. Press of Virginia, Charlottesville. 691 pp.
2. Allison, L. E. 1965. Organic carbon. Pages 1367-1378 *in* C. A. Black, ed. Methods of Soil Analysis, Part 2, Agronomy 9, Am. Soc. Agron., Madison, WI. 802 pp.
3. Beardsell, M. F., Jarvis, P. G., and Davidson, B. 1972. A null-balance diffusion porometer suitable for use with leaves of many shapes. J. Appl. Ecol. 9:677-690.
4. Bremner, J. M. 1965. Inorganic forms of nitrogen. Pages 1179-1237 *in* C. A. Black, ed. Methods of Soil Analysis, Part 2, Agronomy 9, Am. Soc. Agron., Madison, WI. 802 pp.
5. Bremner, J. M. 1965. Total nitrogen. Pages 1149-1178 *in* C. A. Black, ed. Methods of Soil Analysis, Part 2, Agronomy No. 9, Am. Soc. Agron., Madison, WI. 802 pp.
6. Bremner, J. M. and Tabatabai, M. A. 1972. Use of an ammonia electrode for determination of ammonium in Kjeldahl. Comm. Soil Sci. Plant Anal. 3:159-165.
7. Brown, R. W. and Van Haveren, B. P. (eds.). 1972. Psychrometry in water relations research. Proc. Symp. on Thermocouple Psychrometers. Utah Agric. Exp. Stn., Utah State Univ., Logan, UT. 342 pp.
8. Council on Soil Testing and Plant Analysis. 1980. Handbook on reference methods for soil testing. Revised edition. Univ. of Georgia, Athens, GA. 130 pp.
9. Day, P. R. 1965. Particle fractionation and particle-size analysis. Pages 545-567 *in* C. A. Black, ed. Methods of Soil Analysis, Part I, Agronomy 9, Am. Soc. Agron., Madison, WI. 770 pp.
10. Fischer, W. C. and Hardy, C. E. 1976. Fire-weather observer's handbook. USDA For. Serv. Agric. Handb. No. 494. 152 pp.
11. Hillel, D. 1971. Soil and water: physical principles and processes. Academic Press, NY. 288 pp.
12. Kaufmann, M. R. 1968. Evaluation of the pressure chamber technique for estimating plant water potential in forest tree species. Forest Sci. 14:369-374.
13. Kaufmann, M. R. and Eckard, A. N. 1977. A portable instrument for rapidly measuring conductance and transpiration of conifers and other species. Forest Sci. 23:227-237.
14. Kramer, P. J. 1981. Carbon dioxide concentration, photosynthesis, and dry matter production. Bioscience 31:29-33.
15. Lea, R., Ballard, R., and Wells, C. G. 1980. Amounts of nutrients removed from forest soils by two extractants and their relationship to *Pinus taeda* foliage concentrations. Comm. Soil Sci. Plant Anal. 11:957-967.
16. Linacre, E. T. and Harris, W. J. 1970. A thermistor leaf thermometer. Plant Physiol. 46:190-193.
17. McDonald, S. E. and Running, S. W. 1979. Monitoring irrigation in western forest tree nurseries. USDA For. Serv. Gen. Tech. Rep. RM-61, Rocky Mt. For. and Range Exp. Stn., Ft. Collins, CO. 8 pp.
18. Morrow, P. A. and Slatyer, R. O. 1971. Leaf resistance measurements with diffusion porometers: precautions in calibration and use. Agric. Meteorol. 8:223-233.
19. Reifsnyder, W. E. and Lull, H. W. 1965. Radiant energy in relation to forests. USDA For. Serv. Tech. Bull. No. 1344. 111 pp.
20. Richards, L. A. 1965. Physical conditions of water in soil. Pages 128-152 *in* C. A. Black, ed. Methods of Soil Analysis, Part 1, Agronomy 9, Am. Soc. Agron., Madison, WI. 770 pp.

21. Richards, S. J. 1965. Soil suction measurements with tensiometers. Pages 153-163 *in* C. A. Black, ed. Methods of Soil Analysis, Part I, Agronomy 9, Am. Soc. Agron., Madison, WI. 770 pp.
22. Ritchie, G. A. and Hinkley, T. M. 1975. The pressure chamber as an instrument for ecological research. Adv. in Ecol. Res. 9:165-254.
23. Rosenberg, N. J. 1974. Microclimate: the biological environment. John Wiley & Sons, NY. 315 pp.
24. Sands, R. and Reid, C. P. P. 1980. The osmotic potential of soil water in plant/soil systems. Aust. J. Soil Res. 18:13-25.
25. Slavik, B. and Catsky, J. 1965. Colorimetric determination of CO_2 exchanges in field and laboratory. Pages 291-298 *in* F. E. Eckardt, ed. Methodology of Plant Ecophysiology. Proc. Montpellier Symp. 1962, Arid Zone Res. XXV, UNESCO. 531 pp.
26. Soltanpour, P. N. and Schwab, A. P. 1977. A new soil test for simultaneous extraction of macro- and micro-nutrients in alkaline soils. Comm. Soil Sci. Plant Anal. 8:195-207.
27. Soltanpour, P. N. and Workman, S. M. 1981. Soil testing methods used at CSU Soil Testing Laboratory for the evaluation of fertility, salinity, sodicity, and trace element toxicity. Colorado State Univ. Exp. Stn. Tech. Bull. 142, December, 1981. Ft. Collins, CO. 22 pp.
28. Tinus, R. W. and McDonald, S. E. 1979. How to grow tree seedlings in containers in greenhouses. USDA For. Serv. Gen. Tech. Rep. RM-60. Rocky Mt. For. and Range Exp. Stn., Ft. Collins, CO. 256 pp.
29. Watanabe, F. S. and Olsen, S. R. 1965. Test of an ascorbic acid method for determining P in water and $NaHCO_3$ extracts from soil. Soil Sci. Soc. Am. Proc. 29:677-687.
30. West, P. W. and Ramachandran, T. P. 1966. Spectrophotometric determination of nitrate using chromotropic acid. Anal. Chim. Acta. 350:317-324.
31. Wiebe, H. H., Campbell, G. S., Gardner, W. H., Rawlins, S. L., Cary, J. W., and Brown, R. W. 1971. Measurement of plant and soil water status. Utah Agric. Exp. Stn. Bull. 484. 71 pp.

RADIOTRACER METHODS FOR MYCORRHIZAL RESEARCH

L. H. Rhodes and M. C. Hirrel

Introduction

Currently, the study of the physiology of mycorrhizae is largely concerned with the movement of compounds from one point to another: movement in soil, through hyphae, and between fungus and host. Radiotracer methods are ideally suited for the study of this movement. These methods, however, are exacting, and minor flaws in technique can lead to major misinterpretations of results. It is a credit to those conducting research on mycorrhizae that a variety of studies using vastly different radiotracer methods confirm and complement each other, and in addition, support data from studies using non-tracer methods. In this chapter we will present some general information on the use of radionuclides, and on their particular applications to the study of mycorrhizae.

General Considerations of Radiotracer Methodology

Terminology.--As with any area of technology, an essential terminology has developed around the use of radioactive compounds for experimental purposes. Although it would be impossible to provide a comprehensive glossary, the following terms are basic to any discussion of radiotracer methods.

A radionuclide is an element with an unstable nuclear configuration which emits energetic particles from the nucleus in going toward a more stable form. Such elements are said to be radioactive, and detection of particles emitted permits measurement of radioactivity of the element. The term radioisotope is often used interchangeably (although often incorrectly) with radionuclide. Since these elements may be used experimentally to trace the movement or fate of chemical compounds, they are also referred to as radiotracers.

The standard unit of measurement of radioactivity is the curie (Ci). One curie = 3.7×10^{10} disintegrations per second. Most work with mycorrhizae involves work with initial activities in millicurie (mCi) amounts or less.

An alpha particle is a tightly bound unit of two protons and two neutrons (a Helium nucleus) emitted from the nucleus of certain unstable isotopes. Alpha particle emitters are rarely used in biological radiotracer experiments. A beta particle is a negatively charged particle emitted from a nucleus upon conversion of a neutron to a proton. The beta particle is identical to an electron except for its place of origin. A gamma ray is a high energy photon emitted upon elimination of excess excitation energy in a nucleus. Both beta and gamma emitters are frequently used in biological research, including studies on mycorrhizae.

The half-life of a radionuclide is the time required for the activity to decline by one half. This value is constant for any particular radioisotope, e.g., the half-life of ^{32}P is 14.3 days; the half-life of ^{14}C is 5730 years.

One electron-volt is defined as the kinetic energy acquired by an electron passing through a potential gradient of 1 volt. The energy of most particles emitted by radiotracers is given in million electron-volts (MeV).

The shorthand form for designation of a specific nucleus is:

$$^{A}_{Z}\text{Chemical Symbol}_{N}$$

where: A = mass number
Z = proton (atomic) number
N = neutron number

thus:

$$^{32}_{16}P_{16} \quad \text{or, more simply:} \quad ^{32}P$$

(In an older system, the mass number was written as a right superscript.)

Assumptions of radionuclide use.--Radiotracer techniques are based on the fact that certain unstable isotopes of elements emit energetic particles from their nuclei in going toward more stable atomic configurations. These emissions may be detected by various methods, and thus used to trace or quantify the amount of that unstable (radioactive) isotope. In detecting the presence of the radioisotope, one also assumes detection of the stable isotope(s) of the same element. Thus, ^{32}P can be used to trace the movement and chemical reactions of ^{31}P, the predominant, stable isotope of phosphorus.

Here it is important to note that radioisotopes provide a means to trace elements and not compounds. Labeled atoms may be cleaved from the originally administered compound and undergo a multitude of chemical reactions prior to their detection. For example, the presence of ^{14}C activity in one part of a plant after application of a ^{14}C-labeled compound to another part of the plant indicates only that movement of carbon from one place to another has occurred.

It is also assumed that the chemical and physical behavior of the radioisotope is identical to its stable counterpart. In reality, this is not true. Bond strengths may differ considerably between two isotopes. However, for most biological experiments, the commonly-used radioisotopes are considered to behave identically to the stable forms.

Most biological experiments employ extremely low levels of radioactivity. It is assumed that this low level of radioactivity causes no radiation damage to the experimental system. Again, this assumption may not always be valid. Chemical bonds may be broken by ionizing radiation from the tracer. Damage to molecules and thus cells and tissues may result if high levels of activity are concentrated in specific regions.

Safety precautions.--Ultimately, all dangers to humans from radioactive materials result from the interaction of ionizing radiation with living tissue. Thus, any event which would result in excess or unnecessary exposure to radiation must be avoided. Hazards of working with radioactive materials include: 1) Radioactive contamination of the laboratory, e.g., spills, gas escape; 2) Presence of high radiation (although contained) in the work area, e.g., stored materials continually emitting high radiation levels; and 3) Skin contact, ingestion, or inhalation of radioactive materials.

The use of radioactive materials requires that the investigator follow strict procedures to insure maximum safety to all laboratory personnel. Federal regulations govern the acquisition and broad use of radioactive materials. In addition to these regulations, most institutions have prescribed further measures to reduce health hazards. However, the ultimate responsibility for personnel safety rests with the investigator. Thus, a listing of routine safety precautions is provided below:

1. Gloves, lab coats, and safety glasses should always be worn when working with radioactive materials. Handling tools such as tongs and forceps should be used whenever possible.
2. Eating, drinking, or smoking in a laboratory where radioactive materials are being used is not permitted.
3. Whenever possible, work should be done behind adequate shielding.
4. Film badges should be worn and examined regularly when working with high energy radiation or high levels of radioactivity.
5. Radiotracer experiments should be conducted only in designated areas. All radioactive materials, as well as rooms, work surfaces, cabinets, and refrigerators where radioactive materials are used or stored should be clearly labeled.
6. Detailed records of the kinds and amounts of radiotracers used should be kept.
7. Radiation levels of the work area should be monitored regularly.
8. Radioactive spills should be cleaned and reported immediately.

9. Work with radiotracers that may be incorporated into potentially volatile compounds (^{3}H, ^{14}C, ^{35}S, etc.) should be done under a fume hood or in specially-vented rooms.
10. Plastic-backed absorbent toweling should be used to cover all work surfaces. Where possible, enameled metal trays covered with absorbent toweling should be used to transport experimental materials in the laboratory.

Methods for Quantitative Assay

Gas ionization.--One method of quantitatively assaying radiation emitted from a source is by the ionization of gas in a chamber. The most common detector of this type is a Geiger-Muller (G-M) detector. Gas ionization is most effectively used for beta counting. Beta particles enter a chamber containing an ionizable gas and an anode. Ionization of the gas results in a flood of electrons, which rapidly migrate toward the anode in the chamber. This flood of electrons results in an electrical pulse, which is then registered on a scaler.

Liquid scintillation.--With liquid scintillation the sample to be counted is immersed in a 'fluor.' Molecules of the fluor become excited by the energetic particles passing through it and give off photons. These photon emissions (scintillations) are converted into electrical pulses which are registered as individual counts. Liquid scintillation counting is highly efficient for weak beta emitters such as ^{3}H and ^{14}C. Since the sample is intimate contact with and completely surrounded by the fluor, a high proportion of the emissions are detected.

Solid scintillation.--The same principle of interaction of ionizing radiation and a fluor is involved in this method. However, as the name implies, the fluors are solid crystals, one of the most common being sodium iodide. The crystals are much more efficient than liquid fluors in interacting with gamma rays and thus are used almost exclusively for gamma counting.

Autoradiography.--The use of photographic emulsions to detect ionizing radiation can be both qualitative and quantitative. The film is placed in close contact with the radioactive sample. Ionizing radiation emitted from the sample interacts with the emulsion, leaving a 'track' where silver chloride has been converted to metallic silver. These tracks will appear as black spots in the processed film. The greater the density of black spots, the higher the incidence of radiation. An optical scanner may be used to more accurately quantify the number of spots per unit area of the film, giving a fairly accurate picture of the relative amounts of radioactivity in various parts of the sample.

Commonly Used Radionuclides

By far the most frequently-used tracers in mycorrhizal research have been ^{32}P and ^{14}C. The use of ^{32}P is understandable, considering the importance of phosphorus in the mycorrhizal symbiosis. Carbon-14 has also been employed in several studies, and with the increased interest in movement of carbon compounds from host to fungus, the use of ^{14}C will likely continue to increase.

There are several elements for which convenient radiotracer techniques unfortunately do not exist. Most notable of these is nitrogen. Nitrogen-15 (heavy nitrogen) is a stable isotope and thus cannot be detected by radioassay. Other elements which could be studied to great advantage with radiotracer techniques have isotopes which do no readily lend themselves to experiments with mycorrhizae. The best example of these is ^{64}Cu, which has a half-life of 12.8 hr. The importance of copper in the stimulation of plant growth by mycorrhizae has been aptly demonstrated. We have attempted to use ^{64}Cu in translocation experiments, but have been unsuccessful, owing to the disappearance of activity during the period of exposure.

The following radionuclides are those most frequently used in experiments with mycorrhizae.

Phosphorus-32.--(beta, 1.71 MeV; Half-life = 14.3 days). Radiation emitted by ^{32}P is easily measured with a variety of types of instrumentation. The high-energy beta particles can be readily detected with a G-M counter. Liquid scintillation counting is also suitable for assay of ^{32}P, and in fact, water may be substituted for the fluor since the high energy beta particles passing through water cause Cerenkov radiation to be emitted. Solid scintillation detectors can also be used to detect Bremsstrahlung (braking radiation).

Phosphorus-32 can be obtained in a variety of inorganic salts, as phosphoric acid, or in organic compounds. It is one of the least expensive of all radionuclides. Phosphorus-32 requires storage behind brick, metal, or other heavy shielding. Likewise, prolonged work with ^{32}P should be done behind shielding, whenever possible. The 2-week half-life of ^{32}P is suitable for most experiments with mycorrhizae, yet decay to near background levels occurs in a relatively short time so that waste disposal is not a substantial problem. Because of the high-energy beta, ^{32}P autoradiographs usually have only fair resolution, although microautoradiographs of ^{32}P in VA mycorrhizal fungus structures presented by Bowen et al. (4) show extremely fine detail. Phosphorus-32 has been used to study uptake of phosphorus from soil and solution culture; translocation of P from fungus to host; movement of P between two plants linked by mycorrhizal hyphae; localization of P in internal fungal structures; and incorporation of absorbed P into plant and fungal metabolites (2,4,5,6,7,8,9,10,11,14,15,16,17, 18,22,25,26,30,33,34,35,36,39,40,41,44,45,46)

Carbon-14.--(beta, 0.156 MeV; Half-life = 5730 years). Although ^{14}C activity may be measured with a G-M detector, liquid scintillation counters are usually preferred because of their higher counting efficiencies with this radionuclide. Autoradiography can be used successfully to determine the localization of ^{14}C in plant and fungal components of mycorrhizae. The low-energy beta results in relatively high-resolution autoradiographs.

Carbon-14 can be obtained in carbonate salts or in a multitude of organic compounds. Within a given molecule, all carbon atoms may be labeled (uniform labeling) or only specific atoms may be ^{14}C. Since ^{14}C-labeled sugars, amino acids, vitamins, etc. may be readily broken down by microorganisms, they should be stored dry, frozen, or in some way protected from microbial degradation.

Carbon-14 is not particularly hazardous as an external source of radiation since most of its low-energy beta radiation is absorbed by the walls of glass containers, or in a few decimeters of air. Hazards of ^{14}C result from: 1) Its potential liberation as $^{14}CO_2$; 2) Its high rate of incorporation into living tissue; and 3) Its extremely long half-life. Because of the possibility of evolution of $^{14}CO_2$, experimental work with ^{14}C should be done in specially vented rooms or under fume hoods to prevent radioactive contamination of the laboratory atmosphere.

Carbon-14 has been used to study the flow of carbon from hosts to fungal symbionts; incorporation of carbon obtained from hosts into fungal metabolites; decomposition of cellulose by orchid mycorrhizal fungi; hyphal translocation and transfer of organic compounds to mycorrhizal hosts; and interplant transfer of carbon via hyphal connections (1, 2, 9, 13, 19, 23, 24, 32, 33, 38, 47, 48).

Sulfur-35.--(beta, 0.167 MeV; Half-life = 87.2 days). Sulfur has received little attention in studies on mineral nutrition of mycorrhizae, probably because of the difficulties involved in determining sulfur concentration in plant tissue using standard analytical methods. Sulfur-35 thus provides a useful alternative for the study of mycorrhiza-sulfur relationships. Sulfur-35 has been employed in studies on nutrient uptake from soil and solutions, and on hyphal translocation and transfer of S (5,12,29,42,43)

Both liquid scintillation and G-M detectors are satisfactory for ^{35}S assay. The tracer can be obtained in a variety of chemical forms, including sulfate,sulfite, and sulfide salts; as sulfuric acid, in organic compounds, as sulfur dioxide, or as elemental sulfur.

Zinc-65.--(gamma, 1.11 MeV; Half-life = 245 days). A high-energy gamma emitter, ^{65}Zn can be hazardous as an external source of radiation in the laboratory. When working with ^{65}Zn, the handler should be protected by lead or other suitable high-density shielding even if only microcurie amounts are being used. The relatively long half-life also contributes to the problem of waste disposal. For these reasons, ^{65}Zn should be chosen as a research tool only when non-tracer methods are inadequate.

Zinc-65 has been used to study uptake of Zn from solution culture for both VA and ectomycorrhizae (3, 8) and translocation by external hyphae of VA mycorrhizae (5, 44). Although ^{65}Zn activity may be measured with G-M or liquid scintillation detectors, counting efficiencies are extremely low, and thus solid scintillation (NaI crystal) detectors are much preferable for this high energy gamma emitter.

Other radiotracers.--The reactions of calcium and strontium (Group IIA elements) are very similar. Investigations of ^{90}Sr uptake (21) and hyphal translocation of ^{45}Ca (27, 41) indicate the importance of mycorrhizae in the utilization of these ions by plants. Other tracers which have found limited use in mycorrhizal research include ^{86}Rb, as a tracer for K^+ efflux from citrus roots (37); ^{54}Mn, to determine Mn uptake rates from solutions (8); ^{33}P, for autoradiography of the depletion zones around mycorrhizal and nonmycorrhizal onion roots (31); and ^{134}Cs, to determine hyphal translocation of Cs (A.-C. McGraw, pers. comm.)

Experimental Systems and Techniques

Uptake of ions from soil and solutions.--Several researchers have studied short-term uptake of nutrients from soil by applying radiolabeled compounds to the soil after mycorrhizal or nonmycorrhizal root systems have had sufficient time to develop. The root or mycorrhizal systems are exposed to the tracers, usually after several weeks of growth, either by pouring a dilute radiolabeled solution into the soil (11, 29), by injecting the tracers into the soil (10, 12, 43), or by transferring the intact soil and root mass to labeled soil (21). One disadvantage of this type of experiment is the non-uniform exposure of roots and hyphae to the tracer solutions. Also, removal of radioactive soil from the root masses can be difficult. Finally, the procedure is likely to generate large volumes of radioactive soil and rinse water.

A second type of nutrient uptake experiment involves the absorption of ions directly from radiolabeled solutions by mycorrhizal and nonmycorrhizal roots (3, 8, 14, 15, 22, 39, 43). A variety of experiments have been conducted with both excised roots and with intact plants, and have involved both endo- and ectomycorrhizae. Phosphorus-32 has been used most extensively. Nutrient solutions with various concentration of carrier ion are labeled with the appropriate tracer. Roots or mycorrhizae are immersed in the labeled solution for a short period, usually 30 min or less, removed from the radioactive solution, and after an appropriate period of desorption to remove adsorbed ions, analyzed for radioactivity. In nearly all cases ion uptake by mycorrhizal roots has occurred at a more rapid rate than that of nonmycorrhizal roots. However, the use of excised roots and stirred nutrient solutions has prompted some questions on the relevance of these techniques to soil situations. There is now reasonable agreement that much of the response of plants to mycorrhizae is a direct result of the ion-diffusion-limiting properties of soils, and that these effects are negated in stirred nutrient solutions. Also the use of excised roots eliminates the effects of ion transport to the shoot. Nevertheless, the results of numerous such studies provides clear evidence for the more efficient absorption of ions from solution by mycorrhizae.

Some studies have used radiotracers to study the partitioning of absorbed nutrients between fungus and host. The classic studies of Harley and McCready (15), for example, showed that much of the ^{32}P absorbed by excised beech mycorrhizae was concentrated in the fungal mantle prior to its release to host cortical cells. Absorbed ^{32}P is also initially concentrated in internal structures of VA mycorrhizal fungi, as shown by counts of infected and uninfected root segments (11) and by autoradiography (4).

Movement of nutrients from fungus to host.--Several systems have been devised to study the translocation of radiolabeled nutrients through external mycorrhizal hyphae and their ultimate release to the plant symbiont. Melin and Nilsson (25) were the first to devise such a system. Pine seedlings were grown in a flask partially-filled with a sterilized sand medium, and a petri dish culture of an ectomycorrhizal fungus was placed on the surface of the sand. Hyphae grew over the edge of the plate and into the soil, and formed mycorrhizae with the pine seedling. Labeled nutrients were then applied to the petri dish culture, and tracer movement through mycelia and into the pine seedlings was monitored (25, 26, 27, 28). Although the evolution of experimental systems and techniques has been considerable since these early experiments, all subsequent studies on hyphal translocation have involved the same concept, i.e., separation of roots from a portion of the external mycelium, exposure of external mycelium to the radiotracer, and assay of plant parts for presence of the tracer.

To study the movement of ions from some point on the external mycelium into roots, tracers are injected into soils, or applied to an agar medium, in both cases insuring that the tracer does not reach roots by diffusion or mass flow through the soil or agar. Phosphorus has extremely low mobility in soil, and ^{32}P injected into sandy loam soil mixes has been shown to move 1.5 cm or less from the point of injection (16, 40). This degree of movement will vary with soil type, soil moisture conditions and levels of phosphorus and other nutrients in the soil. Sulfur, on the other hand, moves much more readily with the soil solution, and it is considerably more difficult to prevent ^{35}S from reaching roots by mass flow or diffusion. This is an important point, since those nutrients which are easy to work with in this respect are also the ones likely to be important in the mycorrhizal growth response.

Pearson and Read (33) used 'split plates' to study movement of compounds through external hyphae of ericoid mycorrhizae. In this system the plastic divider of the plate and a lanolin coating over this barrier was used to restrict non-translocational movement of nutrients to plant roots. Split plates have also been used successfully with VA mycorrhizae (34, 5, 6). Since the labeled nutrient is applied uniformly over the agar surface containing external hyphae, and since the concentration of this nutrient in the agar is known, the total amount of nutrient taken up and translocated to the host may be calculated. This calculation is based on the assumption of uniform mixing and availability of labeled and unlabeled nutrient in the agar medium.

Englander and Hull (9) have used ^{32}P translocation to obtain evidence for a mycorrhizal relationship between a Clavaria sp. and Rhododendron. The tracer was applied to mature fruiting bodies in the vicinity of Rhododendron plants and subsequently was found in high concentrations in roots, but not in soil immediately beneath the treated fruiting bodies. Although definitive evidence for a mycorrhizal symbiosis between Clavaria and Rhododendron must be provided by experimental synthesis of mycorrhizae, the results of these experiments strongly suggest this possibility. This technique could be particularly useful in determining relationships between basidiocarps found regularly associated with suspected ectomycorrhizal hosts.

Skinner and Bowen (46) studied translocation of ^{32}P in hyphal strands of Rhizopogon luteolus to Pinus radiata roots, and determined that transport of the tracer could occur from as far as 12 cm. Similarly, Rhodes and Gerdemann (40) found that Glomus fasciculatus was capable of translocating ^{32}P over distances of at least 7 cm. The phenomenon of nutrient translocation to plant roots is certainly an important feature of the mycorrhizal symbiosis. and one in which radiotracers will continue to provide new information.

Movement of nutrients from host to fungus.--Radiotracers have also been used to study the flow of carbon and other elements from host to fungus. Lewis and Harley (23) traced the fate of ^{14}C photosynthate from leaves of Fagus sylvatica to its storage as glycogen in the ectomycorrhizal symbiont. Ho and Trappe (20) followed the movement of ^{14}C from mycorrhizal fescue plants to spores of Endogone (Glomus) mosseae. The work of Bevege et al. (1) has confirmed and elaborated on information from these studies for both ecto- and VA mycorrhizae.

Perhaps the simplest method to incorporate ^{14}C into plant compounds has been to apply ^{14}C-labeled glucose or sucrose solutions directly to the foliage of experimental plants (18, 38). Since it is not necessary (or at least has not been to date) to know the fate of any particular carbon atom, uniformly labeled sugars (i.e., all carbon atoms are ^{14}C) are preferable. These will have the highest activity per unit of compound. As an alternative to direct foliar application of ^{14}C-labeled solutions, plant carbon compounds may be labeled by allowing the plant to utilize atmospheric $^{14}CO_2$ to photosynthesize ^{14}C-glucose. Labeling an enclosed atmosphere with $^{14}CO_2$ is usually done through chemical reactions which liberate $^{14}CO_2$ from solid or liquid ^{14}C-labeled compounds. Ho and Trappe (20) used Escherichia coli to evolve $^{14}CO_2$ from liquid cultures containing ^{14}C-sucrose. Labeled CO_2 may also be generated by addition of lactic acid to $Na_2{}^{14}CO_3$ (1) or $Ba^{14}CO_3$ (19) in an enclosed space.

Bevege et al (1) followed the incorporation of ^{14}C into organic compounds after pulse labeling mycorrhizal and non-mycorrhizal Araucaria cunninghamii and Pinus radiata seedlings. The various organic fractions were extracted and separated by chromatographic methods, and the percent of label in each fraction was measured. Using these techniques they determined that C is transferred from host to VA mycorrhizal fungus primarily in the form of sucrose, as is the case with ectomycorrhizae. However they were unable to detect incorporation of ^{14}C into specific fungal metabolites. In the Pinus radiata ectomycorrhiza, however, ^{14}C was rapidly incorporated into the fungal carbohydrates trehalose and mannitol and ultimately into glycogen for storage as had been previously shown for ectomycorrhizae of beech (23).

Lösel and Cooper (24) have conducted studies on the incorporation of ^{14}C photosynthate into organic fractions of mycorrhizal and nonmycorrhizal plants. In addition to the exposure of plant tops to $^{14}CO_2$, excised roots were exposed to labeled acetate, glycerol, or sucrose. In all cases, mycorrhizal roots had significantly higher rates of incorporation of ^{14}C. Of interest was the lack of a significant difference between mycorrhizal and nonmycorrhizal roots in the incorporation of ^{14}C into the lipid fraction. This may have resulted from low activities in the relatively brief exposure to the tracer.

Interplant transfer of nutrients.--Of considerable interest ecologically is the question of movement of both inorganic and organic compounds from one plant to another via interconnecting mycorrhizal hyphae. Bjorkmann (2) appears to have been the first to obtain results indicating the involvement of mycorrhizae in interplant nutrient transfer. Both ^{14}C and ^{32}P were traced from trunks of spruce and pine trees to adjacent Monotropa plants. Woods and Brock (51) performed a similar experiment with maple and associated tree species. Phosphorus-32 and ^{45}Ca were injected into the maple stump and subsequently detected in several tree species in close proximity to the treated stump. They suggested that interconnecting mycorrhizal hyphae may have played a role in movement of the radiotracers between the various species. A close examination of their species list reveals that both VA and ectomycorrhizal species were involved, so that it is likely that root exudation and absorption of exuded ions by adjacent roots probably also occurred.

Controlled laboratory experiments with tracers also indicate that plant-to-plant movement of compounds occurs through mycorrhizal hyphae, although the magnitude and significance of this phenomenon has not yet been determined. Details of the various experimental systems have varied widely; however, the general method has involved the application of a radiotracer to the foliage of one plant (sometimes called the 'donor') followed by the monitoring of an adjacent plant (the 'recipient') for the appearance of the tracer. When potentially volatile radiolabeled compounds are used, the atmosphere surrounding donor and recipient plants may be physically separated. Thus, it is assumed that all transfer of nutrients between the two plants occurs between roots, presumably facilitated by interconnecting hyphae. Hirrel and Gerdemann (19) and Heap and Newman (18) grew donor and recipient plants in the same pot and found that in most cases the presence of mycorrhizae enhanced the movement of ^{14}C (19) and ^{32}P (18) from one plant to another. Tracers were also transferred, although to a lesser extent, in nonmycorrhizal treatments. One problem with this type of system is that the possibility of exuded ions or organic compounds being taken up by adjacent roots (or hyphae) cannot be excluded.

Carbon-14 transfer between two pine seedlings was shown by Reid (38). Mycelial strands of Thelephora terrestris from one seedling were brought into close contact with roots of a second seedling. Labeled glucose was applied to the foliage of the donor, and ^{14}C activity was rapidly detected in the root system of the recipient plant.

Determination of P forms available to mycorrhizal plants.--A technique which has been used to determine whether or not mycorrhizae have access to forms of soil P other than those available to plant roots, i.e., whether mycorrhizae have the ability to solubilize relatively insoluble phosphate compounds, involves the radiolabeling of the 'labile pool' of soil phosphate. The labile pool is considered to be soil solution P and forms in rapid equilibrium with it. Phosphorus-32 is mixed with soil and allowed to equilibrate with soil P over a period of several days. Periodic soil extracts are taken to determine if ^{32}P has reached equilibrium with the labile pool of soil P. Mycorrhizal and nonmycorrhizal plants are established in the soil as soon as the activities of soil extracts become consistent. After a suitable period for growth, plants are harvested and analyzed for total P and ^{32}P. Should phosphates be solubilized and utilized by mycorrhizal plants, the specific activities, i.e., the ratio of ^{32}P to total P, will be lower in those plants. This is because the ^{31}P atoms in the insoluble phosphates do not exchange with ^{32}P added to the soil, and thus remain entirely nonradioactive (or nearly so). Any utilization of these phosphates will consequently result in a higher proportion of ^{31}P in plant tissue.

Using this method, Sanders and Tinker (45), Hayman and Mosse (17), and Powell (36) have determined that VA mycorrhizae do not have access to forms of phosphate other than those readily available to nonmycorrhizal plants. To our knowledge, this technique has not been used with other types of mycorrhizae.

Since the half-life of ^{32}P is only 14.3 days, it is necessary to apply a correction factor to the activities obtained from soil solution extracts during the equilibration period. This is easily done since the activity of any radionuclide after decay over a period of time, t, can be easily calculated from the following equation if the initial activity, A_o, is known:

$$A_t = \frac{A_o}{2^{t/T_{1/2}}}$$

where: A_t = Activity after time, t

A_o = Initial activity (zero time)

t = time over which decay has occurred

$T_{1/2}$ = half-life of the radionuclide

Soil sterilization by gamma irradiation.--Although not specifically a radiotracer technique, soil sterilization by exposure to a gamma source deserves some consideration in a discussion of radionuclide uses. Because there is less alteration of the soil chemistry with gamma irradiation than with autoclaving, steaming, or chemical treatment, irradiation has been advocated for work with mycorrhizae. The method is widely used in Great Britain, but has rarely been employed in the United States. Partial sterilization of the soil is accomplished by placement of the soil beneath a source of gamma radiation such as ^{60}Co. Although there is no published data on the minimum dose required to kill mycorrhizal fungi, a dose of 0.8 Mrad appears to be sufficient to eliminate VA mycorrhizal fungi. Irradiation effects on soil fertility also deserve study. Stribley et al (49) found a considerably greater release of nitrogenous compounds with gamma irradiation of heathland soil than with soil not treated by gamma irradiation. The high levels of available N in one irradiated soil were sufficient to negate effects of ericoid mycorrhizae. With gamma-irradiated soil available nutrient levels may closely approximate autoclaved soil, causing difficulties in determination of the effects of ericoid mycorrhizae on plant nutrition.

Localization of nutrients by autoradiography.--The most frequent purpose for using autoradiography in mycorrhizal research has been to determine the distribution of applied radiotracers in host and fungal structures. One of the simplest and most effective methods for microautoradiography is the stripping-film technique (50, 7). Stained tissue sections are prepared on microscope slides. A prepared emulsion, e.g., Kodak AR-10, is stripped from a glass plate and floated on water. The slides with the radioactive tissue sections are then brought up under the floating pieces of emulsion. Upon contact, the emulsion adheres to the slide and is tightly appressed to the tissue section. After a period of exposure the film is developed using standard solutions. The black spots appearing on the film correspond to the radioactive host and fungal structures immediately beneath them. By successively focusing on the film and the tissue with the microscope, the areas of high tracer density can be determined.

Counting Efficiency and Accuracy

The term 'counting efficiency' takes into account a number of factors which determine how many of the emissions from the sample are actually detected by the instrument. For some instrument-radionuclide combinations, counting efficiencies may be quite low, e.g., only 1 percent of the high-energy gamma rays might be detected by an end-window G-M detector. Conversely, nearly 100 percent of the beta particles emitted by ^{14}C might be detected in a liquid scintillation counter. Some of the factors affecting counting efficiency, such as the electronic capabilities of the instrument, are beyond the control of the investigator. In some instances, however, counting efficiencies can be improved by adjustments in methodology

Accuracy and repeatability of radioactivity measurements depend largely on sound procedures in sample preparation and counting. Most of the factors that affect counting accuracy are specific to the method of assay being used. Space permits mention of only a few of the problems which are particularly relevant to work with mycorrhizae.

Self-absorption.--This occurs in G-M counting when particles emitted from atoms in the inner or lower portions of a sample have insufficient energy to emerge from the sample and enter the detector. Measurement of activity in intact plant parts with a G-M detector can result in considerable loss of counts from low energy particles which cannot penetrate throughthe outer layers of tissue. Self-absorption can be a serious problem when using ^{14}C, ^{35}S, and ^{45}Ca, and all but excludes the use of G-M counters for ^{3}H assay. If all samples are of equal thickness and composition, and one is concerned only with comparisons between samples rather than total activity, then some loss to self-absorption is tolerable. However, if an estimate is to be made of the total activity of a sample, or if samples of different thicknesses are compared, serious errors can result.

Quenching.--This occurs when the sample itself interferes with the light output of the fluor in liquid scintillation counting. Color quenching is one of the most common problems of this type. For example, plant digests may be deeply colored, and when added to the vial containing the fluor, mask the scintillations given off by the interaction of radiation and the fluor. Thus, as sample concentrations are increased, radioactivity will not increase proportionately. Quench correction curves may be constructed by adding increasing contrations of the sample digest to be assayed to known activities of tracer. The appropriate corrections may then be applied to measurements of experimental samples.

Technical contamination.--The high sensitivity of radiotracer techniques also means that even the slightest contamination of equipment or instruments can result in erroneous measurements. For example, we have found that the pressure applied with forceps in handling radioactive root segments is sufficient to cause expression of considerable radioactive sap, which then adheres to the inner surfaces of the forceps. If a non-radioactive root segment is then handled with the same forceps, radioactivity can be transferred to the latter root segment. Obviously, activity recorded for the second sample is erroneous. This is known as 'technical contamination' and undoubtedly occurs to a much greater extent than is realized, particularly in experiments where replicates from different treatments are handled sequentially. If it is not possible to have separate handling instruments for each sample, then it is imperative that forceps, tongs, containers, etc. be continually monitored for radioactivity and if necessary, carefully cleaned prior to use with each new sample.

LITERATURE CITED

1. Bevege, D. I., Bowen, G. D. and Skinner, M. F. 1975. Comparative carbohydrate physiology of ecto- and endomycorrhizas. Pages 149-174 in: Sanders, Mosse, and Tinker, eds., Endomycorrhizas. Academic Press, London. 626 pp.
2. Bjorkman, E. 1960. *Monotropa hypopitys* L. - an epiparasite on tree roots. Physiol. Plant. 13:308-329.
3. Bowen, G. D., Skinner, M. F. and Bevege, D. I. 1974. Zinc uptake by mycorrhizal and uninfected roots of *Pinus radiata* and *Araucaria cunninghamii*. Soil Biol. Biochem. 6:141-144.
4. Bowen, G. D., Bevege, D. I. and Mosse, B. 1975. Phosphate physiology of vesicular-arbuscular mycorrhizas. Pages 241-260 in: Sanders, Mosse, and Tinker, eds., Academic Press, London. 626 pp.
5. Cooper, K. M. and Tinker, P. B. 1978. Translocation and transfer of nutrients in vesicular-arbuscular mycorrhizas. II. Uptake and translocation of phosphorus, zinc and sulphur. New Phytol. 81:43-52.
6. Cooper, K. M. and Tinker, P. B. 1981. Translocation and transfer of nutrients in vesicular-arbuscular mycorrhizas. IV. Effect of environmental variables on movement of phosphorus. New Phytol. 88:327-339.
7. Cox, G., Sanders, F. E., Tinker, P. B., and Wild, J. A. 1975. Ultrastructural evidence relating to host-endophyte transfer in a vesicular-arbuscular mycorrhiza. Pages 289-312 in: Sanders, Mosse, and Tinker, eds., Endomycorrhizas. Academic Press, London. 626 pp.
8. Cress, W. A., Throneberry, G. O., and Lindsey, D. L. 1979. Kinetics of phosphorus absorption by mycorrhizal and nonmycorrhizal tomato roots. Plant Physiol. 64:484-487.
9. Englander, L. and Hull, R. J. 1980. Reciprocal transfer of nutrients between ericaceous plants and a *Clavaria* sp. New Phytol. 84:661-667.
10. Gray, L. E., and Gerdemann, J. W. 1967. Influence of vesicular-arbuscular mycorrhizas on the uptake of phosphorus-32 by *Liriodendron tulipifera* and *Liquidambar styraciflua*. Nature 213:106-107.
11. Gray, L. E. and Gerdemann, J. W. 1969. Uptake of phosphorus-32 by vesicular-arbuscular mycorrhizae. Plant Soil 30:415-422.

12. Gray, L. E. and Gerdemann, J. W. 1973. Uptake of sulphur-35 by vesicular-arbuscular mycorrhizae. Plant Soil 39:687-689.
13. Hadley, G. and Purves, S. 1974. Movement of ^{14}Carbon from host to fungus in orchid mycorrhiza. New Phytol. 73:475-482.
14. Harley, J. L. and McCready, C. C. 1950. Uptake of phosphate by excised mycorrhizal roots of the beech. New Phytol. 49:388.
15. Harley, J. L. and McCready, C. C. 1952. The uptake of phosphate by excised mycorrhizal roots of the beech. III. The effect of the fungal sheath on the availability of phosphate to the core. New Phytol. 51: 342-348.
16. Hattingh, M. J., Gray, L. E. and Gerdemann, J. W. 1973. Uptake and translocation of ^{32}P-labeled phosphate to onion roots by endomycorrhizal fungi. Soil Sci. 116:383-387.
17. Hayman, D. S. and Mosse, B. 1972. Plant growth responses to vesicular-arbuscular mycorrhiza. III. Increased uptake of labile P from soil. New Phytol. 71:41-47.
18. Heap, A. J. and Newman, E. I. 1980. The influence of vesicular-arbuscular mycorrhizas on phosphorus transfer between plants. New Phytol. 85:173-179.
19. Hirrel, M. C. and Gerdemann, J. W. 1979. Enhanced carbon transfer between onions infected with a vesicular-arbuscular mycorrhizal fungus. New Phytol. 83:731-738.
20. Ho, I. and Trappe, J. M. 1973. Translocation of ^{14}C from Festuca plants to their endomycorrhizal fungi. Nature 244:30-31.
21. Jackson, N. E., Miller, R. H. and R. E. Franklin. 1973. The influence of vesicular-arbuscular mycorrhizae on uptake of ^{90}Sr from soil by soybeans. Soil Biol. Biochem. 5:205-212.
22. Kramer, P. J. and Wilbur, K. M. 1949. Absorption of radioactive phosphorus by mycorrhizal roots of pine. Science 110:8-9.
23. Lewis, D. H. and Harley, J. L. 1965. Carbohydrate physiology of mycorrhizal roots of beech. III. Movement of sugars between host and fungus. New Phytol. 64:256-269.
24. Lösel, D. M. and Cooper, K. M. 1979. Incorporation of ^{14}C-labelled substrates by uninfected and VA mycorrhizal roots of onion. New Phytol. 83:415-426.
25. Melin, E. and Nilsson, H. 1950. Transfer of radioactive phosphorus to pine seedlings by means of mycorrhizal hyphae. Physiol. Plant. 3:88-92.
26. Melin, E. and Nilsson, H. 1954. Transport of labelled phosphorus to pine seedlings through the mycelium of Cortinarius glaucopus (Schaeff. ex Fr.) Fr. Svensk Bot. Tidskr. 48:555-558.
27. Melin, E. and Nilsson, H. 1955. Ca^{45} used as an indicator of transport of cations to pine seedlings by means of mycorrhizal mycelia. Svensk Bot. Tidskr. 49:119-122.
28. Melin, E., Nilsson, H. and Hacskaylo, E. 1958. Translocation of cations to seedlings of Pinus virginiana through mycorrhizal mycelium. Bot. Gaz. 119:243-246.
29. Morrison, T. M. 1962. Uptake of sulphur by mycorrhizal plants. New Phytol. 61:21-27.
30. Mosse, B., Hayman, D. S. and Arnold, D. J. 1973. Plant growth responses to vesicular-arbuscular mycorrhiza. V. Phosphate uptake by three plant species from P-deficient soils labelled with ^{32}P. New Phytol. 72:809-815.
31. Owusu-Bennoah, E. and Wild, A. 1979. Autoradiography of the depletion zone of phosphate around onion roots in the presence of vesicular-arbuscular mycorrhiza. New Phytol. 82:133-140.
32. Paul, E. A. and Kucey, R. M. N. 1981. Carbon flow in plant microbial associations. Science 213:473-474.
33. Pearson, V. and Read, D. J. 1973. The biology of mycorrhiza in the Ericaceae. II. The transport of carbon and phosphorus by the endophyte and the mycorrhiza. New Phytol. 72:1325-1339.

34. Pearson, V. and Tinker, P. B. 1975. Measurement of phosphorus fluxes in the external hyphae of endomycorrhizas. Pages 277-287 in: Sanders, Mosse, and Tinker, eds., Endomycorrhizas. Academic Press, London. 626 pp.
35. Pichot, J. and Binh, T. 1976. Action of endomycorrhizae on growth and phosphorus nutrition of Agrostis in pots and on isotopically exchangeable phosphorus in soil. Agronomia Tropical 31:375.
36. Powell, C. L. 1975. Plant growth responses to vesicular-arbuscular mycorrhiza. VII. Uptake of P by onion and clover infected with different Endogone spore types in ^{32}P labelled soils. New Phytol. 75:563-566.
37. Ratnayake, M., Leonard, R. T., and Menge, J. A. 1978. Root exudation in relation to supply of phosphorus and its possible relevance to mycorrhizal formation. New Phytol. 81:543-552.
38. Reid, C. P. P. 1971. Transport of C^{14}-labeled substances in mycelial strands of Thelephora terrestris. Pages 222-227 in: Hacskaylo, ed., Mycorrhizae, Proceedings of the First North American Conference on Mycorrhizae. U. S. Dept. of Agric. Forest Service Misc. Pub 1189, Washington, D. C. 255 pp.
39. Reid, C. P. P. and Bowen, G. D. 1979. Effect of water stress on phosphorus uptake by mycorrhizas of Pinus radiata. New Phytol. 83:103-107.
40. Rhodes, L. H. and Gerdemann, J. W. 1975. Phosphate uptake zones of mycorrhizal and non-mycorrhizal onions. New Phytol. 75:555-561.
41. Rhodes, L. H. and Gerdemann, J. W. 1978. Translocation of calcium and phosphate by external hyphae of vesicular-arbuscular mycorrhizae. Soil Sci. 126:125-126.
42. Rhodes, L. H. and Gerdemann, J. W. 1978. Hyphal translocation and uptake of sulfur by vesicular-arbuscular mycorrhizae of onion. Soil Biol. Biochem. 10:355-360.
43. Rhodes, L. H. and Gerdemann, J. W. 1978. Influence of phosphorus nutrition on sulfur uptake by vesicular-arbuscular mycorrhizae of onion. Soil Biol. Biochem. 10:361-364.
44. Rhodes, L. H., Hirrel, M. C. and Gerdemann, J. W. 1978. Influence of soil phosphorus on translocation of ^{65}Zn and ^{32}P by external hyphae of vesicular-arbuscular mycorrhizae of onion. Phytopathology News 12(9):197 (Abstr).
45. Sanders, F. E. and Tinker, P. B. 1971. Mechanism of absorption of phosphate from soil by Endogone mycorrhizas. Nature 233:278-279.
46. Skinner, M. F. and Bowen, G. D. 1974. The uptake and translocation of phosphate by mycelial strands of pine mycorrhizas. Soil Biol. Biochem. 6: 53-56.
47. Smith, S. E. 1967. Carbohydrate translocation in orchid mycorrhizas. New Phytol. 66:371-378.
48. Stribley, D. P. and Read, D. J. 1974. The biology of mycorrhiza in the Ericaceae. III. Movement of carbon-14 from host to fungus. New Phytol. 73:731-741.
49. Stribley, D. P., Read, D. J. and Hunt, R. 1975. The biology of mycorrhiza in the Ericaceae. V. The effects of mycorrhizal infection, soil type and partial soil-sterilization (by gamma irradiation) on growth of cranberry (Vaccinium macrocarpon). New Phytol. 75:119-130.
50. Wang, C. H., Willis, D. L. and Loveland, W. D. 1975. Radiotracer Methodology in the Biological, Environmental, and Physical Sciences. Prentice Hall, Englewood Cliffs, N. J. 480 pp.
51. Woods, F. W. and Brock, K. 1964. Interspecific transfer of Ca-45 and P-32 by root systems. Ecology 45:886-889.

ELECTRON MICROSCOPY OF MYCORRHIZAE

M. F. Brown and E. J. King

INTRODUCTION

The primary reason for utilizing electron microscopy is to visualize structures of interest at higher magnifications and enormously greater resolution than can be achieved by light microscopy. It has been used increasingly over the past 15 years in descriptive studies of numerous mycorrhizae and, more recently, it has been employed to identify certain chemical constituents of these associations. The greatest challenge in electron microscopy is the initial preservation of structural characteristics of the living system and successful retention of them through the entire preparative sequence for the particular type of microscopic investigation undertaken. The following discussion provides a general overview of the principal factors involved in the preparation of mycorrhizal specimens for transmission electron microscopy (TEM), scanning electron microscopy (SEM) and certain analytical procedures employed at the ultrastructural level.

Transmission Electron Microscopy

Mycorrhizae rank among the most difficult of biological subjects to prepare successfully for TEM. Host tissues are often highly vacuolated and acutely sensitive to osmotic imbalances while the fungal components are frequently densely cytoplasmic and may be aggregated in compact tissues or located within vacuolated host cells. Achieving the goal of optimal preservation in both organisms may require judicious compromise in the establishment of an appropriate preparative procedure, and will require exercise of extreme care in each phase of the protocol. While the basic procedures employed for mycorrhizae are similar to those used for other biological materials (15,17), sample collection, fixation, dehydration and embedding of mycorrhizae merit special consideration.

<u>Collection of specimens</u>.--The initial steps in achieving life-like cytological preservation are to first ensure that the mycorrhizae being collected are in a normal, 'healthy' condition representative of the circumstances being studied and, then to remove them from the growth medium in a manner which minimizes structural damage to the roots and concomitant autolytic changes within the tissues. Since the rate at which autolytic changes occur is temperature dependent, they can be reduced by rinsing the soil from the root system with cold (0^oC) water. These rinses should be performed quickly and gently, but must be thorough enough to remove adherent soil particles which would interfere with ultramicrotomy later. The cleaned roots should be exposed immediately to the chosen fixative and reduced to pieces no larger than 1 mm in the largest dimension. The use of even smaller pieces may be advantageous with mycorrhizae which have a compact mantle or are more than 1 mm in diameter. Small samples are essential for high quality TEM since sample size exerts a profound effect upon the quality of fixation achieved, and is a major factor influencing the length of exposure intervals required during dehydration and embedding sequences.

<u>Fixation techniques</u>.--Fixation is the most important step in TEM protocols. The quality of fixation is greatly influenced by numerous factors including: the type, concentration and pH of fixative(s) used, the type and molarity of buffers, and the duration and temperature of fixation. These variables must be defined experimentally since no single fixation procedure produces optimal results with all mycorrhizae. Most mycorrhizal studies employ glutaraldehyde, in concentrations from 1-6% in 0.05M-0.1M phosphate or cacodylate buffers at

pH 6.8-7.4, as the initial fixative (3,8,11,29,30,34,39,46,47). Fixation is generally performed at 4°C for 1-4 hrs. and is followed by post-fixation for 1-4 hrs. at 4°C in 1-2% OsO_4 in the same buffer used with the aldehyde. Figures 1-3 illustrate three different vesicular-arbuscular (VA) mycorrhizal associations fixed with glutaraldehyde and post-fixed with OsO_4.

Since mycorrhizae are inherently difficult to penetrate and since many structures of interest occur internally to a mantle of fungal tissue or are located deep within the host tissue, both the rate of fixative penetration and the rate at which cytoplasmic stabilization occurs are important considerations in the selection of initial fixatives to be used. Although glutaraldehyde stabilizes cytoplasm rapidly, it penetrates tissues slowly increasing the possibility, with samples such as mycorrhizae, that structural alterations may occur prior to stabilization by the fixative. Glutaraldehyde has been used effectively in combination with other aldehydes (eg., formaldehyde and acrolein) which penetrate tissues more rapidly (10,23,48, J. A. Duddridge, personal communication). The objective with combined aldehyde fixatives is to achieve partial stabilization of cytoplasmic components rapidly with formaldehyde or acrolein, and more complete fixation as glutaraldehyde penetrates the tissue. The excellent quality of cytological preservation which can be achieved with combined formaldehyde-glutaraldehyde fixatives is shown in Figs. 7-10. Since

---

Fig. 1-3. VA mycorrhizae fixed in glutaraldehyde and post-fixed with OsO_4. Fig. 1. An infection in Taxus baccata showing typical preservation of host cytoplasm around the arbuscular branches (A), numerous vacuoles (V) in both symbionts and the host plasmalemma (P) surrounding the endophyte. The interfacial zone between the dense fungal wall and host plasmalemma varies in width and contains varying amounts of moderately dense interfacial materials (arrows). Uniform preservation of the host tonoplast (T) without disruption or displacement (*) is difficult to achieve with glutaraldehyde fixation. x6,500. Micrograph courtesy of D. G. Strullu (47). Fig. 2. Glomus fasciculatus/Ornithogalum umbellatum showing several arbuscular branches (A), some in states of normal deterioration and collapse (CB), and host membrane systems. The interfacial material (IM) is uniformly dispersed in some portions of the interfacial zone while in others (arrows) clear areas are present in samples prepared with this fixative. Host plasmalemma (P). x18,000. Micrograph courtesy of S. Scannerini and P. Bonfante-Fasolo (39). Fig. 3. Glomus mosseae/Allium cepa. With the exception of some outward displacement of the interfacial material (IM) around one arbuscular branch (A), the appearance of the interfacial zone is comparable to Figs. 1 and 2. The preservation of host organelles shown here is representative of glutaraldehyde fixed samples. Plasmalemma (P), mitochondrion (M), dictyosome (D), plastid (PL). x30,500. Micrograph courtesy of J. Dexheimer. Fig. 4, 5. VA mycorrhizae of Glycine max fixed with a mixture of glutaraldehyde and OsO_4 and post-fixed in OsO_4. Fig. 4. The arbuscular branches (A) of Gigaspora margarita are enclosed by host cytoplasm which is appressed quite uniformly around them. Host and fungal organelles are well preserved and clear areas within the interfacial zone are infrequent. Host plasmalemma (P), mitochondrion (M), plastid (PL), vacuole (V), fungal nucleus (N). x6,750. Fig. 5. Small arbuscular branches (A) of Glomus caledonius illustrating the dense material which completely fills the interfacial zone (arrows) in this association in samples prepared by simulaneous glutaraldehyde-OsO_4 fixation. Vacuole (V), mitochondrion (M), plastid (PL). x16,200. M. F. Brown, unpublished micrographs. Fig. 6. Glomus mosseae/Allium cepa fixed solely with a mixture of glutaraldehyde and OsO_4. The arbuscular branches (A) show well preserved fungal organelles but the interfacial material (IM) is dispersed more uniformly, and the host ground plasm and plastid (PL) matrix are less dense than in samples of the same association fixed sequentially with glutaraldehyde and OsO_4 (see Fig. 3). Host plasmalemma (P), mitochondrion (M), dictyosome (D). x24,000. Micrograph courtesy of J. Dexheimer.

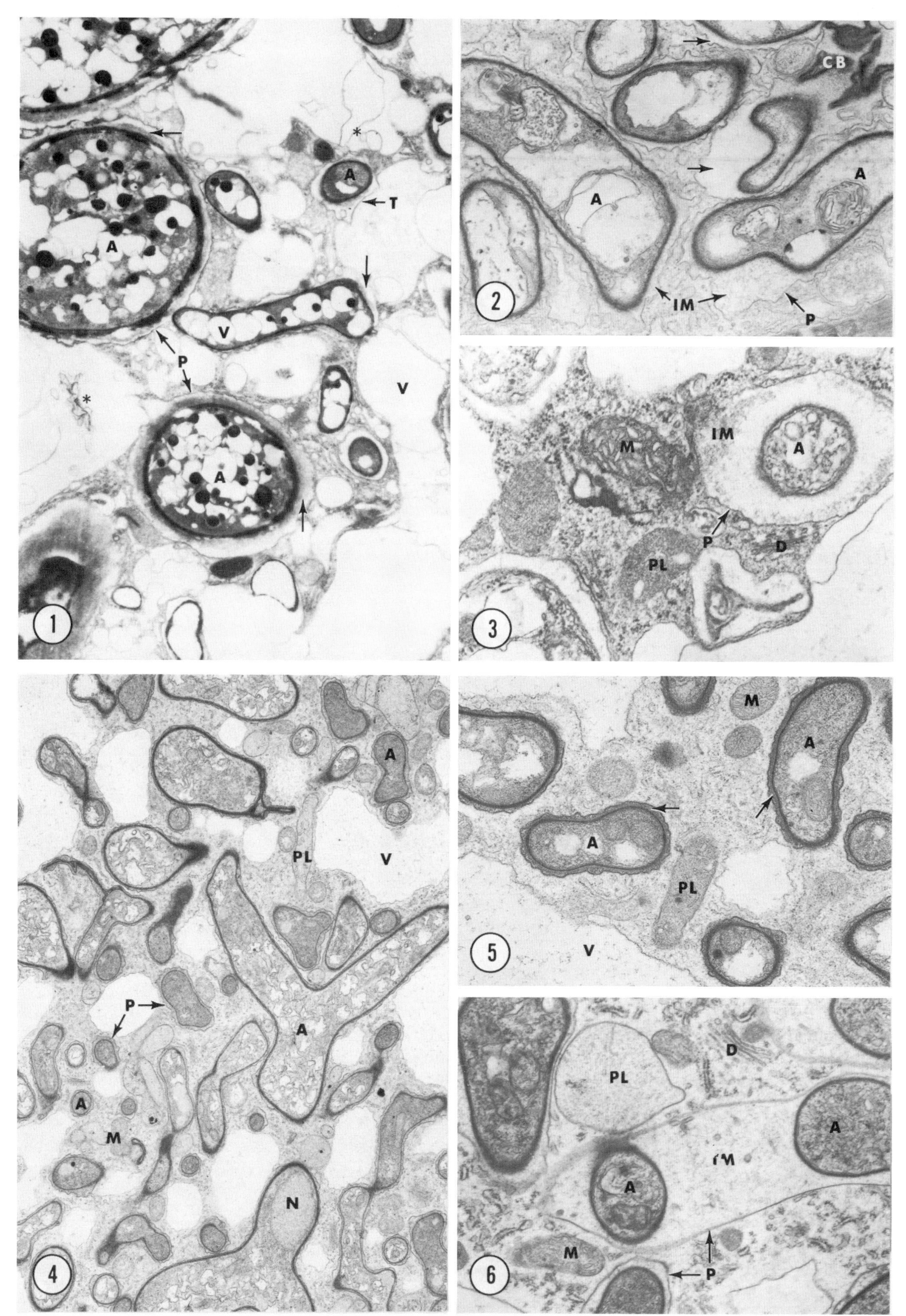
A
T
A
P
V
V
A
1
CB
A
A
2
IM
P
M
IM
A
P
D
PL
3
A
PL
V
P
A
A
M
N
4
M
A
A
PL
V
5
D
PL
A
IM
A
M
P
6

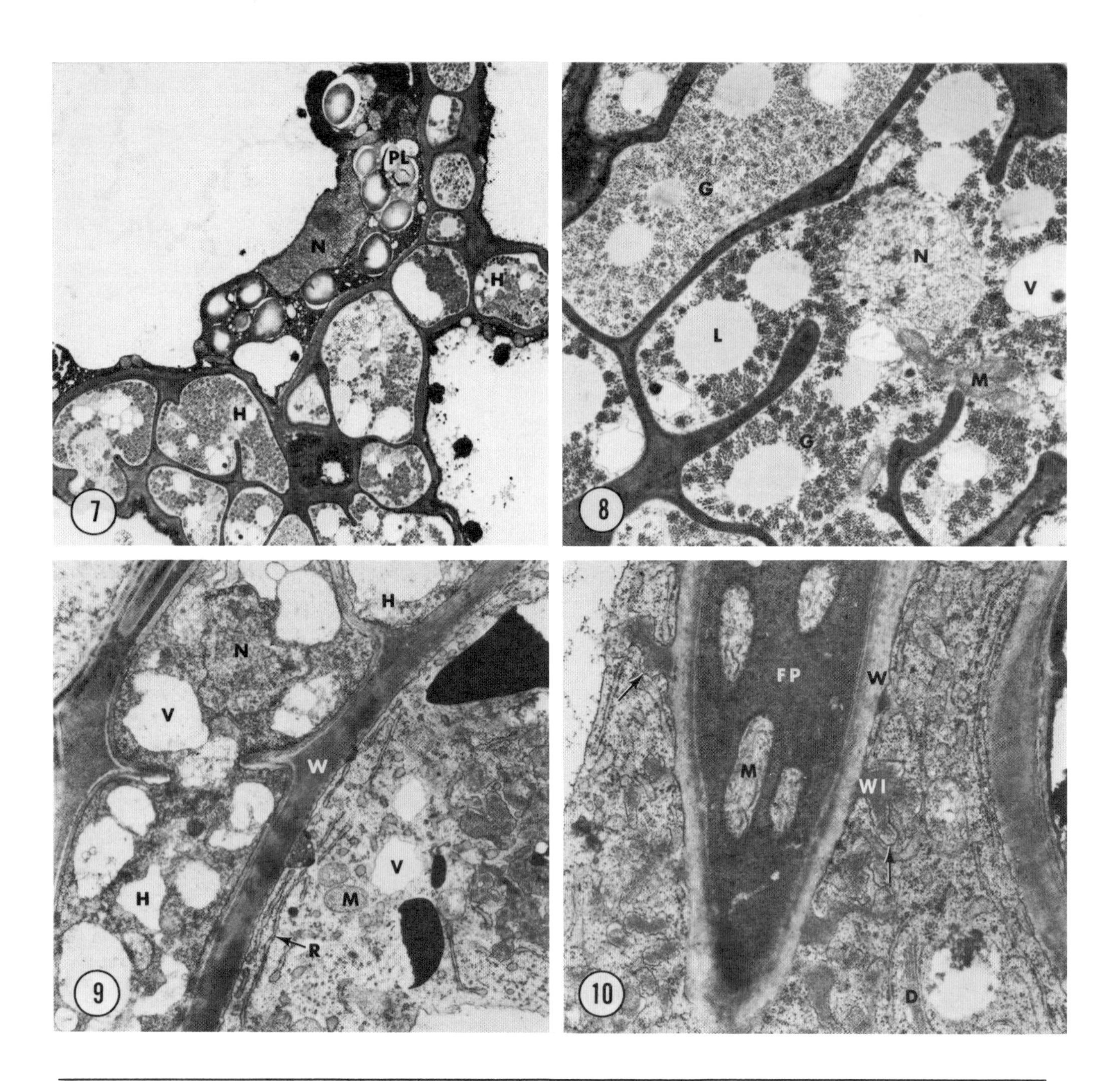

Fig. 7-10. Mycorrhizae fixed with formaldehyde and glutaraldehyde, post-fixed with OsO_4. Fig. 7, 8. Hebeloma alpinum/Dryas octopetala ectomycorrhizae. Fig. 7. Excellent cytological preservation of the cortical cell in the upper left and cells of the Hartig net (H) is illustrated. Unconfined host cytoplasmic components (lower right) indicates loss of tonoplast integrity in that cell demonstrating that the quality of preservation of highly vaculolated components of mycorrhizae may vary from cell to cell even in well fixed specimens. Host nucleus (N), plastid with starch grains (PL). x8,700. Fig. 8. Cells of the Hartig net showing well preserved fungal organelles and cytoplasmic inclusions. Nucleus (N), mitochondrion (M), lipid droplet (L), glycogen (G), vacuole (V). x23,300. Micrographs courtesy of J. C. Debaud (10). Fig. 9, 10. Mycorrhiza of Monotropa hypopitys showing excellent preservation of both symbionts. Fig. 9. Portion of an outer cortical cell (right) and cells of the Hartig net (H). Mitochondrion (M), rough endoplasmic reticulum (R), vacuole (V), host cell wall (W), fungal nucleus (N). x13,500. Fig. 10. Longitudinal section of a mature fungal peg (FP) surrounded by the invaginated cortical cell wall (W). Fungal mitochondrion (M), host wall ingrowth (WI), host plasmalemma (arrows), dictyosome (D). x21,500. Micrographs courtesy of J. A. Duddridge.

these micrographs were obtained using two very different fixation procedures (see TEM protocols) with dissimilar mycorrhizal associations, the quality of preservation attained indicates that formaldehyde-glutaraldehyde fixatives should be seriously considered and evaluated for TEM studies of other mycorrhizae. It should be noted, however, that formaldehyde used for TEM purposes is prepared from dry paraformaldehyde just prior to use. Commercial formalin, which contains methanol, should not be used. While all fixatives should be handled with care in a fume hood, acrolein is especially toxic and must be handled with extreme caution. We do not recommend its use.

In addition to combinations of aldehydes, glutaraldehyde-OsO_4 mixtures have been used as initial fixatives for VA mycorrhizae (5,19,20). Simultaneous glutaraldehyde-OsO_4 fixation, followed by OsO_4 post-fixation (5), has yielded excellent preservation of soybean mycorrhizae (Figs. 4, 5) produced by all VA mycorrhizal (VAM) endophytes examined. While this procedure has been reported to provide no advantages over sequential glutaraldehyde and OsO_4 fixation with VAM infections of *Allium* and *Ornithogalum* (11,39), in our view it merits critical appraisal with other mycorrhizal associations, especially those formed by 'fine rooted' host species. Combined glutaraldehyde-OsO_4 fixation, with no OsO_4 post-fixation, has also been employed with VA mycorrhizae of onion (Fig. 6). However, the quality of cytological preservation achieved with this protocol was considered no better than that obtained with conventional glutaraldehyde, OsO_4 preparations (J. Dexheimer, personal communication).

Dehydration.--The primary concerns of the dehydration sequence are to remove water from the sample fairly rapidly to minimize extraction of cytoplasmic components in the lower concentrations of the dehydrant, and to ultimately achieve complete dehydration. Either acetone or ethanol can be used for this purpose but acetone is preferred by many since it causes fewer shrinkage artifacts and it is miscible with the epoxy resins used for embedding. Ideally, exposure intervals in the lower dehydrant concentrations should be shorter than those above 40-50%. The dehydration series for mycorrhizae should include 3 changes of 100% (absolute) dehydrant no less than 20 minutes each to ensure total dehydration. Some workers exchange the last absolute dehydrant with propylene oxide (PO) and use PO as the diluent in infiltration mixtures.

Infiltration and embedding.--Many different embedding media have been employed with mycorrhizae including Durcopan, Epon (Epikote) 812, Epon-Araldite mixtures, Vestopal and Spurr's resin. Mycorrhizae are difficult to infiltrate thoroughly with resins and, for that reason, those with lower formula weight components and lower viscosities are recommended. In our experience, Spurr's low viscosity resin (44) is the best choice of those media indicated for mycorrhizal samples. However, two problems encountered with Spurr's medium deserve comment: sectioning difficulties may be experienced unless suitable precautions are taken to protect the unpolymerized resin from atmospheric moisture; and longer exposures to uranyl and lead solutions than those used with other embedding media are typically required to achieve comparable staining intensity. Infiltration of resins can be improved by employing a mild vacuum or, preferably, by using longer exposures in each solvent-resin mixture and by slowly rotating (1-2 rpm) the samples continuously throughout the infiltration sequence. A high pressure infiltration technique has been reported to overcome embedding difficulties with pine ectomycorrhizae (50). Infiltration of spores of VA fungi has been achieved by puncturing the spore wall (27) or by treating them briefly with dilute sodium hypochlorite before fixation (48).

Criteria for evaluating sample quality.--While there are no precisely comparable controls for evaluating the quality of preservation achieved with mycorrhizal specimens, it should be expected that they would exhibit ultrastructural characteristics comparable to those of well preserved plant tissues and fungal specimens. Each plasmalemma should be continuous and closely appressed to its respective cell wall. Tonoplasts in host and fungal cells should be intact and continuous. The ground plasm in both symbionts should be uniform in density and should show no voids lacking content. Cytoplasmic organelles should be well preserved in both organisms. Nuclei should have closely spaced, parallel double membranes; mitochondria should have parallel outer membranes and cristae which are more or less parallel in profile. The matrices of mitochondria and plastids should be uniform and, typically, more dense than the ground plasm in the respective symbiont. This quality of preservation is not easily achieved with mycorrhizae but, as demonstrated by the figures included here, it can be attained through the use of appropriate techniques.

TEM protocols.--The following protocols are presented with respect to the initial fixative used since this treatment exerts the greatest influence on the ultimate quality of cytoplasmic preservation achieved. They should serve only as potential 'starting points' for individual determinations of those conditions best suited to one's own materials.

1. Glutaraldehyde fixation. This procedure yields preservation of VA mycorrhizae comparable to that shown in Figs. 1-3. Aside from the buffer pH indicated, it is representative of procedures used in the majority of mycorrhizal TEM publications. Clean soil from roots and cut mycorrhizae into small pieces in cold fixative. Fix for 4 hrs. at 4°C in 3% glutaraldehyde in 0.075M Sorensen's phosphate buffer, pH 6.8. Rinse in three 15-min. changes of buffer at 4°C. Post-fix in 2% OsO_4 (same buffer) for 4 hrs. at 4°C. Dehydrate with acetone at room temperature (RT): 20, 40, 60, 80, 100, 100, 100% (30 min. each). Infiltrate with Spurr's 'standard' resin mixture (44) at RT with continuous rotation using acetone as the diluent: 25, 50, 75, 100% resin (4 hrs. each), 100% resin (overnight). Transfer to flat embedding molds or aluminum weighing dishes in fresh resin and cure 24 hrs. at 65°C.

2. Combined paraformaldehyde-glutaraldehyde fixation.

A. *Monotropa* (J. A. Duddridge, personal communication) (Figs. 9, 10). Fix 2 hrs. in modified Karnovsky's fixative (15, p. 47-48) containing 2% paraformaldehyde and 2.5% glutaraldehyde in 0.1M Sorensen's phosphate buffer, pH 7.2. Rinse in three-20 min. changes of buffer at 4°C. Post-fix in 1% OsO_4 (same buffer) for 1½ hrs. at 4°C. Rinse in three-20 min. changes of buffer. Dehydrate in ethanol: 25% (15 min.), 50, 75, 90, 100, 100, 100% (20 min. each), propylene oxide (PO) (30 min.). Infiltrate with Spurr's 'standard' resin in PO at RT with continuous rotation: 25% (8 hrs.), 50%, 75% resin (24 hrs. each), 100% resin (7 days, changed daily). Flat embed and cure 16 hrs. at 70°C.

B. Ectomycorrhiza (10) after (33) (Figs. 7, 8). Fix 4-8 hrs. under vacuum at RT in 0.5% paraformaldehyde and 2% glutaraldehyde in 0.1M phosphate-citrate (P-C) buffer, pH 6.8. Rinse in 0.2M P-C buffer (pH 6.8) 2 hrs. to overnight at 4°C. Post-fix 8-14 hrs. at RT with 0.5% OsO_4 in 0.1M P-C buffer, pH 6.8 (time inversely proportional to duration of aldehyde fixation). Dehydrate in ethanol and embed in Spurr's resin.

3. Simultaneous glutaraldehyde-OsO_4 fixation. Soybean VA mycorrhizae, after (5) (Figs. 4,5). Remove soil in ice water. Fix in a fresh 1:1 mixture of 2% glutaraldehyde and 2% OsO_4 in 0.075M cacodylate buffer, pH 6.8, for 45 min. in an ice bath. Rinse in three changes of buffer and post-fix in 2% OsO_4 (same buffer) for 4 hrs. at 4°C. Dehydrate in acetone: 20, 40, 60, 80, 100, 100, 100% (30 min. each). Infiltrate with Spurr's 'standard' resin at RT with continuous rotation using acetone as the diluent: 20, 40, 60, 80, 100% resin (4 hrs. each), 100% resin (overnight). Transfer to aluminum weighing dishes in fresh resin and cure 24 hrs. at 65°C.

Scanning Electron Microscopy

Scanning microscopy is employed to study surface morphology, and can be applied to naturally occurring external surfaces of mycorrhizae or, by using suitable preparative techniques, it can also be used to examine their internal characteristics. Since preservation of the sample in a realistic and life-like condition is a prerequisite for high quality SEM studies, many of the considerations discussed in relation to TEM also apply to SEM preparations. However, SEM preparations differ in several respects from those used for TEM. Typically, SEM samples are larger, and therefore require longer exposures to fixatives and dehydration solutions. They are generally prepared without prior infiltration of supporting media and must be absolutely dry when they are examined in the microscope, so water contained in the fixed samples must be removed by dehydration and drying procedures which prevent morphological alterations. In addition, the samples must be made electrically conductive before they are examined in the microscope. In this presentation, we will consider fixation procedures, techniques for obtaining sectional views of mycorrhizae, dehydration and critical point drying, methods for mounting dried mycorrhizal specimens, and criteria for evaluating sample quality. More comprehensive discussions of preparative techniques for SEM and equipment employed for sample preparation than can be included here are available in other sources (4,16,35).

Fixation procedures.--The same fixatives used for TEM are also employed for SEM preparations. However, fixation periods ranging from several hours to several days are preferred in order to achieve maximum stabilization of the sample. Samples may be fixed with aldehydes alone (9,41) (Fig. 11-13) or may also be post-fixed with OsO_4 (3,19,22,45,46) (Fig. 14, 15). The use of aldehyde fixatives, particularly those incorporating formaldehyde or acrolein with glutaraldehyde, is preferred for samples which are to be sectioned for examination of internal structures. Extended OsO_4 post-fixation (ie., 12-48 hrs. at 4°C) in many instances is desirable for SEM preparations since it substantially hardens the samples, rendering them less susceptible to drying artifacts, and simultaneously increases their electrical conductivity. The fixative must be changed if the solution darkens, however, to avoid precipitation of osmium black particles over the sample surface. Conductivity of both whole and sectioned mycorrhizal samples can be enhanced, and SEM image quality improved, by employing techniques which increase the quantity of OsO_4 bound to constituents of exposed surfaces of the sample (4,28). These techniques are especially useful with samples which have abundant aerial components and/or extremely irregular surfaces. Fixative solutions employed for light microscopic study of mycorrhizae are not recommended for samples to be examined by SEM although, occasionally, useful preparations can be obtained with them, provided other facets of SEM protocols are suitably executed. The samples shown in Figs. 12 and 13, for example, were fixed in FAA (formalin, ethyl alcohol, acetic acid) and were stored in this solution at room temperature for two years before they were processed for SEM (37)., We have also had good success with a variety of plant tissues infected by fungal pathogens, sent to us from other laboratories, after fixation in 5% commercial formalin. These samples were rinsed in distilled water, fixed for 1-2 days in OsO_4 at 4°C, and were subsequently processed by the usual SEM procedures.

Techniques for obtaining sectional views.--Internal views of mycorrhizae can be obtained by cutting the fixed specimens in the desired plane with a sharp razor blade before they are dehydrated or in one of the dehydration solutions (Fig. 14, 15, 18). Samples fixed in OsO_4 are most suitable for manual sectioning since they are reasonably hard and can be sectioned with less crushing or distortion host tissues. Materials fixed with aldehydes are often quite soft, but can be sectioned successfully with a freezing microtome (37) or freeze fractured in absolute ethanol or acetone after they have been dehydrated (9). Arbuscules of VA mycorrhizae may be exposed for SEM observation by treating sectioned samples with reagents which remove host cytoplasm. Fig. 15 illustrates a sample from which host cytoplasm was removed by treatment with dilute base prior to dehydration and critical point drying (21). An alternative technique, more suitable for small rooted species such as soybean, involves a milder treatment performed on dehydrated specimens (25). Nylund (30) has employed pronase to remove host cytoplasm in studies of pitting characteristics of cortical cell walls in spruce ectomycorrhizae.

Dehydration and critical point drying.--Most SEM studies of mycorrhizal associations have been performed with samples which have been critical point dried. This drying procedure circumvents the effects of surface tension stresses, associated with other drying methods, which produce structural alterations of surface morphology. Samples to be critical point dried must be thoroughly dehydrated and this is generally accomplished with a graded ethanol or acetone series similar to that used for TEM. We routinely employ 30-40 minute intervals in each of the following concentrations of ethanol: 20, 40, 60, 80, 100, 100, 100%. Structural damage or drying artifacts, which are easily introduced as the solution changes are made, can be minimized by leaving the sample immersed in a small quantity of the previous solution and adding 2-3 changes of the next solution in the sequence. Dehydration may also be

---

Fig. 11. *Hebeloma alpinum*/*Dryas octopetala* ectomycorrhiza fixed with a formaldehyde-glutaraldehyde mixture, dehydrated in acetone and critical point dried. Excellent preservation of the mycorrhizal rootlets and external mycelium is illustrated. x100. Micrograph courtesy of J. C. Debaud (9). Figs. 12, 13. *Pisolithus tinctorius*/*Eucalyptus nova-angelica* ectomycorrhiza fixed in FAA dehydrated in ethanol and critical point dried. Fig. 12. Freezing microtome section showing surprisingly good morphological preservation of host cells and the fungal mantle (MT). Cytoplasm (C) in the cortical cells is poorly preserved, as expected with the use of a coagulant type of fixative, but does not detract from the usefulness of such preparations for low magnification study. x800. Fig. 13. Hyphae of the outer mantle with clamp connections (arrows). The minor distortion of hyphae, which should not be present in conventionally fixed specimens, may have been produced by the high alcohol concentration (43%) in the initial fixative solution. x2,600. Micrographs courtesy of C. G. Van Dyke (37). Figs. 14, 15. *Glomus mosseae*/*Liriodendron tulipifera* VA mycorrhizae fixed with a glutaraldehyde-acrolein mixture, post-fixed in OsO_4 and sectioned by hand. Fig. 14. A section dehydrated in ethanol and critical point dried. Intracellular hyphae (H) and arbuscules (A) enclosed by host cytoplasm are well preserved. Compare the appearance of host cytoplasm stabilized by the non-coagulative fixatives used here with that shown in Fig. 12. x700. D. A. Kinden and M. F. Brown, unpublished micrograph. Fig. 15. A sample prepared as in Fig. 14 but de-osmicated with periodic acid after sectioning, treated with dilute base to remove host cytoplasm, treated with OsO_4, dehydrated in ethanol and critical point dried. Host cytoplasm is absent, while cortical cell walls and arbuscular branches are well preserved. x1,800. Micrograph courtesy of D. A. Kinden (24). Fig. 16. Chlamydospores of *Glomus fasciculatus* fixed with OsO_4, dehydrated in ethanol and critical point dried. The subtending hyphae are preserved well but critical point drying often causes radial fractures in chlamydospore walls. While the preservation achieved is unacceptable, the fractures permit examination of spore wall structure. Lipid droplet (arrow). x475. M. F. Brown, unpublished micrograph. Fig. 17. Azygospore of *Gigaspora coralloidea* fixed with OsO_4 dehydrated and air-dried from acetone. Normal morphology of the spore and basal suspensor-like cells (arrows) has been preserved. x95. Micrograph by E. J. King (4). Fig. 18. Cross section of a resin impregnated mycorrhiza of *Monotropa uniflora*. Following fixation with formaldehyde-glutaraldehyde and post-fixation in OsO_4, the sample was dehydrated and infiltrated with Spurr's resin as for TEM. It was partially polymerized, washed with propylene oxide to remove resin from the surface, and then cured to achieve complete polymerization before examination with the SEM (26). Several nuclei (N) within host cortical cells and components of the mantle (MT) are clearly shown. x530. R. W. Lutz, unpublished micrograph.

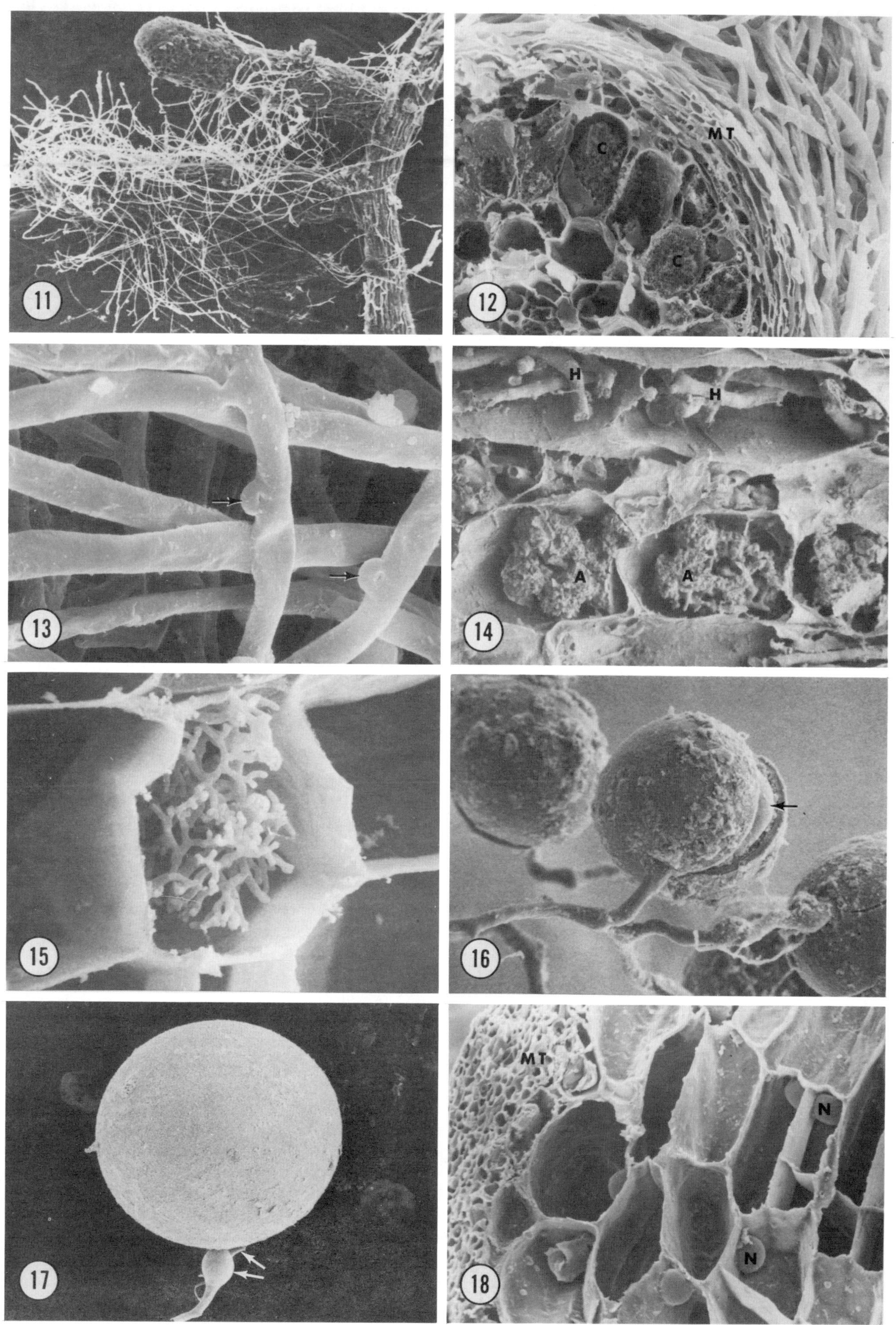
11
12
C
C
MT
13
14
H
H
A
A
15
16
17
18
MT
N
N

accomplished with an easily assembled cold distillation apparatus (43). This apparatus, which achieves dehydration gradually by vapor diffusion, has been successfully employed with mycorrhizae of Monotropa (J. A. Duddridge, personal communication).

Thoroughly dehydrated specimens are placed in the chamber of the critical point dryer and immersed, under pressure, in either liquid CO_2 or various Freons. Liquid CO_2 is most frequently used since it is inexpensive, readily available, and is miscible with both acetone and ethanol. Formerly, these dehydrants were replaced with amyl acetate before exposure to liquid CO_2 but it is now recognized that comparable or better preparations can be achieved without amyl acetate substitution. One should adhere to the manufacturer's instructions pertaining to the safe operation of the critical point dryer but the following procedural modifications of their recommendations are suggested for drying mycorrhizal specimens: 1) place the samples in the dryer chamber immersed in a small quantity of the dehydrant used to preclude any air drying or structural damage which may occur during this transfer, and introduce the CO_2 slowly; 2) substantially increase the number and length of exposures to CO_2 beyond manufacturer's recommendations to ensure total removal of the dehydrant from the sample and; 3) after the specimen chamber is heated to convert the liquid CO_2 to a gas, release the gas pressure slowly over a 20-30 minute period. Figs. 11-15 illustrate several types of mycorrhizal preparations dried successfully by the critical point method. In our experience, however, this technique has not produced optimal results with OsO_4-fixed chlamydospores or azygospores of VAM fungi (Fig. 16). Better morphological preservation of these materials has been obtained by dehydrating the fixed samples in an ethanol or acetone series and allowing them to air dry slowly from the pure solvent (Fig. 17).

An additional approach for preserving morphological characteristics of Monotropa mycorrhizae has been employed by Lutz and Sjolund (26). These workers infiltrated fixed and dehydrated mycorrhizae with an epoxy embedding medium, partially polymerized the resin, then washed the resin from the surface of the samples with propylene oxide and subsequently cured the residual resin within the specimens as with TEM preparations. A cross sectional view of a mycorrhiza prepared in this manner is shown in Fig. 18.

Mounting specimens for SEM.--Electrically conductive liquid adhesives are generally recommended for attaching samples to SEM specimen mounts. However, dried biological samples rapidly absorb the solvents used in these adhesives and distortion of the sample occurs as the solvent dries. We have found foil backed tape, available from EM suppliers, to be a more satisfactory means of fastening dried specimens to the specimen mount. The tape, which has a protective paper backing over the adhesive, is cut into pieces no more than 1.5 times the diameter of the specimen mount used. The ends of the tape are folded backward, with paper backing attached, so that the adhesive side faces outward. The paper backing is peeled from the folded end tabs and the tape is positioned firmly on the specimen mount. The paper is peeled from the remainder of the tape and the dried sample is placed on the exposed adhesive surface with a fine needle or forceps. Regardless of the mounting technique employed, SEM samples are coated with a thin film of carbon and/or a heavy metal to provide uniform electrical conductivity from all parts of the sample surface to the specimen mount. Most scanning microscopists now employ sputter-coating techniques with gold or gold alloys for this purpose.

Criteria for evaluating sample quality.--The surface morphology of biological samples can be altered at any stage of the preparative sequence as well as during observation in the SEM. An impressive array of artifacts and other types of damage which may be encountered with SEM preparations has been presented by Goldstein and coworkers (16). Foremost among preparation artifacts are those caused by shrinkage during dehydration, and surface tension stresses during drying of the specimen. The severity of these artifacts ranges from obvious fractures and widespread collapse of external components of the sample to less conspicuous distortions such as partial depression of convex structures, wrinkles in surfaces, or alterations in surface texture. Correlative studies, employing other microscopic techniques, are essential for evaluating the quality of structural preservation achieved with SEM preparations, as well as for establishing the validity of observations derived from them.

Analytical Electron Microscopy

Analytical EM can be broadly defined to include all techniques that provide information about the identity, chemical composition or function of cells or subcellular structures beyond that provided by those procedures employed in descriptive EM. Analytical EM studies differ from conventional descriptive studies in several important respects. Analytical techniques are employed as experimental procedures, each designed to obtain a specific bit of information about the specimen. Therefore, one requirement of analytical procedures is that appropriate control specimens be examined to insure that the results obtained with the analytical procedure are reliable. Because of the focus of each analytical technique on obtaining a specific type of information about the specimen, compromises in specimen preparation are often required to balance the need for adequate overall preservation and observation of ultrastructure with those requirements imposed by a particular analytical procedure. Further compromises may be necessary in applying certain analytical techniques to the study of mycorrhizae, because many procedures were developed in studies of specific animal tissues, and may not perform well on mycorrhizae without modification. Before any analytical procedure is applied to the study of mycorrhizae, the most recent applications of the method to plants or fungi should be consulted. The techniques considered briefly here include several from the areas of ultrahistochemistry, enzyme cytochemistry and x-ray microanalysis.

Ultrahistochemistry.--Ultrahistochemistry includes EM techniques in which specific staining procedures are used to determine the identity or chemical nature of cellular structures, and also techniques in which specific enzymatic or chemical removal of cellular components permits inferences to be made concerning their composition (18,31,32). Both types of procedures have been used to study mycorrhizae.

The simplest histochemical procedures may require little modification of protocols used to prepare specimens for descriptive study. The use of phosphotungstic acid (PTA) as a specific stain for the plasma membrane (2,11) requires only that osmium be removed from the ultrathin sections by periodic acid oxidation prior to staining with PTA in a chromic acid solution. Similarly, ruthenium red may be employed as a stain for pectic substances with only minor alteration in specimen preparation. Staining is usually accomplished during fixation by addition of ruthenium red to cacodylate-buffered aldehyde and osmium tetroxide fixatives (Fig. 19) (39).

Another type of histochemical staining procedure takes advantage of the fact that tissue components such as proteins and polysaccharides have reactive groups that may be used to specifically stain them in ultrathin sections. The Swift procedure for demonstration of proteins and the Thiery procedure for polysaccharides have both been applied to the study of mycorrhizae (2,11,14,39). The Swift procedure is based on the oxidation of disulfide linkages in cystine-rich proteins by treatment with silver methenamine. This specific oxidation is accompanied by reduction of the silver, and deposition of metallic silver grains at the reactive sites (Fig. 20). In the Thiery procedure, periodic acid treatment of the sections oxidizes vicinal glycol groups in the polysaccharides to produce aldehydes. The sections are then reacted with thiocarbohydrazide, and finally with silver proteinate, to form metallic silver deposits over the polysaccharides (Fig. 21). Controls for both of these procedures include chemical treatments designed to block the specific reactive groups and thus prevent the specific staining reactions.

Both chemical and enzymatic extractions of aldehyde-fixed specimens have been used to study the interfacial zone in VA mycorrhizae (2,11,39). These techniques may result in severe disruption of cytoplasm, and the enzymatic treatments may not always be as specific as expected, but they do offer additional approaches to determination of the chemical nature of subcellular structures. Extraction of non-cellulosic material may be followed by potassium permanganate staining to reveal the fibrillar component of the cell wall structure, and permanganate staining has also been used for this purpose without prior extraction (Fig. 22) (39).

Enzyme Cytochemistry.--EM techniques may be used to determine the subcellular locations of many enzyme activities by the observation of electron dense reaction products deposited in the tissue at the specific sites occupied by the enzymes (42). Because most enzyme reaction products are soluble, a "simultaneous capture" procedure is commonly used to locate the sites of enzyme activity. The simultaneous capture method for acid phosphatase activities, for example, involves exposure of the tissue to a buffered medium containing a suitable substrate, often β-glycerophosphate, and a capture reagent, usually

---

Fig. 19-22. Ultrahistochemistry of the interfacial zone in VA mycorrhizae. Fig. 19. Ruthenium red staining of the interfacial material (IM) surrounding a collapsed hypha (H). x27,000. Micrograph courtesy of S. Scannerini and P. Bonfante-Fasolo (39). Fig. 20. Swift procedure for demonstration of proteins, showing deposition of silver grains over the interfacial zone (IZ) and heavy deposition over the fungal wall (W) and cytoplasm (C). x89,000. Unpublished micrograph courtesy of P. Bonfante-Fasolo and S. Scannerini. Fig. 21. Thiery procedure for demonstration of polysaccharides. Silver grains are deposited over the interfacial material (IM) as well as the fungal wall (W) and cytoplasm (C). x64,000. Micrograph courtesy of S. Scannerini and P. Bonfante-Fasolo (39). Fig. 22. Potassium permanganate-stained section showing a fibrillar component in the interfacial material (IM) surrounding a collapsed hypha (H). Host plasmalemma (arrows). x97,000. Micrograph courtesy of S. Scannerini and P. Bonfante-Fasolo (39). Fig. 23, 24. Acid phosphatase localization in a soybean mycorrhiza. Fig. 23. Lead phosphate deposits (arrows) indicative of phosphatase activity are located in the interfacial zone surrounding arbuscular branches (A). x24,500. Fig. 24. Section of a control specimen for acid phosphatase localization incubated in a medium lacking the enzyme substrate. No lead phosphate deposits are present in the interfacial zone. x13,600. E. J. King, unpublished micrographs. Fig. 25-27. Energy dispersive x-ray microanalysis of granules in hyphae of Glomus mosseae. Fig. 25. Micrograph of an unstained intercellular hypha containing phosphorus granules (PG) in small vacuoles, and an unidentified cytoplasmic inclusion (CI) of similar appearance. x21,300. Fig. 26. X-ray sprectrum of a phosphorus granule showing the presence of both phosphorus and calcium. The copper peaks originated from the specimen support grid. Fig. 27. X-ray spectrum of an unidentified cytoplasmic inclusion in which neither phosphorus nor calcium was detected. Micrograph and spectra courtesy of J. A. White (51). Fig. 28-30. Wavelength dispersive x-ray analysis of phosphorus granules contained in the endophyte of a Taxus mycorrhiza. Fig. 28. The secondary electron image of a 1300Å epoxy section of the fungus containing phosphorus granules (PG). x15,000. Fig. 29. An x-ray 'dot map' of the same section in which the sites of phosphorus x-ray emission appear as white dots. x15,000. Fig. 30. A line scan analysis for phosphorus superimposed on the secondary electron image. The horizontal scan line (SL) is the path over which the electron beam was moved. The peaks in the phosphorus profile line indicate high relative phosphorus concentrations at specific points (arrows) along the scan line. x18,000. Micrographs courtesy of D. G. Strullu (47).

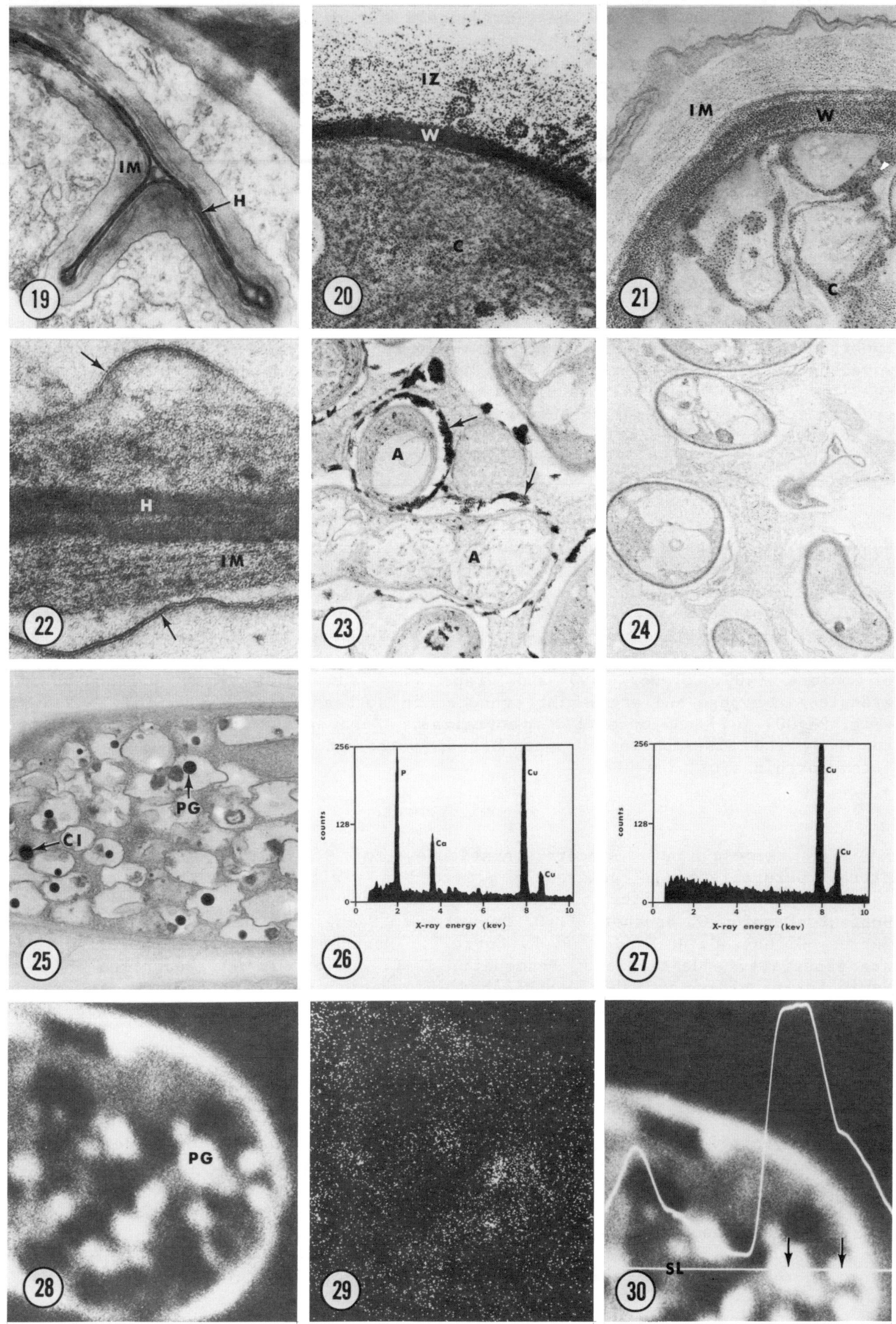
19
IM
H
20
IZ
W
C
21
IM
W
C
22
H
IM
23
A
A
24
25
PG
CI
26
256
128
counts
P
Ca
Cu
Cu
X-ray energy (kev)
27
256
128
counts
Cu
Cu
X-ray energy (kev)
28
PG
29
30
SL

lead nitrate. In theory, enzymatic hydrolysis of the substrate will release phosphate ions which will be immediately "captured" and precipitated as insoluble lead phosphate at the site of the enzyme activity (Fig. 23). In practice, the results may be greatly influenced by the preparation of specimen pieces, the fixation procedure employed, the composition of the enzyme reaction medium, the conditions of tissue incubation in the reaction medium, and the amount of activity actually present at particular sites in the tissue. A trial-and-error approach is usually necessary to determine the best procedure for a particular specimen, and appropriate controls are essential for proper interpretation of the results (Fig. 24). Enzyme cytochemistry at the EM level has been used to study only phosphatases in VA mycorrhizae (13,36,38).

X-ray analysis.--X-ray analysis yields information about the elemental composition of cellular structures by detection and recording of the x-rays emitted by a specimen during irradiation in the electron beam of the microscope (6,16). Identification of elements is possible because each element emits characteristic x-rays when irradiated, and these x-rays may be detected with spectrometers designed to distinguish them on the basis of either their wavelengths or their energies. Wavelength dispersive spectrometers are more sensitive, but permit analysis of only one element at a time. Energy dispersive spectrometers permit simultaneous detection and identification of many elements. X-ray analysis may be performed in a scanning electron microscope, but scanning transmission electron microscopes are more useful for many biological applications.

Although various cryotechniques are available for the study of diffusable ions (1,12,16), x-ray analysis is more easily applied to structures that can be subjected, without loss or alteration, to the conventional specimen preparation procedures used for transmission electron microscopy. It is often preferable to examine the sections without heavy metal post-staining, and some alterations in specimen preparation may also be required. Fixation with osmium tetroxide, for instance, may interfere with later detection of phosphorus by energy dispersive analysis. X-ray analysis has been used to detect phosphorus in VA mycorrhizae of onions (40), clover (49) and grape (1); and to identify polyphosphate granules in hyphae and arbuscular branches in soybean (Fig. 25-27) (51), Taxus (Fig. 28-30) (47) and onion (7) mycorrhizae. X-ray analysis has also been used to study the distribution of phosphorus, potassium, calcium and magnesium in ectomycorrhizae of pine (12).

Acknowledgements

We express our sincere gratitude to the following mycorrhizal ultrastructuralists who generously provided a wealth of electron micrographs and/or x-ray spectra for our use in this presentation: H. Bartschi, P. Bonfante-Fasolo, G. Bruchet, J. C. Debaud, J. Dexheimer, J. A. Duddridge, J. P. Garrec, G. Gay, D. A. Kinden, R. W. Lutz, S. Scannerini and D. G. Strullu. We are especially endebted to J. Boumendil, J. L. Duclos, J. A. Duddridge and R. Pepin for providing valuable unpublished information and/or manuscripts in press for our use. We also thank D. L Pinkerton for performing the photographic work and J. A. White for reviewing the manuscript.

LITERATURE CITED

1. Bartschi, H., and Garrec, J. P. 1980. Étude comparative de la répartition cytologique de quelques éléments minéreaux dans l'écorce de racines saines et d'endomycorhizes de Vitis vinifera L. C. R. Acad. Sc. Paris, Series D. 290:919-922.
2. Bonfonte-Fasolo, P., Dexheimer, J., Gianinazzi, S., Gianinazzi-Pearson, V., and Scannerini, S. 1981. Cytochemical modifications in the host-fungus interface during intracellular interactions in vesicular-arbuscular mycorrhizae. Plant Sci. Letters 22:13-21.

3. Bonfante-Fasolo, P., and Gianinazzi-Pearson, V. 1979. Ultrastructural aspects of endomycorrhiza in the Ericaceae. 1. Naturally infected hair roots of Calluna vulgaris L. Hull. New Phytol. 83:739-744.
4. Brown, M. F., and Brotzman, H. G. 1979. Phytopathogenic fungi: a scanning electron stereoscopic survey. Extension Division, University of Missouri, Columbia. 355 pp.
5. Carling, D. E., White, J. A., and Brown, M. F. 1977. The influence of fixation procedure on the ultrastructure of the host endophyte interface of vesicular-arbuscular mycorrhizae. Can. J. Bot. 55:48-51.
6. Chandler, J. A. 1977. X-ray microanalysis in the electron microscope. Pages 317-547 in: Glauert, A. U. ed. Practical methods in electron microscopy, Volume 5. North Holland Press, Amsterdam and New York. 547 pp.
7. Cox, G., Moran, K.J., Sanders, F., Nickolds, C., and Tinker, P.B. 1980. Translocation and transfer of nutrients in vesicular-arbuscular mycorrhizas. III. Polyphosphate granules and phosphorus translocation. New Phytol. 84:649-659.
8. Cox, G., and Sanders, F. E. 1974. Ultrastructure of the host fungus interface in a vesicular-arbuscular mycorrhiza. New Phytol. 73:901-912.
9. Debaud, J. C., Pepin, R., and Bruchet, G. 1981. Etude des ectomycorhizes de Dryas octopetala. Obtention de synthèses mycorhiziennes et de carpophores d'Hebeloma alpinum et H. marginatulum. Can. J. Bot. 59:1014-1020.
10. Debaud, J. C., Pepin, R., and Bruchet, G. 1981. Ultrastructure des ectomycorhizes synthetiques a Hebeloma alpinum et Hebeloma marginatulum de Dryas octopetala. Can. J. Bot. 59:2160-2166.
11. Dexheimer, J., Gianinazzi, S., and Gianinazzi-Pearson, V. 1979. Ultrastructural cytochemistry of the host-fungus interfaces in the endomycorrhizal association Glomus/Allium cepa. Z. Pflanzenphysiol. 92:191-206.
12. Gay, G., and Garrec, J. P. 1980. Premiers essais de microlocalisation de quelques éléments minéreaux dans les racines courtes et les mycorhizes de Pinus halpenensus Mill. C. R. Acad. Sc. Paris, Series D. 290:69-71.
13. Gianinazzi, S., and Gianinazzi-Pearson, V. 1979. Enzymatic studies on the metabolism of vesicular-arbuscular mycorrhiza. III. Ultrastructural localization of acid and alkaline phosphatase in onion roots infected by Glomus mosseae (Nicol. and Gerd.). New Phytol. 82:127-132.
14. Gianinazzi-Pearson, V., Morandi, D., Dexheimer, J., and Gianinazzi, S. 1981. Ultrastructural and ultracytochemical features of a Glomus tenuis mycorrhiza. New Phytol. 88:633-639.
15. Glauert, A. M. 1975. Fixation, dehydration and embedding of biological specimens. North Holland Publishing Co., Amsterdam and New York. 207 pp.
16. Goldstein, J. I., Newbury, D. E., Echlin, P., Joy, D. C., Fiori, C., and Lifshin, E. 1981. Pages 534-539 in: Scanning electron microscopy and x-ray microanalysis. Plenum Press, New York and London. 673 pp.
17. Hayat, M. A. 1970. Principles and techniques of electron microscopy: biological applications. Vol. I. Van Nostrand Reinhold Co., New York. 412 pp.
18. Hayat, M. A. 1975. Positive staining for electron microscopy. Van Nostrand Reinhold, New York. 361 pp.
19. Holley, J. D., and Peterson, R. L. 1979. Development of a vesicular-arbuscular mycorrhiza in bean roots. Can. J. Bot. 57:1960-1978.
20. Kaspari, H. 1973. Elektronenmikroskopische Untersuchung zur Feinstruktur der endotrophen Tabakmykorrhiza. Arch. Microbiol. 92:201-207.
21. Kinden, D. A., and Brown, M. F. 1975. Technique for scanning electron microscopy of fungal structures within plant cells. Phytopathology 65:74-76.

22. Kinden, D. A., and Brown, M. F. 1975. Electron microscopy of vesicular-arbuscular mycorrhizae of yellow poplar. I. Characterization of endophytic structures by scanning electron stereoscopy. Can. J. Microbiol. 21:989-993.
23. Kinden, D. A., and Brown, M. F. 1975. Electron microscopy of vesicular-arbuscular mycorrhizae of yellow poplar. II. Intracellular hyphae and vesicles. Can. J. Microbiol. 21:1768-1780.
24. Kinden, D. A., and Brown, M. F. 1975. Electron microscopy of vesicular-arbuscular mycorrhizae of yellow poplar. III. Host-endophyte interactions during arbuscular development. Can. J. Microbiol. 21:1930-1939.
25. King, E. J., Schubert, T. S., and Brown, M. F. 1981. Techniques for developmental studies of VA mycorrhizae. Page 46 in: Program and Abstracts, Fifth North American Conference on Mycorrhizae. University Laval, Quebec. 83 pp.
26. Lutz, R. W., and Sjolund, R. D. 1973. Monotropa uniflora: ultrastructural details of its mycorrhizal habit. Amer. J. Bot. 60:339-345.
27. Mosse, B. 1970. Honey-coloured, sessile Endogone spores. II. Changes in fine structure during spore development. Arch. Mikrobiol. 74:129-145.
28. Murphy, J. A. 1978. Non-coating techniques to render biological specimens conductive. Pages 175-193 in: SEM/1978/II, SEM Inc. AMF O'Hare, IL. 1134 pp.
29. Nieuwdorp, P.J. 1972. Some observations with light and electron microscope on the endotrophic mycorrhiza of orchids. Acta. Bot. Neerl. 21:128-144.
30. Nylund, J. E. 1980. Symplastic continuity during Hartig net formation in Norway spruce ectomycorrhizae. New Phytol. 86:373-378.
31. Pearse, A. G. E. 1968. Histochemistry, Volume 1. Third edition. Churchill Livingstone, Edinburgh and London. 759 pp.
32. Pearse, A. G. E. 1972. Histochemistry, Volume 2. Third edition. Churchill Livingstone, Edinburgh and London. 757 pp.
33. Pepin, R., and Boumendil, J. 1982. Préservation de l'ultrastructure du sclérote de Sclerotinia tuberosa (champignon discomycete). Un modèle pour la préparation des échantillons imperméables et hétérogènes. Cytologia 47. (In press).
34. Peterson, T. A, Mueller, W. C., and Englander, L. 1980. Anatomy and ultrastructure of a Rhododendron root-fungus association. Can J. Bot. 58:2421-2433.
35. Postek, M. T., Howard, K. S., Johnson, A. H., and McMichael, K. L. 1980. Scanning electron microscopy. A student's handbook. Ladd Research Industries, Inc. Burlington, VT. 305 pp.
36. Protsenko, M. A. 1973. Elektronnomikroskopisheskoe izushemie lokalizatsii kisloi fosfatazi v perevarivaiushih griv kletkah mikorizii goroha. Dokl. Akad. Nauk SSSR 211:213-215.
37. Rose, R. W., Jr., Van Dyke, C. G., and Davey, C. B. 1981. Scanning electron microscopy of three types of ectomycorrhizae formed on Eucalyptus nova-angelica in the southeastern United States. Can. J. Bot. 59:683-688.
38. Scannerini, S. 1975. Le ultrastrutture delle micorrize. Giorn. Bot. Ital. 109:109-144.
39. Scannerini, S., and Bonfonte-Fasolo, P. 1979. Ultrastructural cytochemical demonstration of polysaccharides and proteins within the host-arbuscule interfacial matrix in an endomycorrhiza. New Phytol. 83:87-94.
40. Schoknecht, J. D., and Hattingh, M. J. 1976. X-ray microanalysis of elements in cells of VA mycorrhizal and nonmycorrhizal onions. Mycologia 68:296-303.
41. Seviour, R. J., Hamilton, D., and Chilvers, G. A. 1978. Scanning electron microscopy of surface features of eucalypt mycorrhizas. New Phytol. 80:153-156.

42. Sexton, R., and Hall, J. L. 1978. Enzyme cytochemistry. Pages 63-147 in: Hall, J. L. ed. Electron microscopy and cytochemistry of plant cells. Elsevier/North Holland Biomedical Press, Amsterdam and New York. 444 pp.
43. Sitte, P. 1962. Einfaches Verfahren zur stufenlosen Gewebe-Entwässerung für die elektronenmikroskopische Präparation. Naturwissenschaften 17:402-403.
44. Spurr, A. R. 1969. A low-viscosity epoxy resin embedding medium for electron microscopy. J. Ultrastruct. Res. 26:31-43.
45. Strullu, D. G. 1976. Contribution à l'étude ultrastructurale des ectomycorrhizes à basidiomycètes de Pseudotsuga menziesii (Mirb.). Bull. Soc. Bot. Fr. 123:5-16.
46. Strullu, D. G. 1979. Ultrastructure et représentation spatiale du manteau fongique des ectomycorhizes. Can. J. Bot. 57:2319-2324.
47. Strullu, D. G., and Gourret, J. P. 1981. Ultrastructure and electron-probe microanalysis of metachromatic vacuolar granules occurring in Taxus mycorrhizas. New Phytol. 87:537-545.
48. Sward, R. J. 1981. The structure of the spores of Gigaspora margarita. I. The dormant spore. New Phytol. 87:761-768.
49. Walker, G. D., and Powell, C. Ll. 1979. Vesicular-arbuscular mycorrhizas in white clover: a scanning electron microscope and x-ray microanalytical study. New Zealand J. Bot. 17:55-59.
50. Warmbrodt, R. D., and Fritz, E. 1981. Embedding plant tissue with plastic using high pressure: a new method for light and electron microscopy. Stain Tech. 56:299-305.
51. White, J. A., and Brown, M. F. 1979. Ultrastructure and x-ray analysis of phosphorus granules in a vesicular-arbuscular mycorrhizal fungus. Can. J. Bot. 57:2812-2818.

MYCORRHIZAE IN INTERACTIONS WITH OTHER MICROORGANISMS

A. Endomycorrhizae

R. W. Roncadori and R. S. Hussey

INTRODUCTION

Vesicular-arbuscular (VA) mycorrhizal fungi are the most common endomycorrhizal symbionts and, therefore, will be emphasized in this discussion. Frequently, mycorrhizal plant root systems are infected by various pathogenic or non-pathogenic organisms. Since VA mycorrhizae and plant pathogens usually have opposite effects upon plant health, these interactions have been extensively studied and recently reviewed by Hussey and Roncadori (11), Schenck and Kellam (18), and Schönbeck (19). The interaction studies have emphasized relationships between VA mycorrhizae and soil-borne fungal pathogens causing root rots and vascular wilts and plant-parasitic nematodes. There are relatively few reports of interactions between VA mycorrhizae and pathogens attacking aerial parts of plants, viruses, or nonpathogenic organisms; therefore these groups will be excluded from further discussion.

The outcome of VA mycorrhizae-plant pathogen challenges will vary considerably depending upon the fungal symbiont, host plant, and pathogen involved (11,18,19). An interaction can be neutral if the mycorrhiza does not affect disease development or symbiosis or other activities of the two microorganisms. If the mycorrhiza offsets pathogen damage or sporulation, the interaction is regarded as positive. However, should the endophyte enhance disease development or pathogen reproduction, or the pathogen suppress mycorrhizal development, the interaction is considered negative.

Although VA mycorrhizal plants have been shown to react to disease differently than nonmycorrhizal plants, the mechanisms responsible for the interactions are poorly understood. Unlike ectomycorrhizae, VA mycorrhizae have neither an external mechanical barrier, such as a fungus mantle, nor do they produce any apparent antibiotics. Instead, most studies indicate that changes within the root tissue influence disease development more than alterations in the rhizosphere (11,17,19). Suggested protective mechanisms in mycorrhizal plants are cell wall thickenings, changes in amino acids and reducing sugars, increased chitinase activity, and a general change in plant physiology. The methods used most often in challenge studies with soil-borne plant pathogens in the greenhouse and several specialized techniques to elucidate the mechanisms of protection are discussed herein.

Challenge Studies

Soil-borne Fungal Pathogens.--Most challenge studies have involved the root rot fungi, *Pyrenochaeta terrestris*, *Pythium* spp., *Phytophthora* spp., *Rhizoctonia solani*, and *Thielaviopsis basicola* and the vascular wilt pathogens, *Fusarium oxysporum* f. sp. *lycopersici* and *Verticillium dahliae*. The plant is inoculated with an appropriate fungal symbiont using one of the methods described elsewhere in this manual to establish VA mycorrhizae. Preferred inocula from pot cultures are a mixture of soil, roots, and spores or a spore suspension along with appropriate controls. The latter may consist of a filtrate from the pot culture, a mixture of autoclaved soil, roots, and spores, or soil and roots from a mycorrhizal-free plant of the same species grown under similar conditions used to increase inoculum. Crops which are propagated directly from seed or cuttings can be inoculated by placing the inoculum with the seed or cutting in the planting hole (14,21), adding the inoculum just below the seed in a layer prior to planting (4,7,21), or infesting the soil throughout with the inoculum (6,15). Crops which are propagated from transplants, such as citrus (7) or onion (2,4,16), or cuttings can be grown for a specified time in flats, pots, or cups containing a low phosphorus soil mix to promote maximum mycorrhizal development prior to transplanting. The extent of colonization by the fungal symbiont should be determined at the time of inoculation with the pathogen by clearing and staining roots using an appropriate method for quantification described earlier in this mannual.

The VA mycorrhizal plant may be challenged with the pathogen using appropriate methods of inoculation or varying the inoculation sequence. Before or at planting, the pathogen may be introduced by using naturally infested field soil (7,22), broadcast and incorporation into treated or fumigated soil (4,6,15), placed at various depths under the seed or transplant (21), or injected or poured onto the soil (7,16). Mycorrhizal citrus (4) and tomato (8) have been challenged with Phytophthora parasitica and F. oxysporum f. sp. lycopersici, respectively, by washing the roots of seedlings and immersing them in an inoculum suspension prior to transplanting to soil in a pot. The challenge sequence can be arranged to allow mycorrhizal development to occur first or can be simultaneous by adding both the fungal symbiont and pathogen at planting or during transplanting.

Soil mixes for challenge studies usually contain low concentrations of available phosphorus to promote maximum mycorrhizal development. A mix supplemented with a high rate of available phosphorus should be included to permit challenge comparisons between mycorrhizal and nonmycorrhizal plants of similar size (6). Changes in symptom expression and the disease cycle associated with high phosphorus fertilization can provide information as to the general nature of the interaction between the endophyte and pathogen (Plate 8E).

After a suitable incubation period, depending upon the challenge pathogen, plant growth and development and the interacting effect of the microorganisms are ascertained. Plant growth data usually include root and shoot weights (dry or fresh), shoot height, stem diameter, reproduction, and chemical analysis of tissue. Disease symptoms may be rated visually or the degree of infection or reproduction determined by isolation of the pathogen from plant tissue or by soil assay. Likewise, roots are cleared and stained to detect the extent of mycorrhizal development and spore populations determined by centrifugal-flotation (12) or other methods described in this manual.

Plant Parasitic Nematodes.--Plant-parasitic nematodes are obligate parasites and often placed into three groups based on the manner in which they feed on roots. These feeding habits may influence how these pathogens interact with VA mycorrhizal fungi. Species feeding as sedentary endoparasites (Meloidogyne, Heterodera, Tylenchulus, Rotylenchulus) penetrate roots as vermiform juveniles; after feeding commences, the body swells and the somatic musculature system degenerates, resulting in the nematode becoming immobile. Further development and reproduction by the nematode is dependent upon the plant cells adjacent to its head providing nourishment. These cells are usually modified by the nematode into elaborate feeding cells which become the permanent feeding site for the parasite. Migratory endoparasites (Pratylenchus, Radopholus) remain vermiform throughout their life cycle. They bodily penetrate roots and migrate throughout the root tissue, feeding on different cells without establishing a localized or permanent feeding site. Considerable destruction of root tissue usually occurs with this type of feeding. Ectoparasites (Belonolaimus, Macroposthonia, Paratrichordorus, Xiphinema) remain outside the root, using their stylet to feed on internal cells. With the exception of a few species that feed at root tips, nematodes with this type of feeding habit generally cause the least amount of tissue damage. Selection of the type (stage in life cycle of nematode) of inoculum and the methods of determining the final nematode population at the end of an interaction experiment will vary depending on the nematode species used and its feeding habit.

Initial experiments are usually designed to determine if an interaction occurs between the two groups of organisms. In these challenge studies, plant performance and population densities of both organisms are determined at the end of the experiment. The nematode population data are best expressed as nematodes per gram of root rather than nematodes or galls (for Meloidogyne) per plant. Expression of the data in this manner allows comparisons to be made among different studies and provides an index indicating the suitability of a root system as a substrate for the nematode. However, a lower nematode population density on a mycorrhizal plant than on the nonmycorrhizal plant could be interpreted as a dilution of the nematode population as a result of the larger root system usually associated with mycorrhizal plants. Therefore, an appropriate check such as an additional fertility treatment of supplemental phosphorus should be included to ensure that a comparison can be made of nematode populations occurring on mycorrhizal and nonmycorrhizal plants with similar size root systems. Mycorrhizal fungi should be assayed by determining the percentage of roots colonized using the gridline intersect (9) or other appropriate methods. The fungal spore population in the soil, determined as previously mentioned, is also useful as an indication of mycorrhizae development.

If an interaction occurs in the previous experiments then the next step is to determine if it's the result of either altered nematode attraction to roots, nematode penetration into roots (for endoparasites), or nematode development. For these studies, endoparasitic nematodes in the root systems are stained in situ and counted using a stereoscopic microscope. An acid fuchsin staining procedure (13) permits the assessment of the physical relationship between the nematode and fungus. With this procedure, however, the nematodes are stained very well whereas the mycorrhizal fungal hyphae, arbuscules, and vesicles inside root tissue stain less intensely. Plants with split-root systems are also useful for studying interactions and will be discussed later.

The sequence in which plant roots become colonized by the mycorrhizal symbiont and infected by the plant-parasitic nematode should also be considered. Inoculation of plants with the mycorrhizal fungus prior to adding the nematode inoculum will permit the slow growing symbiont to become established before the roots are attacked by the nematode. Usually a symbiont requires 3 wk or longer to become established in plant roots. At the time of inoculation with nematodes, the percentage of the root system colonized by the fungus should be determined. For the delayed nematode inoculation, the inoculum can be added in a shallow trench excavated around the base of the plant or via glass tubes inserted at planting in the soil around the seed or seedling.

Methods that have been used with specific plant-parasitic nematodes in interaction studies are described below:

a. Sedentary Endoparasites.--The infective stage of Meloidogyne spp. is the second-stage juvenile which hatches from an egg. Either freshly hatched second-stage juveniles or eggs can be used as inoculum. Freshly hatched juveniles are collected by placing galled tomato roots in a mist chamber (20). The juveniles collected during the first 24 hr are discarded and those collected over the next 48 hr are used as the inoculum. Root-knot nematode eggs can be collected from galled tomato roots using 0.53% sodium hypochlorite (Clorox) according to the method of Hussey and Barker (10). Although the inoculum density used will vary depending on the type of study, pot size, etc., the low percentage of egg hatch requires that the number of eggs used for inoculum should be several times greater than if juveniles are used as inoculum. Also juveniles will continue to hatch from the eggs for 3 to 4 wk whereas juveniles used directly as inoculum will be infective for a much shorter time.

For measuring the final nematode reproduction, Meloidogyne eggs produced per root system can be collected or the number of second-stage juveniles per volume of soil can be determined. Eggs are collected from the root systems using 1.0% sodium hypochlorite and laboratory stirrers (10) whereas the juveniles are collected from the soil with a centrifugal-flotation procedure (12). The eggs may be stained with acid fuchsin to facilitate counting (3). The affect on female fecundity can be assessed by determining number of eggs per egg mass. The influence of mycorrhizae on penetration of roots by second-stage juveniles and subsequent development is observed by staining the nematodes in the root tissue at weekly intervals after inoculation and counting the nematodes in the various life stages under a stereoscopic microscope.

b. Migratory Endoparasites.--Inoculum of these nematodes (e.g. Pratylenchus) is collected by incubating infected root systems in a mist chamber. All stages of the nematode are infective and the inoculum will consist of a mixture of the different life stages. At the end of an experiment the final nematode population is determined by placing the root systems in a mist chamber and collecting and counting the nematodes every 3 days for 9-12 days. Nematodes inside the roots can also be stained with acid fuchsin as described above.

c. Ectoparasites.--Ectoparasitic species are more difficult to work with than the sedentary and migratory endoparasites. Inoculum is collected from stock cultures by repeated decanting and sieving (1,20). A major problem encountered in working with ectoparasitic nematodes is obtaining inoculum free of spores of VA mycorrhizal fungi. The final nematode population in an experiment is determined by extracting the nematodes from an aliquant of soil by centrifugal-flotation.

Specialized Techniques

Should an interaction be evident in challenge tests, additional experiments are required to determine the nature of the mechanisms involved. The following techniques can indicate whether the nature of the interaction has either a physical or physiological basis.

Split Root System.--This method involves the development of a plant with equal size half-root systems growing in two adjacent containers. The mycorrhizal fungus or the challenge pathogen may be added to one or both half-root systems as needed. A high phosphorus soil treatment should be included to indicate the effect of phosphorus nutrition on colonization and reproduction by both microorganisms.

Split root systems may be developed in a variety of crops such as peanuts, peach, sweet orange, and tomato. The method of promoting growth of a split root system will vary depending upon the crop; therefore, the procedure with sweet orange (5) is used as an example. Seed are germinated in an autoclaved sandy loam soil. After primary lateral roots are produced, the tap root is severed to promote lateral root growth. Seven wk later seedlings with divided half root systems are transplanted into two adjacent 10-cm-diameter clay pots filled with an autoclaved, low phosphorus sandy loam. The desired type of inoculum of either microorganism can be introduced at this time or inoculation with the pathogen may be delayed. Following an appropriate incubation period, evaluation of plant and microorganism performance is made as previously described. Results will indicate whether the nature of the interaction is local when both microorganisms co-inhabit the same half root system or systemic when they colonize opposite half-root systems.

Direct Root Challenge.--This method involves inoculation of specific sections of a root system on a local basis and facilitates selection of co-inhabited tissue for more detailed study such as histology. Plants with a small, slow growing root system are best suited for this technique. A plastic petri dish (14.5 diameter x 1.8 cm deep) with a 4 cm segment removed in the top of the chamber and a drainage slit at the bottom is used (2). The chamber is filled with an appropriate autoclaved medium and, for example, onion seed germinated for 5 days on moist filter paper are transplanted one per vessel. The soil chamber should be taped shut and the upper two-thirds covered with aluminum foil. The chambers are then inserted at a 60° angle from the horizon into trays filled with autoclaved quartz sand. The seedlings may be placed in a specific environment to promote root growth and the sand kept moist. After the seedlings are established, the chambers are removed, opened, and the VA mycorrhizal fungus inoculum (spore suspension) added in a pattern to match anticipated root system growth. Suitable controls should be included. The chambers are closed, taped and wrapped, and placed back into the sand. The desired level of mycorrhizal development may be determined by periodically sampling a population of plants intended solely for this purpose using an appropriate assay such as the grid line intersect method. The development of a yellow pigment in the mycorrhizal plant tissue (eg. onion) may preclude the destructive sampling needed to assess the degree of mycorrhizal development in nonpigmented root systems. The challenge organism is added to specifically chosen sites within the root system (yellow pigmented tissue when present) of the remaining plants. Agar plugs from an actively growing colony of a pathogen, or other desirable inoculum may be placed near a segment of root and the chamber incubated long enough to permit disease development. At this time, the challenged root sections may be removed and processed for further study.

Other techniques commonly used to demonstrate pathogen-host interactions are likely to be adopted to study the nature of the VA mycorrhizae-plant pathogen interaction. These should include methods to determine pathogen attraction to the host and physical and biochemical changes induced in VA mycorrhizal plants.

Most of the challenge studies conducted to date have been carried out in the greenhouse. Therefore, the outcome of the interaction may be influenced by environmental conditions unique to greenhouse culture, such as light intensity and quality and physical properties of the soil mix in pots. Furthermore, due to the limited period that many plant species may be grown in containers, study results are derived from only a portion of the time required for crop development and maturity and often do not include effects upon yield. If mycorrhizal technology is to be incorporated into crop management, it is imperative that the interactions also be evaluated in microplots or field plots to overcome these obstacles.

LITERATURE CITED

1. Barker, K. R. 1978. Determining nematode population responses to control agents. Pages 114-125 in: E. I. Zehr, ed. Methods for evaluating plant fungicides, nematicides, and bactericides. The Am. Phytopathol. Soc., St. Paul, Minnesota. 256 p.
2. Becker, W. N. 1976. Quantification of onion vesicular-arbuscular mycorrhizae and their resistance to *Pyrenochaeta terrestris*. Ph.D. Dissertation, Univ. of Illinois, Urbana. 72 p.
3. Byrd, D. W., Jr., Ferris, H. and Nusbaum, C. J. 1972. A method for estimating number of eggs of *Meloidogyne* spp. in soil. J. Nematol. 4:266-269.

4. Davis, R. M. 1980. Influence of Glomus fasciculatus on Thielaviopsis basicola root rot of citrus. Plant Dis. 64:839-840.
5. Davis, R. M. and Menge, J. A. 1980. Influence of Glomus fasciculatus and soil phosphorus on Phytophthora parasitica root rot of citrus. Phytopathology 70:447-452.
6. Davis, R. M., Menge, J. A. and Erwin, D. C. 1979. Influence of Glomus fasciculatus and soil phosphorus on Verticillium wilt of cotton. Phytopathology 69:453-456.
7. Davis, R. M., Menge, J. A. and Zentmyer, G. A. 1978. Influence of vesicular-arbuscular mycorrhizae on Phytophthora root rot of three crop plants. Phytopathology 68:1614-1617.
8. Dehne, H. W. and Schonbeck, F. 1975. The influence of the endotrophic mycorrhiza on the fusarial wilt of tomato. Z. Pflanzenkr. and Pflanzenschutz. 82:630-632.
9. Giovannetti, M. and Mosse, B. 1980. An evaluation of techniques for measuring vesicular arbuscular mycorrhizal infection in roots. New Phytol. 84:489-500.
10. Hussey, R. S. and Barker, K. R. 1973. A comparison of methods of collecting inocula of Meloidogyne spp., including a new technique. Plant Dis. Rep. 57:1025-1028.
11. Hussey, R. S. and Roncadori, R. W. 1982. Vesicular-arbuscular mycorrhizal fungi may limit nematode activity and improve plant growth. Plant Dis. 66: 9-14
12. Jenkins, W. R. 1964. A rapid centrifugal-flotation technique for separating nematodes from soil. Plant Dis. Rep. 48:692.
13. McBryde, M. C. 1936. A method of demonstrating rust hyphae and haustoria in unsectioned leaf tissue. Am. J. Bot. 23:686-689.
14. Roncadori, R. W. and Hussey, R. S. 1977. Interaction of the endomycorrhizal fungus Gigaspora margarita and root-knot nematode on cotton Phytopathology 67:1507-1511.
15. Ross, J. P. 1972. Influence of Endogone mycorrhiza on Phytophthora root rot of soybean. Phytopathology 62:896-897.
16. Safir, C. 1968. The influence of vesicular-arbuscular mycorrhiza on the resistance of onion to Pyrenochaeta terrestris. M. S. Thesis, Univ. of Illinois, Urbana. 36 p.
17. Schenck, N. C. 1983. Can mycorrhizae control root disease? Plant Dis. 65:230-234.
18. Schenck, N. C. and Kellam, M. K. 1978. The influence of vesicular arbuscular mycorrhize on disease development. Bull. 799. Florida Agr. Expt. Sta. 16 p.
19. Schönbeck, F. 1979. Endomycorrhiza in relation to plant diseases. Pages 271- 280. in: B. Schippers and W. Gams, eds., Soil-borne plant pathogens. Int. Symp. Soil-borne Pl. Pathogens, 4th, Munich, 1978. Academic Press, New York.
20. Southey, J. F. ed. 1970. Laboratory methods for work with plant and soil nematodes. Page 36. Tech. Bull. 2. Her Majesty's Stationery Office, London, 148 p.
21. Stewart, E. L. and Pfleger, F. L. 1977. Development of poinsettia as influenced by endomycorrhizae, fertilizer and root rot pathogens Pythium ultimum and Rhizoctonia solani. Florist's Review 159:37, 79-81.
22. Timmer, L. W. and Leyden, R. F. 1978. Relationship of seedbed fertilization and fumigation to inoculation of sour orange seedlings by mycorrhizal fungi and Phytophthora cinnamomi. Amer. Soc. Hort. Sci. 103:537-541.

MYCORRHIZAE IN INTERACTIONS WITH OTHER MICROORGANISMS

B. Ectomycorrhizae

D. H. Marx

INTRODUCTION

Ectomycorrhizae and feeder root diseases of plants have many traits in common, even though their effects on plants are opposite. Both types of infection intimately involve the succulent fine feeder roots of their hosts. Infective propagules (hyphae or spores) of ectomycorrhizal fungi in soil are stimulated by host roots to symbiotically infect and eventually transform feeder roots into dual organs in which the cortical cells are wrapped in the Hartig net-hyphae matrix and isolated from direct contact with soil by the fungal mantle. During synthesis and maintenance of ectomycorrhizae, the plant host and fungal symbiont respond physiologically to the symbiotic infection. Similarly, infective propagules of feeder root pathogens (mainly species of *Phytophthora*, *Pythium*, *Rhizoctonia*, and *Fusarium*) are chemically stimulated by feeder roots and pathogenically infect these tissues by ramifying into meristematic, primary cortex, and occasionally vascular tissues, causing limited or extensive necrosis.

A pathogen infecting a nonmycorrhizal feeder root is confronted first with succulent, thin-walled epidermal cells. In most instances, the cortical cells beneath have no secondary thickening. If a pathogen infects and destroys this feeder root prior to infection by an ectomycorrhizal fungus, an ectomycorrhiza cannot be formed. However, if the ectomycorrhizal fungus infects and transforms the feeder root into an ectomycorrhiza prior to pathogenic infection, the tissues of this transformed root are no longer vulnerable to attack by the pathogen. The pathogen attacking an ectomycorrhiza is confronted with a tightly interwoven network of hyphae (fungal mantle) and with cortical cells surrounded by the hyphal matrix of the Hartig net. Research has shown that ectomycorrhizae are so physically and chemically altered that they resist infection by certain fungal pathogens and, therefore, function as biological deterrents to feeder root infections (10). Several mechanisms of resistance have been demonstrated by methods that are briefly described here.

Antibiotic Production in Pure Culture

Antifungal, antibacterial, or antiviral compounds are produced by over 100 species of ectomycorrhizal fungi. Since fungi cause most of the feeder root diseases of ectomycorrhizal tree hosts, some methods used to detect antifungal compounds are described. In the simplest method, the ectomycorrhizal fungus is grown at the edge of a petri dish containing a standardized volume of an appropriate agar medium. The inoculum disc containing the fungus is allowed to grow from 1 to 2 cm into the agar. Test fungal pathogens are placed 1 to 6 cm away from the edge of the symbiont colony. Inoculum of the test pathogen is placed in a similar location on petri dishes of the same agar medium, but without the symbiont (control plates); and the dual fungal cultures are incubated long enough for the test pathogen in the control plates to grow the distance initially separating the two fungi in the dual cultures. Incubation time for *Pythium* and *Phytophthora* species is from 3 to 6 days (7). If the ectomycorrhizal fungus produces an agar-diffusible growth inhibitor that affects the test pathogen, a sharp and measurable zone of inhibition should separate the two colonies. Different agar media should be tested since some are not conducive to antibiotic production. Some test pathogens may grow into the mycelium of the ectomycorrhizal fungus and morphologically change both organisms. These changes can be observed by direct microscopic examination. One of the limitations of this technique is that the pH of the agar medium for 0.5 to 1 cm in advance of the ectomycorrhizal fungus frequently becomes acidic. This pH change can affect growth of the test pathogen and can be misinterpreted as an antibiotic effect.

Antibiotics produced by ectomycorrhizal fungi can also be detected in various liquid media. If the antibiotic is agar diffusable, a volume of liquid medium in which the ectomycorrhizal fungus has grown can be placed in a diffusion well (6 to 8 mm diameter disc removed from agar) of an appropriate agar medium (7). The liquid medium should be allowed to diffuse for various times before placing the test pathogen on the agar at various distances from the well as mentioned above. After incubation, zones of inhibition can be measured.

These techniques can be modified in various ways. For example, in the agar plate culture test, instead of an inoculum disc of the test pathogen a diffusion well can be inoculated with a zoospore suspension of a test pathogen. Zones of inhibition of the mycelial growth of the pathogen in the agar can be measured, or zoospores in the well can be examined under a microscope to detect possible inhibition to zoospore germination. Since considerable variation is normally encountered in growth of the various organisms used in the above tests, five to eight replicate plates of each treatment should be used.

Inhibitors in Ectomycorrhizae

Inhibitors present in ectomycorrhizae may be antibiotics produced directly by the fungal symbiont (12) or antibiotics produced by the host as a result of stimulation by fungus infection (5). If its identity is known, the inhibitor can be extracted from ectomycorrhizae with the same solvents and techniques used to identify it (8). General inhibitors in ectomycorrhizae can be determined by macerating the ectomycorrhizae and then extracting with various organic solvents or water. If organic solvents are used, they are evaporated and the residues are suspended or dissolved in water. Such extracts can be tested at different concentrations by placing them in agar diffusion wells or directly in liquid or agar cultures of the test pathogen. Subsequent growth of the test pathogen is measured and compared with that on extract-free control cultures (6).

Ectomycorrhizal development causes an increase in quantities of volatile terpenes produced on roots (3). Certain volatile chemicals have been tested for effects on ectomycorrhizal and pathogenic fungi (4, 5). Tests are performed in deep petri dishes (25 mm x 100 mm) with 30 ml of agar medium. Three agar discs of the test fungi are placed in a triangle on the edges of the plates and incubated long enough to obtain colony diameters of 3 to 4 cm. The under sides of the plates are etched with a scribe to mark the leading edge of each colony prior to testing. A sterile pyrex glass vial (1 cm deep x 2.5 cm diameter) containing 0.5 ml of sterile water is placed in the center of each plate. Test terpenes at various concentrations are then added to the vials with pipettes. Control plates contain only sterile water. Plates are double wrapped in parafilm to reduce loss of the volatile compounds and each treatment should be replicated at least five times. Incubation time depends on the growth rate of the test organisms on control plates. Growth is measured at the end of the test from the etched mark to the edge of new growth and degree of inhibition is expressed as percent growth relative to growth on control plates.

Individual Root Inoculations

A root inoculation cell technique has been used to determine the susceptibility of individual root types to pathogenic infection (1, 9, 12, 13). This technique will work for pine seedlings with specific ectomycorrhizae formed in aseptic culture, with naturally occurring ectomycorrhizae, with specific ectomycorrhizae formed in fumigated nursery soil, or with container seedlings in the greenhouse. Seedling roots are carefully washed of soil and organic matter, laid on a flat surface, and separated. Individual types of roots (ectomycorrhizal or short roots on lateral roots and lateral root tips) still intact on the root system are selected and inserted through slits (Plate 8F-H) in glass cells (25 mm OD x 1 cm deep with slits 1 mm wide x 5 mm deep to accommodate the root). After the root and glass cell interfaces are sealed with a nontoxic adhesive (1, 12), the remaining root system is covered with a layer of soil or other suitable material. The foliage of the seedlings should be exposed to sufficient light to maintain active photosynthesis. Cells of different root types have been inoculated with an aqueous suspension of zoospores or mycelial mats of *Phytophthora cinnamomi*. Sterile, premoistened sand is added to the cells if nematodes and fungal pathogens are to be studied in concert (1). Length of incubation varies with pathogen tested. Incubation for *P. cinnamomi* was from 3 to 10 days. After incubation, the roots are removed from the cells, prepared histologically, and examined microscopically for infection.

Seedling Response in Inoculated Soil

Pine seedlings with and without ectomycorrhizae are produced as mentioned earlier and transplanted into soil containing the test pathogen. Seedlings are then grown for several months. Fertility, moisture and light conditions are maintained at levels comparable to those encountered in the field. Results from these tests have shown that ectomycorrhizal pine seedlings have greater tolerance to *P. cinnamomi* infections than nonmycorrhizal seedlings (6, 11, 14). In these tests, the degree of ectomycorrhizal development (percent short roots ectomycorrhizal) must be assessed before and after exposure to the pathogen to determine whether the degree of protection is correlated with the degree of ectomycorrhizal development and whether pathogen infection of nonmycorrhizal roots on ectomycorrhizal seedlings influences the incidence of new ectomycorrhizal development. Populations of the test pathogen should be monitored before transplanting of test seedlings and after seedling removal so that the effects of ectomycorrhizae on pathogen populations can be determined (11, 14). Seedlings with different amounts of ectomycorrhizae should be grown in soil containing varying population densities of a feeder root pathogen. Results from this test define quantitative relationships between amounts of resistant ectomycorrhizae and susceptible short and lateral roots, and effects of population densities of the pathogen on seedling growth.

Fortin and others (2) recently developed a technique of growing pine seedlings in flat, transparent polyester growth pouches and inoculating these seedlings with ectomycorrhizal fungi. Using this technique, the development of ectomycorrhizae can be macroscopically assessed at desired intervals of time. This pouch technique could be useful in studying relationships between symbiotic and pathogenic organisms. It would be useful for studying the effects of extra-matrical mycelium of ectomycorrhizal fungi with specific traits (antibiotic production, rapid growth rate, etc.) on the growth and development of pathogenic organisms.

These are but a few of the methods that can be employed to study the relationships between ectomycorrhizal fungi, fungal pathogens or nematodes on feeder roots. Each method should be modified to conform to the traits and habits of the fungal symbiont, pathogen, and plant host and to the environmental conditions normally encountered by the tripartite association.

LITERATURE CITED

1. Barham, R. O., Marx, D. H. and Ruehle, J. L. 1974. Infection of ectomycorrhizal and nonmycorrhizal roots of shortleaf pine by nematodes and *Phytophthora cinnamomi*. Phytopathology 64:1260-1264.

2. Fortin, J. A., Piche, Y. and Lalonde, M. 1980. Techniques for the observation of early morphological changes during ectomycorrhiza formation. Can. J. Bot. 58:361-365.
3. Krupa, S. and Fries, N. 1971. Studies on ectomycorrhizae of pine. I. Production of volatile organic compounds. Can. J. Bot. 49:1425-1431.
4. Krupa, S. and Nylund, J. E. 1972. Studies on ectomycorrhizae of pine. III. Growth inhibition of two root pathogenic fungi by volatile organic constituents of ectomycorrhizal root systems of *Pinus sylvestris* L. Eur. J. For. Path. 2:88-94.
5. Krupa, S., Andersson, J. and Marx, D. H. 1973. Studies on ectomycorrhizae of pine. IV. Volatile organic compounds in mycorrhizal and nonmycorrhizal root systems of *Pinus echinata* Mill. Eur. J. For. Path. 3:194-200.
6. Marais, L. J. and Kotze, J. M. 1976. Ectomycorrhizae of *Pinus patula* as biological deterrents to *Phytophthora cinnamomi*. S. Afr. For. J. 99:35-39.
7. Marx, D. H. 1969. The influence of ectotrophic mycorrhizal fungi on the resistance of pine roots to pathogenic infections. I. Antagonism of mycorrhizal fungi to root pathogenic fungi and soil bacteria. Phytopathology 59:153-163.
8. Marx, D. H. 1969. The influence of ectotrophic mycorrhizal fungi on the resistance of pine roots to pathogenic infections. II. Production, identification, and biological activity of antibiotics produced by *Leucopaxillus cerealis* var. *piceina*. Phytopathology 59:411-417.
9. Marx, D. H. 1970. The influence of ectotrophic mycorrhizal fungi on the resistance of pine roots to pathogenic infections. V. Resistance of mycorrhizae to infection by vegetative mycelium of *Phytophthora cinnamomi*. Phytopathology 60:1472-1473.
10. Marx, D. H. 1972. Ectomycorrhizae as biological deterrents to pathogenic root infections. Ann. Rev. Phytopathology 10:429-454.
11. Marx, D. H. 1973. Growth of ectomycorrhizal and nonmycorrhizal shortleaf pine seedlings in soil with *Phytophthora cinnamomi*. Phytopathology 63:18-23.
12. Marx, D. H. and Davey, C. B. 1969. The influence of ectotrophic mycorrhizal fungi on the resistance of pine roots to pathogenic infections. III. Resistance of aseptically formed mycorrhizae to infection by *Phytophthora cinnamomi*. Phytopathology 59:549-558.
13. Marx, D. H. and Davey, C. B. 1969. The influence of ectotrophic mycorrhizal fungi on the resistance of pine roots to pathogenic infections. IV. Resistance of naturally occurring mycorrhizae to infection by *Phytophthora cinnamomi*. Phytopathology 59:559-565.
14. Ross, E. W. and Marx, D. H. 1972. Susceptibility of sand pine to *Phytophthora cinnamomi*. Phytopathology 62:1197-1200.

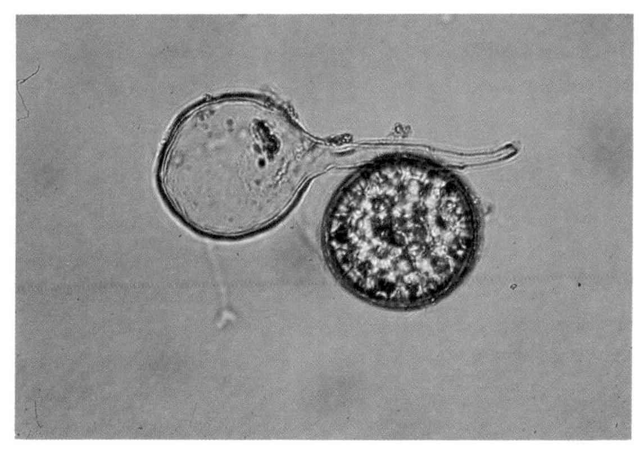

A. *Acaulospora trappei* Ames and Lind.; spore budded from tapered stem of parent cell, × 250 (R. Ames photo).

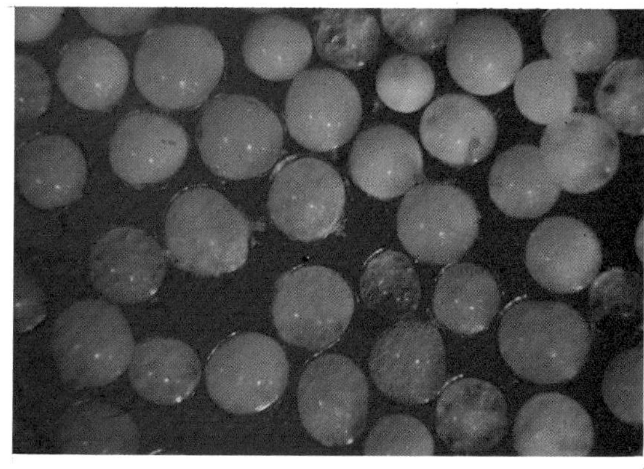

B. *Gigaspora* species; × 8 (V. Furlan photo).

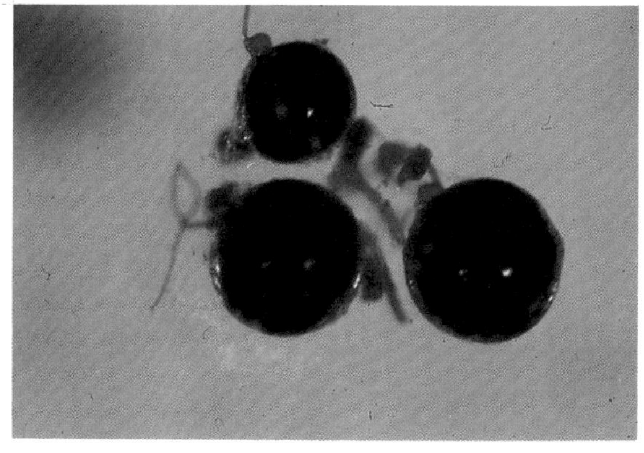

C. *Gigaspora nigra* Redh.; × 20.

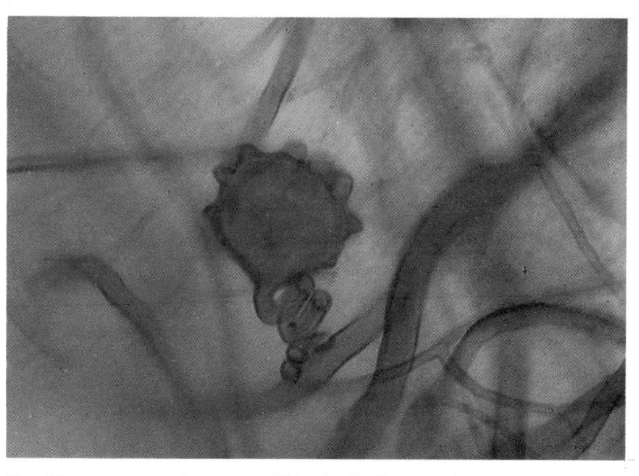

D. *Gigaspora calospora* (Nicol. & Gerd.) Gerd. & Trappe; soil-borne vesicle stained with acid fuchsin; × 200.

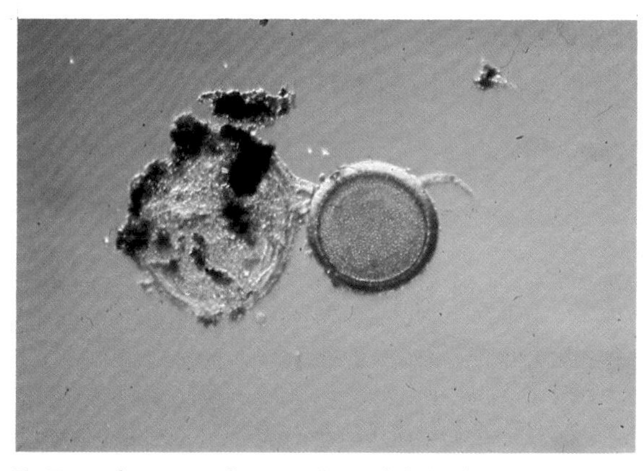

E. *Entrophospora infrequens* Ames & Schneid. Spore formed inside and bulging out basal portion of parent cell; × 50.

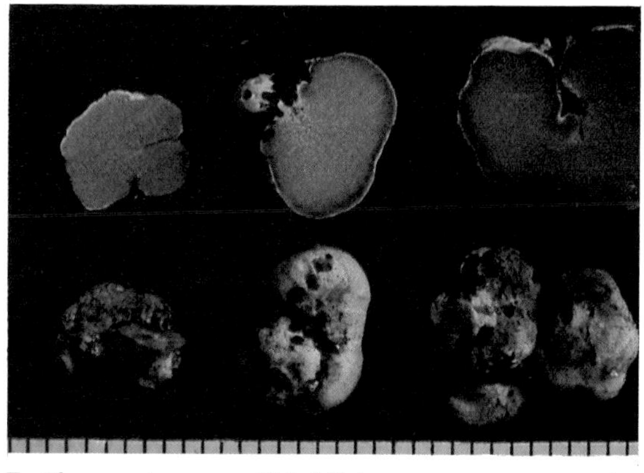

F. *Glomus microcarpus* Tul. & Tul.; sporocarps, cross-section above, surface with peridium below; × 1.

G. *Glomus convolutus* Gerd. & Trappe; sporocarps; × 1.25.

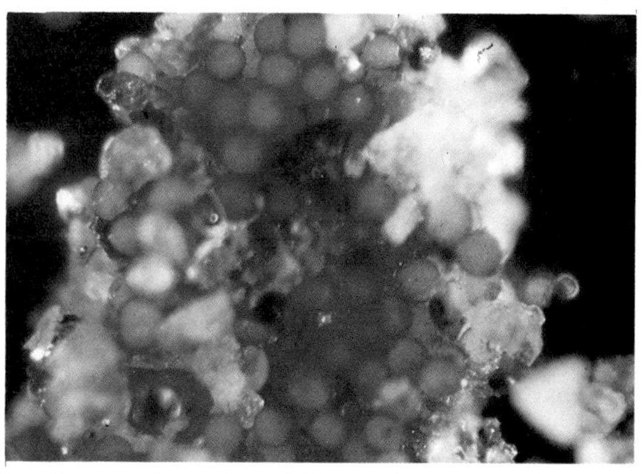

H. *Glomus epigaeus* Dan. & Trappe; Sporocarp; × 20.

PLATE 2

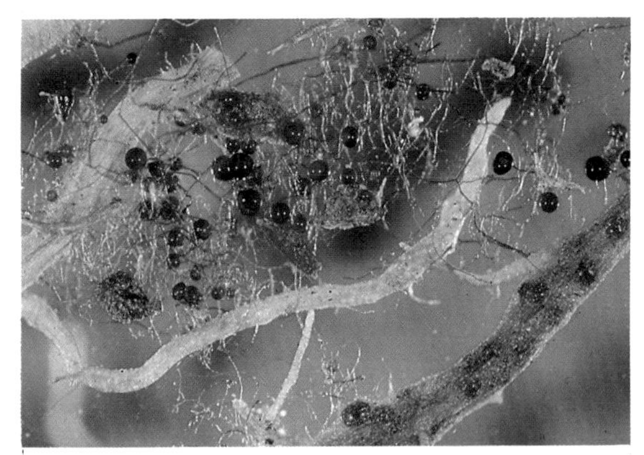

A. *Glomus deserticola* Trappe, Bloss & Menge; spores among roots; × 40.

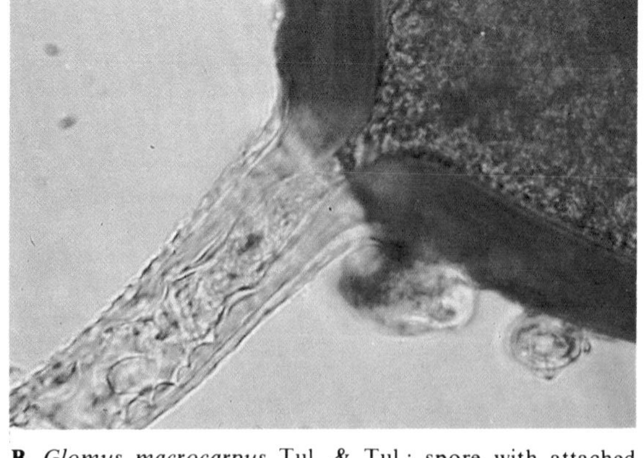

B. *Glomus macrocarpus* Tul. & Tul.; spore with attached hypha; × 200.

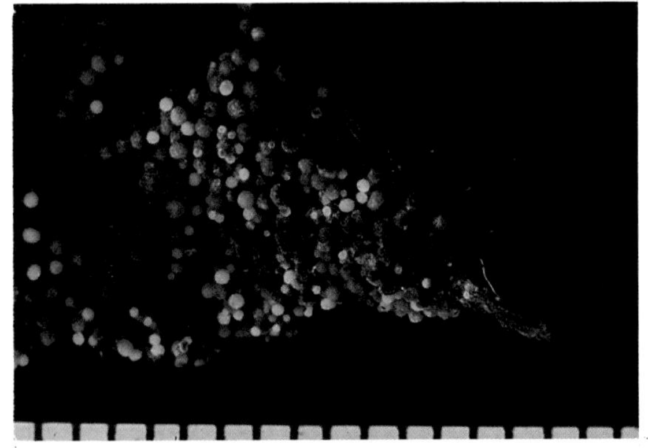

C. *Sclerocystis coremioides* Bk. & Br.; massed sporocarps—white are young, brown are mature; × 2.

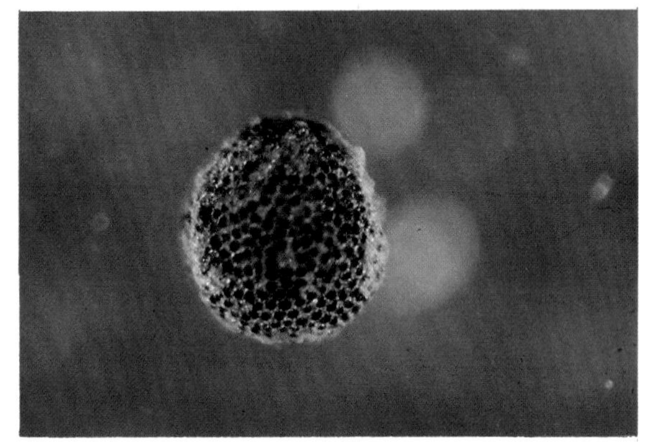

D. *Sclerocystis clàvispora* Trappe; sporocarp; × 20.

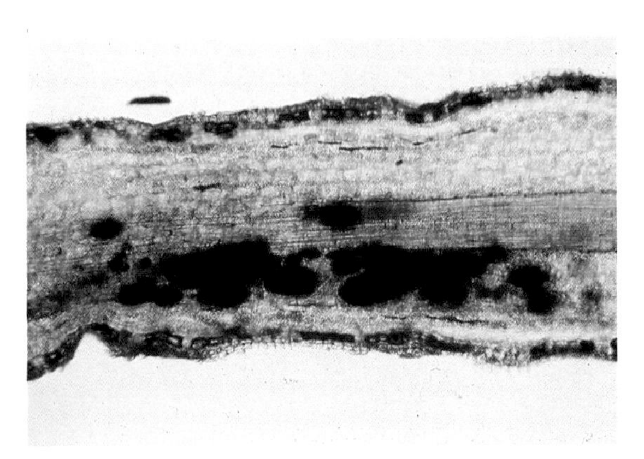

E. OsO_4 detects unsaturated lipid in vesicle of *Glomus etunicatus* in rough lemon roots.

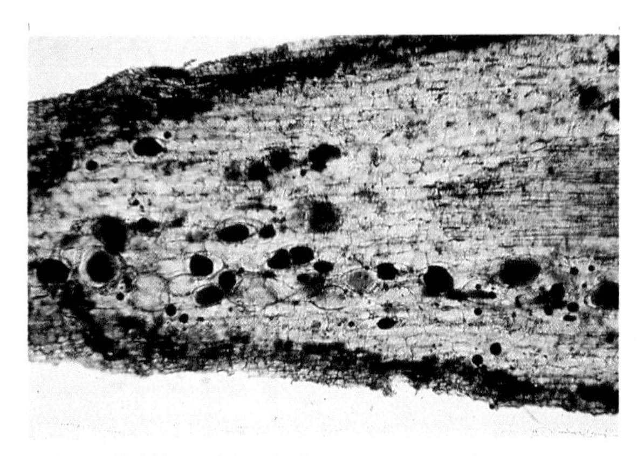

F. Total lipid in vesicle of *Glomus etunicatus* in rough lemon roots stains blue-black with Sudan black B.

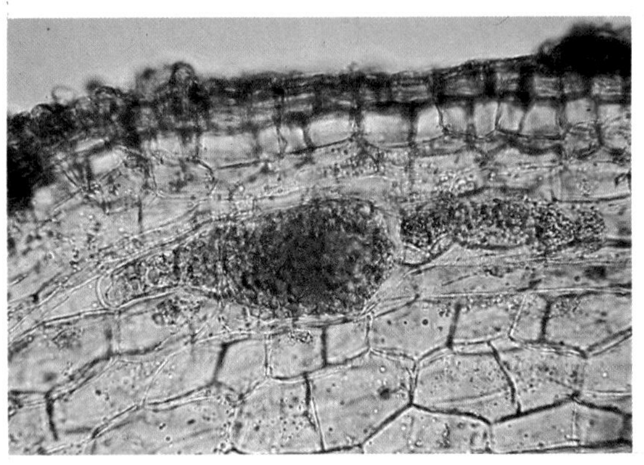

G. Sudan IV stains neutral lipid orange in vesicle of *Glomus etunicatus* in rough lemon roots.

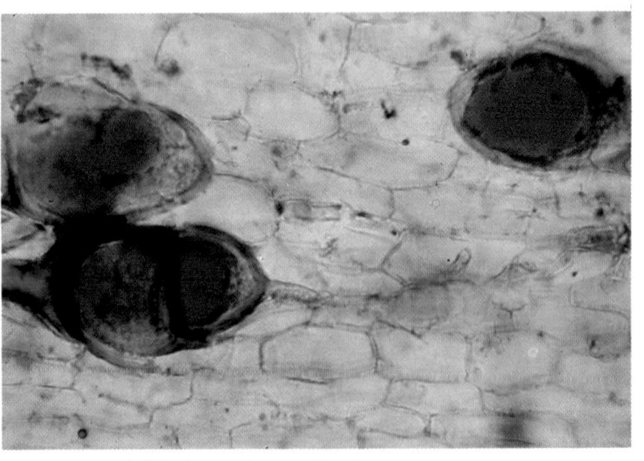

H. Neutral lipids in vesicles of *Glomus etunicatus* in rough lemon roots react pink with oxazone component of Nile Blue.

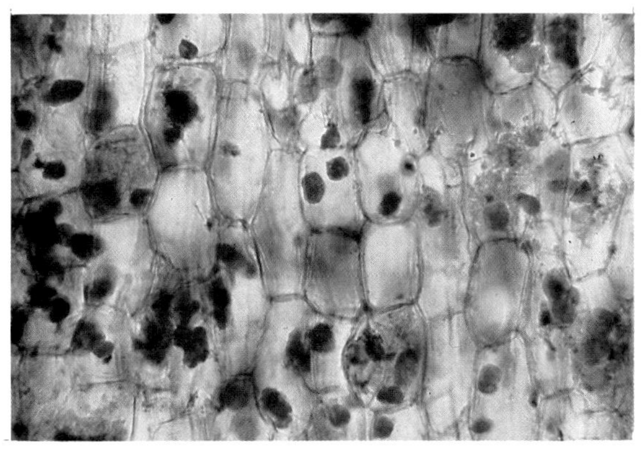

A. Arbuscules of *Glomus etunicatus* in rough lemon roots stain blue for phospholipids with the oxazine component of Nile blue.

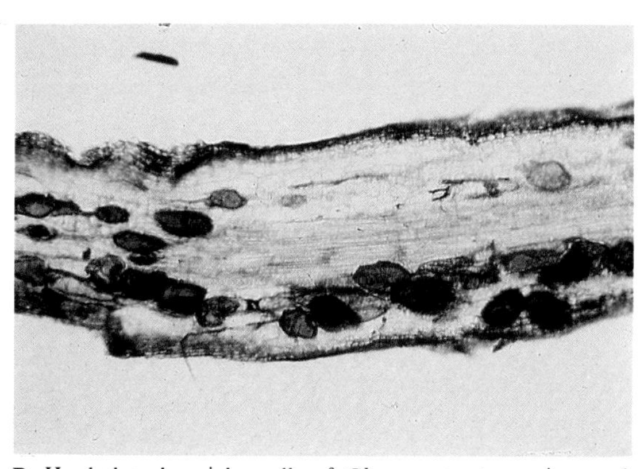

B. Hyphal and vesicle walls of *Glomus etunicatus* in rough lemon roots react positively for phosophoglycerides with the gold hydroxamic reaction.

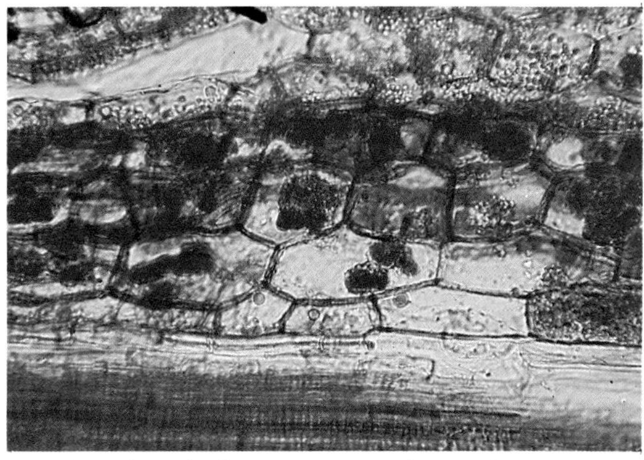

C. Arbuscules of *Glomus etunicatus* in rough lemon roots stain positive for carbohydrate with the periodic acid-Schiff reagent.

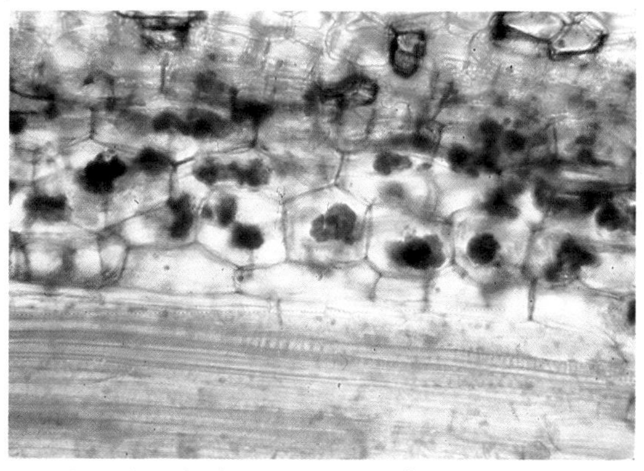

D. Arbuscules of *Glomus etunicatus* in rough lemon roots stain positive for RNA with Azure B.

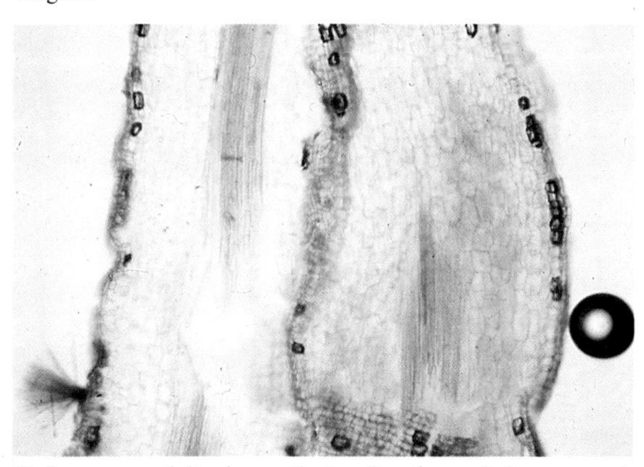

E. Laccase activity in cortical cells of rough lemon roots containing arbuscules of *Glomus etunicatus* (root on right) and no activity in noninfected root on left.

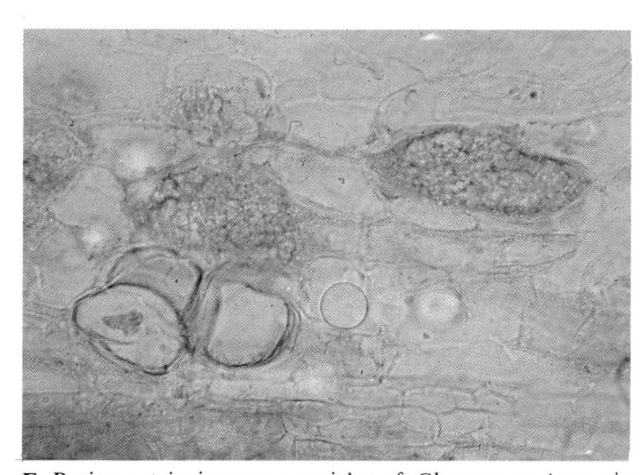

F. Basic protein in young vesicles of *Glomus etunicatus* in rough lemon roots with fast green.

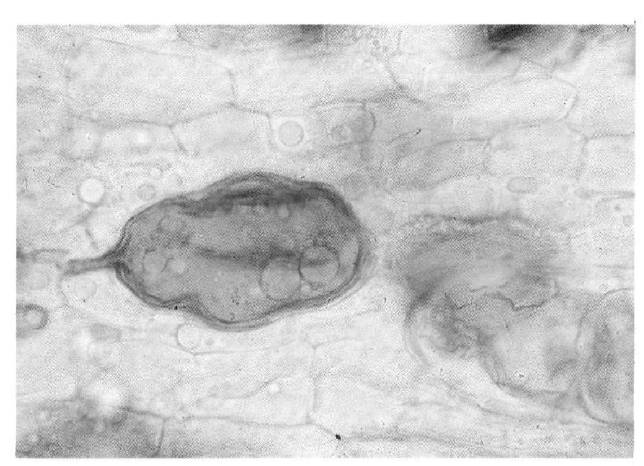

G. Vesicle walls of *Glomus etunicatus* in rough lemon roots stain gold-orange for tannins with the Nitroso test.

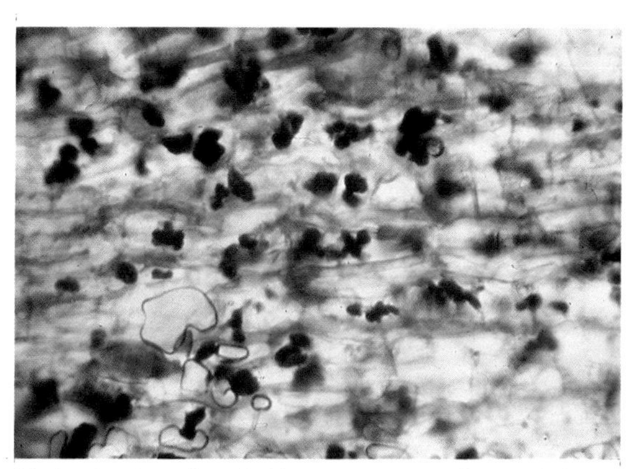

H. Aged arbuscules of *Glomus etunicatus* in rough lemon roots stain positive for peroxidase with the p-phenylene diamine test.

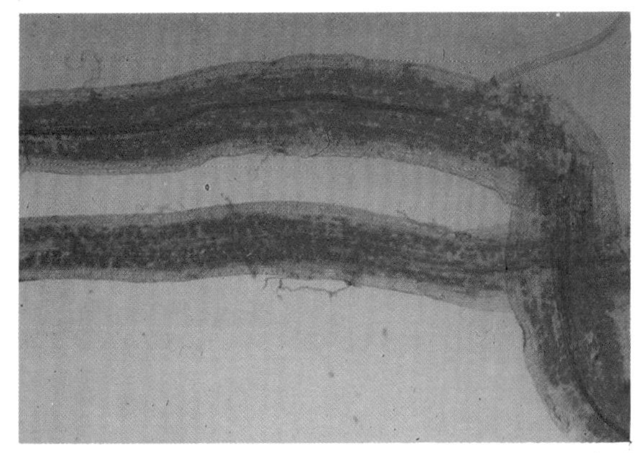

A. Cleared root segments showing abundant VA mycorrhizal infection; × 50.

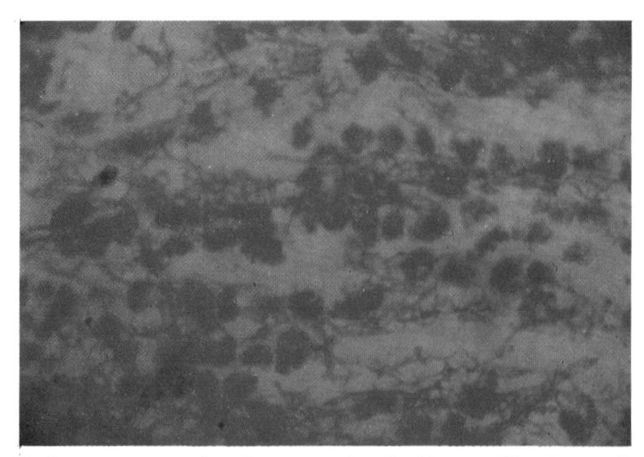

B. Root segment showing more detail of internal hyphae and arbuscules; ca. × 200.

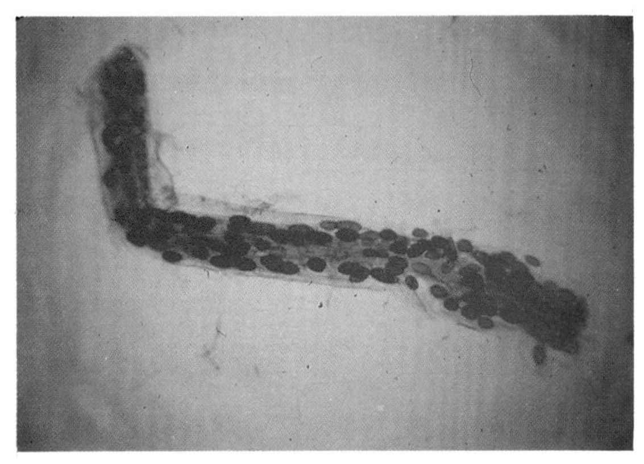

C. Root segment with well developed internal vesicles formed by a *Glomus* species.

D. *Liquidambar styraciflua* seedling inoculated with *Glomus fasciculatus* (R) compared to noninoculated seedling (L) showing stunted growth and phosphorus deficiency symptoms.

E. *Rhizopogon roseolus.* Peridium white, staining yellow, becoming red where handled; gleba chambered.

F. *Truncocolumella citrina.* Yellow stipe-columella with branches which extend among the glebal chambers.

G. *Alpova diplophloeus.* Orange, gel-filled chamber becomes red upon exposure to air.

H. *Leucogaster rebuscens.* White peridium bruises pink to red in age; gleba with pure white, gel-filled chambers.

A. *Macowanites* sp. Stipe-columella connects to peridium; gleba labyrinthoid to sublamellate.

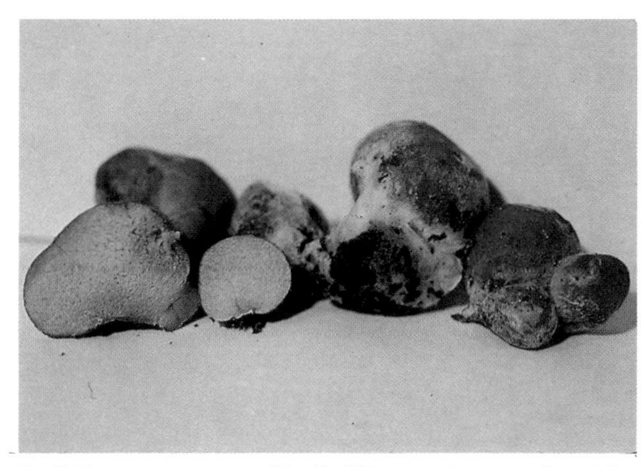

B. *Zelleromyces ravenelii.* Peridium red, smooth; gleba chambered and exuding a latex when cut.

C. *Hysterangium separabile.* Peridium white, bruising pink; gleba green, mucilaginous.

D. *Endogone lactiflua.* Old specimen left has collapsed peridium and exposed spores; rest in prime condition (J. Trappe photo).

E. Unidentified *Pinus resinosa* ectomycorrhiza from Heiberg Forest, Tully, N.Y.; young mycorrhizae irregularly forked, white, smooth, without rhizomorphs.

F. Unidentified *Pinus elliottii* ectomycorrhiza from Gainesville, Fla.; massive corralloid structure of active and inactive mycorrhizae.

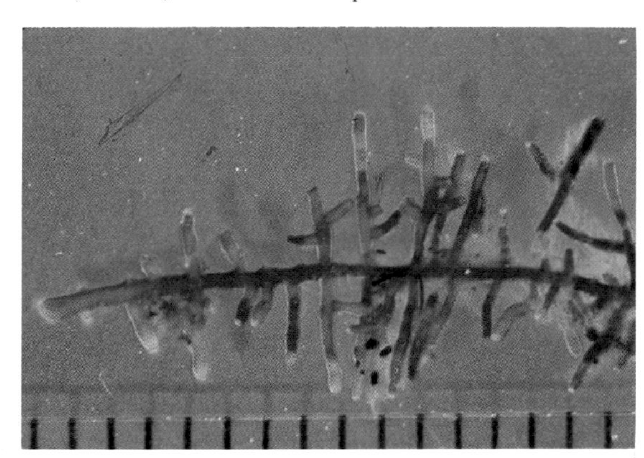

G. Unidentified pinnate *Picea abies* ectomycorrhiza from Highland Forest, Tully, N.Y.

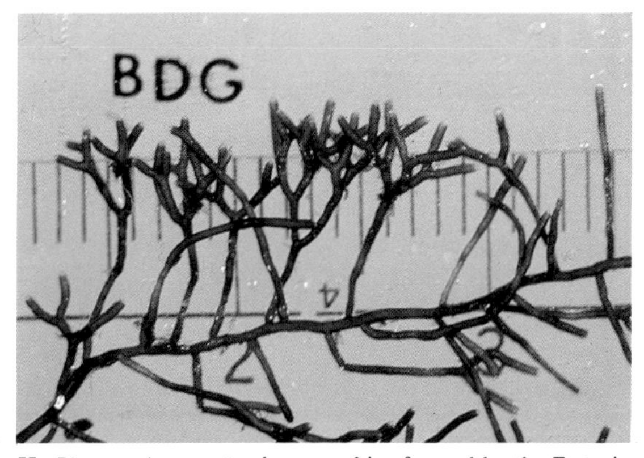

H. *Pinus resinosa* ectendomycorrhiza formed by the E-strain fungus BDG.

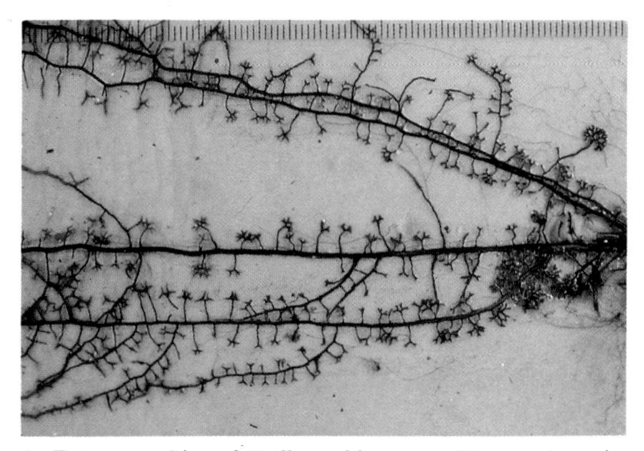

A. Ectomycorrhiza of *Suillus subluteus* on *Pinus resinosa* in tube culture on filter paper; rhizomorphs traverse the filter paper between mycorrhizal clusters.

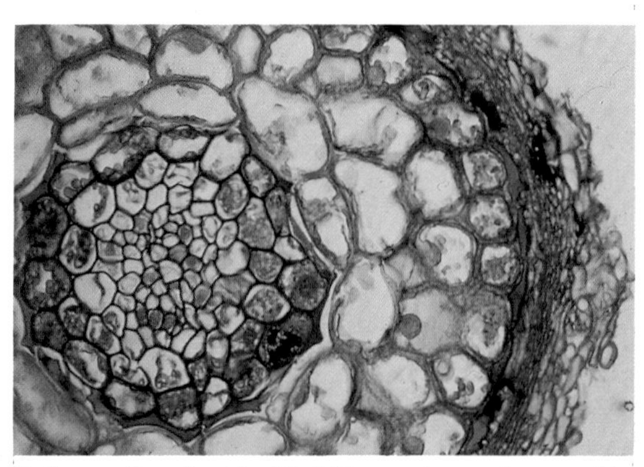

B. Transection of unidentified *Pinus resinosa* ectomycorrhiza from a forest nursery stained with chlorazol black E and Pianese III-B.

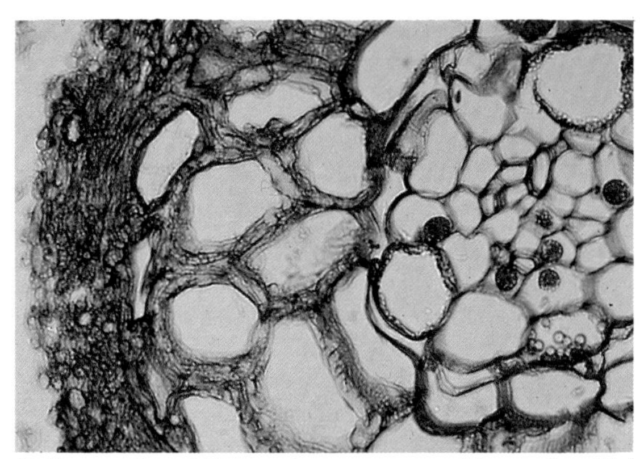

C. Transection of *Pinus resinosa* ectomycorrhiza produced by *Pisolithus tinctorius* in mycorrhizal synthesis culture stained with Conant's quadruple stain.

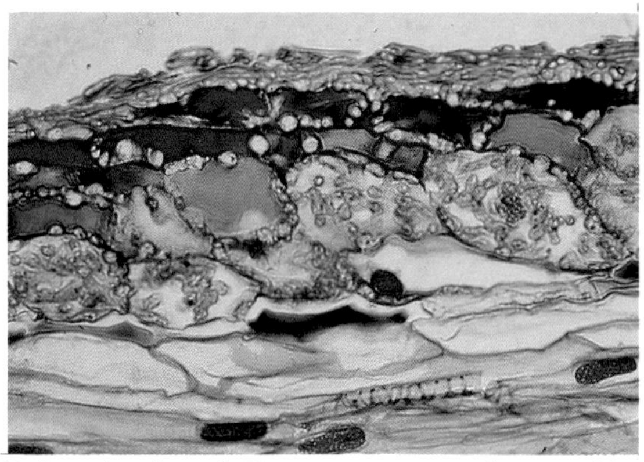

D. Longisection of *Pinus resinosa* ectendomycorrhiza produced by E-strain fungus BDG in mycorrhizal synthesis culture stained with Conant's quadruple stain.

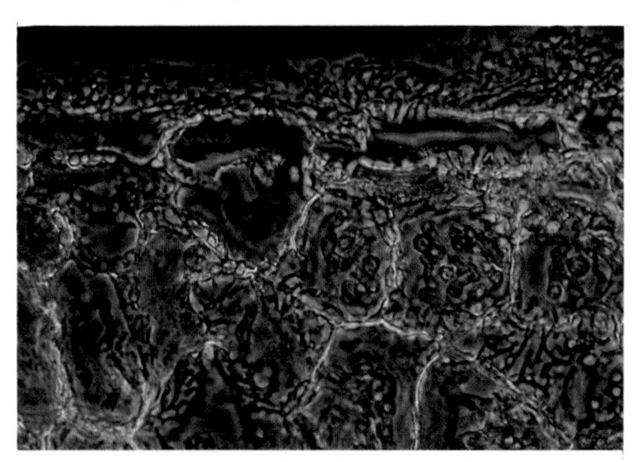

E. Freehand longisection of terminal rootlet of *Pinus resinosa*; prominent intracellular infection confirms the structure as an ectendomycorrhiza.

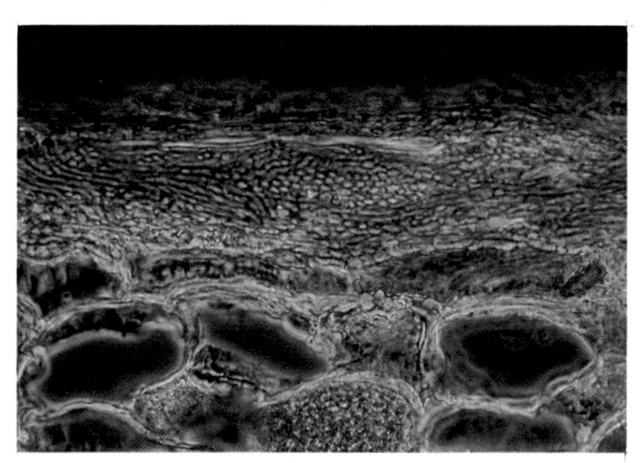

F. Freehand longisection of terminal rootlet of *Pinus resinosa* showing exclusively intercellular infection of the cortex caused by an ectomycorrhizal fungus.

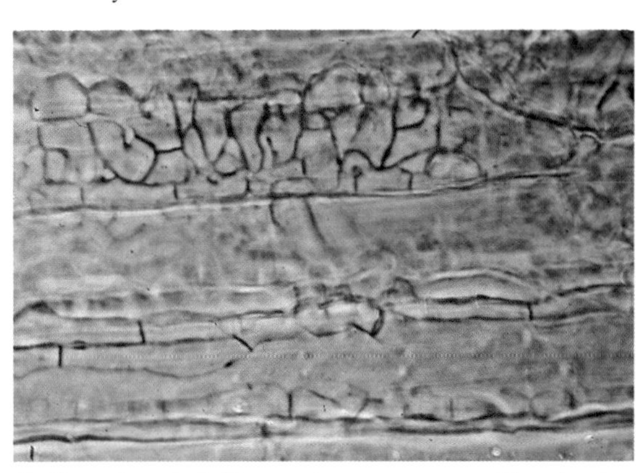

G. Cleared root of *Pinus resinosa* nursery seedling showing coarse Hartig net of E-strain fungus in a small-diameter long root.

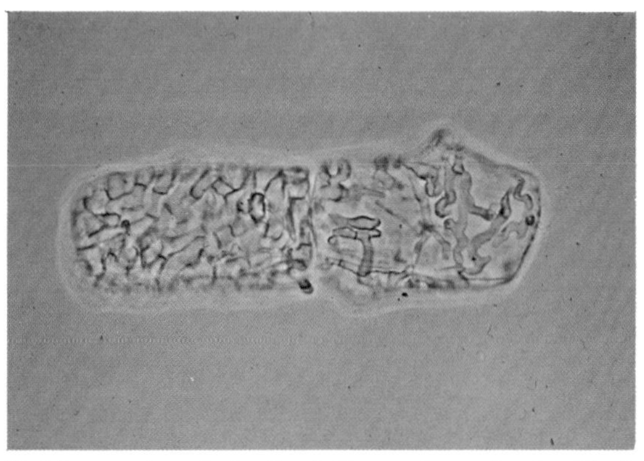

H. Macerated cortical cells of *Pinus resinosa* showing Hartig net and intracellular infection characteristic of an ectendomycorrhiza.

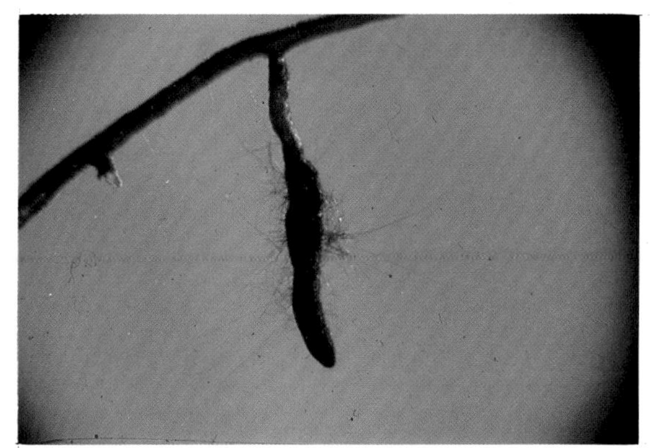

A. Branching habit of ectomycorrhizae on Fraser fir, *Abies fraseri.*

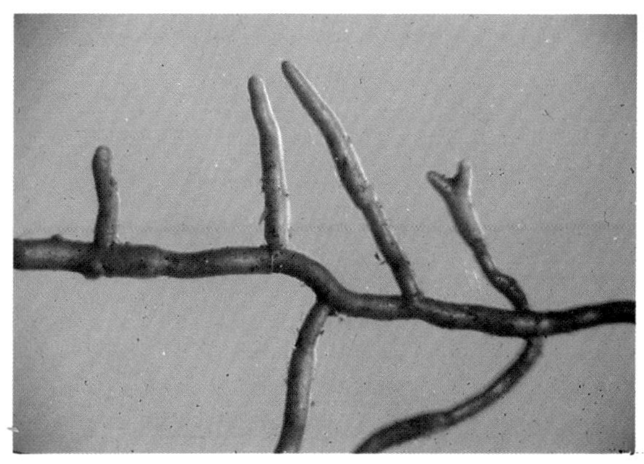

B. Branching habit of ectomycorrhizae on Fraser fir, *Abies fraseri.*

C. Branching habit of ectomycorrhizae on loblolly pine, *Pinus taeda.*

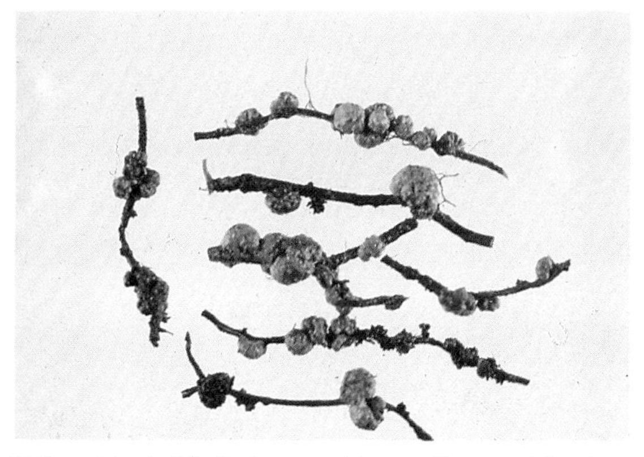

D. Branching habit of ectomycorrhizae on Eastern white pine, *Pinus strobus.*

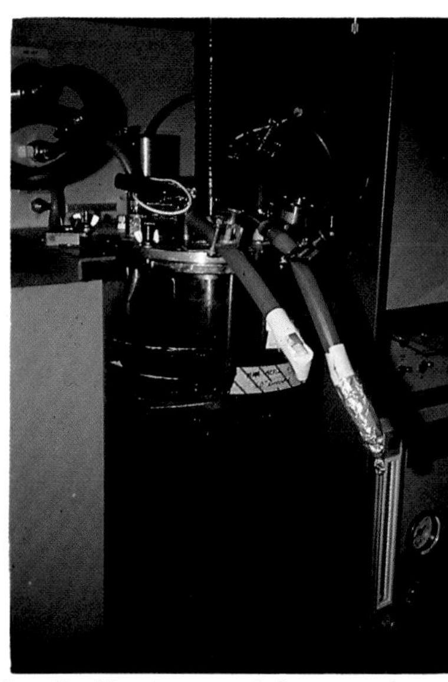

E. A 14-liter liquid culture used for research on industrial production of *Pisolithus tinctorius* vegetative inoculum (MycoRhiz®).

F. Bank of 55-liter fermentors for liquid culture "starter inoculum" of *Pisolithus tinctorius.*

G. A 500-liter rotary drum fermentor for culturing *Pisolithus tinctorius* in vermiculite-peat moss.

H. Filling a 20,000-liter fermentor for MycoRhiz® production.

A. Nursery planter adapted for simultaneous application of vegetative mycorrhizal inoculum and tree seeds.

B. Ectomycorrhizal inoculum bander for containerized seedlings; top: dibble platform with plexiglass cylinders and attached copper tubbing; below: holding tray for inoculum.

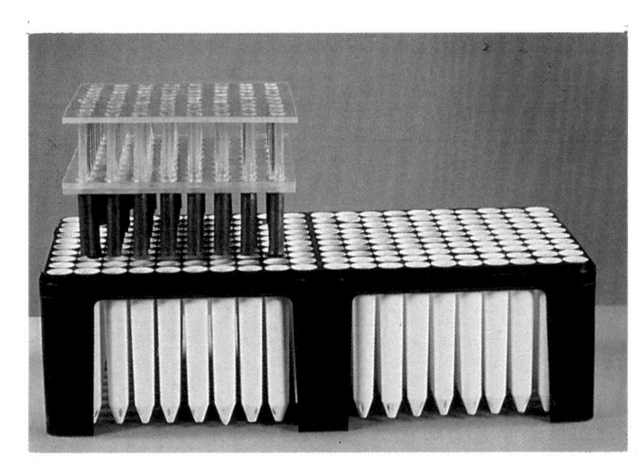

C. Ectomycorrhizal inoculum bander in use; inoculum released into containers by inserting plexiglass cylinders into the copper tubing.

D. *Pisolithus tinctorius* sporocarp on a 5-month old, container-grown short leaf pine seedling, *Pinus echinata*.

E. Phosphorus and mycorrhizae in plant pathogen interactions; L. to R. *Meloidogyne incognita* (MI), MI + phosphorus, MI + *Glomus etunicatus*, MI + *Glomus mosseae*.

F. Intact ectomycorrhizal roots from pine seedling sealed in inoculator cells (25 mm diam.) prior to pathogen inoculation.

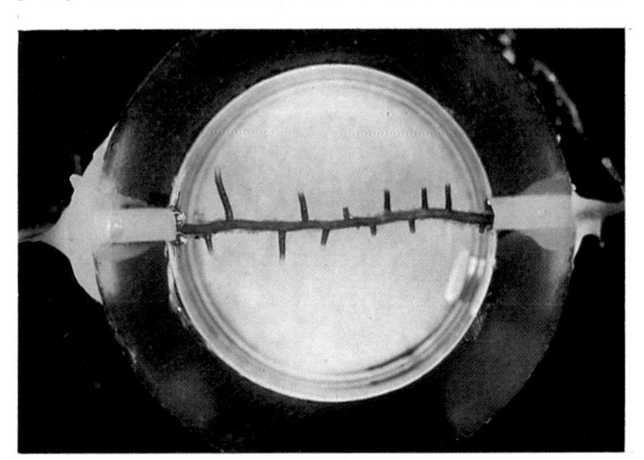

G. Intact nonmycorrhizal short roots from pine seedling in inoculator cells.

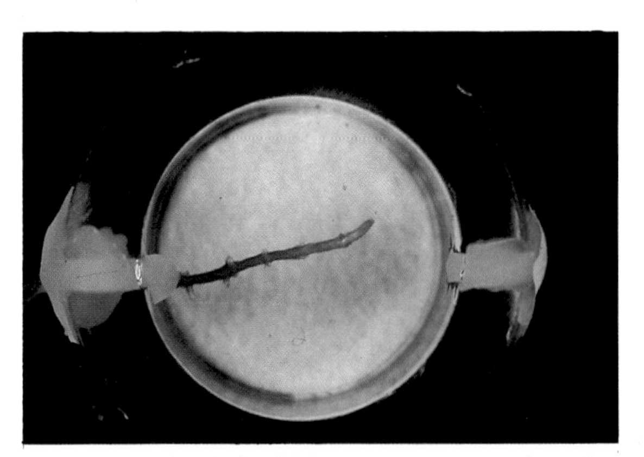

H. Intact lateral root tip with immature short roots from pine seedling in inoculation cell.